Origin of the Earth and Moon

A. E. Ringwood

Research School of Earth Sciences
Australian National University

Origin of the Earth and Moon

Springer-Verlag
New York   Heidelberg   Berlin

A. E. Ringwood

Director, Research School of Earth Sciences
Australian National University
Canberra A.C.T. 2600
Australia

Library of Congress Cataloging in Publication Data

Ringwood, A   E   1930–
  Origin of the Earth and Moon.

  Bibliography: p.
  Includes index.
  1. Earth—Origin. 2. Moon—Origin. I. Title.
QB631.R55   525′.01   78-25974

With 58 illustrations.

Printed in the United States of America.

9 8 7 6 5 4 3 2 1

ISBN 0-387-90369-0 Springer-Verlag New York
ISBN 3-540-90369-0 Springer-Verlag Berlin Heidelberg

Preface

Since the beginning of civilization, the origins of the Earth and Moon have been the subjects of continuing interest, speculation, and enquiry. These are also among the most challenging of all scientific problems. They are, perhaps to a unique degree, interdisciplinary, having attracted the attention of philosophers, astronomers, mathematicians, geologists, chemists, and physicists. A large and diverse literature has developed, far beyond the capacity of individuals to assimilate adequately. Consequently, most of those who attempt to present review–syntheses in the area tend to reflect the perspectives of their own particular disciplines. The present author's approach is that of a geochemist, strongly influenced by the basic philosophy of Harold Urey. Whereas most astronomical phenomena are controlled by gravitational and magnetic fields, and by nuclear interactions, Urey (1952) emphasized that the formation of the solar system occurred in a pressure–temperature regime wherein the *chemical* properties of matter were at least as important as those of gravitational and magnetic fields. This was the principal theme of his 1952 book, "*The Planets*," which revolutionized our approach to this subject.

In many subsequent papers, Urey strongly emphasized the importance of *meteorites* in providing critical evidence of chemical conditions in the primordial solar nebula, and of the chemical fractionation processes which occurred during formation of the terrestrial planets. This approach has been followed by most subsequent geochemists and cosmochemists. Numerous detailed physical, chemical, and mineralogical studies of meteorites have yielded a bonanza of critical evidence concerning events which occurred in the solar system during and soon after its formation, about 4.5 billion years ago. The evidence has provided many crucial boundary conditions which must be explained by any acceptable theory of the origin of the solar system. Tremendous effort has been expended in studying the fractionation processes responsible for the differing chemical compositions of the various classes of meteorites. However, although important progress has been made, many essential aspects of these processes remain obscure and controversial.

It has been widely assumed that these chemical fractionations were representative of large-scale physicochemical processes occurring throughout the inner regions of the solar nebula, and that the formation of the terrestrial planets was governed by this same set of physicochemical fractionations. For reasons which will appear subsequently, I have become sceptical of the validity of this latter assumption. I suspect that the study of meteorites may be revealing details of important but often localized fractionation processes which occurred on a small scale in particular regions of the nebula, and doubt whether these processes were necessarily identical to those which occurred on a far larger scale during the formation of the terrestrial planets. Accordingly, the geochemical–cosmochemical approach adopted in this book differs in its emphasis from that of Urey in his later papers, and also from those of many subsequent workers, such as the Chicago school. In some ways, it is closer to the approach adopted by Urey in his 1952 book.

Our knowledge of the constitution and composition of the Earth's interior has advanced enormously during the last 30 years. For the mantle, it is now possible to provide a reasonably satisfactory explanation of the variation of density and seismic velocities with depth in terms of the nature and thermodynamic stability fields of mineral phases present in the various regions. These advances, combined with extensive geochemical and petrological studies of rocks and magmas derived from the upper mantle, have made it possible to estimate the bulk chemical composition of the mantle and to place constraints upon the extent of possible chemical zoning within this region. Corresponding advances have also been made in our understanding of the structure and composition of the Earth's core.

As a result of these developments, many new and important boundary conditions for the origin of the Earth have emerged. I do not believe that the significance of these boundary conditions, mainly of a geochemical nature, have been adequately recognized in many recent discussions of the origins of terrestrial planets in general, and of the Earth in particular. The primary emphasis in this book will be the utilization of geochemical evidence obtained by direct study of the Earth itself in an attempt to understand its origin.

Likewise, as a result of the Apollo project, our knowledge of the composition and constitution of the Moon has increased enormously over the past decade. Much of this new information is relevant to the venerable problem of the origin of the Moon and we shall also explore this topic.

With these objectives in mind, the book has been divided into three parts. Part I seeks to present an overview of knowledge and interpretations concerning the present constitution and composition of the Earth, and is aimed at setting out the relevant background, particularly for non-Earth scientists. This part concludes with a set of geochemical boundary conditions which are believed to be applicable to the Earth's origin. The condensation of such a complex topic may lead, in some instances, to oversimplifications. More detailed justifications for many of the geochemical generalizations concerning the Earth's mantle are given in an earlier book (Ringwood, 1975a), whilst the topics relating to the composition of the core are discussed more specifically in Ringwood (1977b).

Part II commences with a discussion of the development and properties of the solar nebula and of general problems relating to the formation of planets therein. Current theories of the origin of the Earth are discussed in the light of boundary conditions and evidence considered in Part I, as well as from cosmochemistry and related fields. An attempt is made to present a synthesis which, hopefully, reconciles much of this evidence.

In Part III, we first explore some of the chemical relationships between the Earth and the terrestrial planets. The insights obtained into the constitution and formation of the Earth in Part II are extended to interpret the compositions of Mars, Venus, and (with less success) Mercury. These comparisons highlight the totally anomalous composition of the Moon, if viewed as a "normal" terrestrial planet. In two subsequent chapters, the composition and constitution of the Moon are discussed, and certain key aspects of lunar geochemistry, relevant to the Moon's origin, are described in some detail. In the final chapter, a review of current hypotheses of lunar origin is presented, concluding with the presentation of a preferred working hypothesis.

Except, perhaps, for some of the material in Chapters 10 and 11, the book is written at a level which, it is hoped, will be intelligible to a wide range of undergraduate and graduate students in the earth and space sciences. Extensive references to the primary literature are provided for the use of graduate students and workers in this field.

At the same time, the prospective reader should be aware that this is not a textbook in the accepted sense. The subject matter includes many areas which have been, and still are, considered controversial. In many of these, I have advocated a particular point of view. In such cases, I have attempted to justify these preferred views, and to provide adequate references dealing with both sides of the controversy, so that the interested reader can follow up the issue himself. This is not to state that large parts of the book are unsuitable for textbook use. I would hope that the sections dealing with the composition and constitution of Earth, Moon, and planets present a reasonably objective and comprehensive overview of their subject matter. As might be expected, the more controversial discussions are mainly (but not exclusively) in the sections dealing more specifically with the origins of the Earth and Moon.

The incentive to write this book was provided in the first instance by an invitation to deliver the Vernadsky Lecture in Moscow in March, 1975. Most of the chapters of Parts I and II will be published separately in the U.S.S.R. during 1978 as the Vernadsky Lecture. I wish to acknowledge the hospitality extended by the Vernadsky Institute and by the late Academician A. P. Vinogradov on that occasion.

This book is dedicated to Professor Harold Urey and to Academician Vinogradov, whose monumental contributions in cosmochemistry provided entirely new and fundamental perspectives on the origin and evolution of the Earth and solar system.

Acknowledgments

I wish to express my sincere appreciation to many colleagues who have read and commented on one or more chapters in this volume, particularly Dr. R. J. Arculus, Dr. S. E. Kesson, Dr. J. W. Delano, Dr. D. J. Stevenson, Dr. M. W. McElhinny, and Professor K. Lambeck of the Australian National University; Dr. A. R. Duncan of the University of Capetown; and Dr. M. G. Langseth of Lamont–Doherty Geological Observatory.

Chapter 11 is based on papers which I have written jointly with Dr. Kesson and Dr. Delano. I am grateful for the collaboration of these colleagues and their willingness for this material to be so included. A condensed account of the subject matter covered in Parts I and II will appear as a chapter in *The Earth: It's Origin, Structure, and Evolution*, edited by Dr. M. W. McElhinny (Academic Press, London and New York). Permission to use some diagrams and other material from this chapter in the present book is gratefully acknowledged.

Three of the chapters were written at Cornell University during periods when the author was present as Andrew D. White Professor-at-large. I wish to express my gratitude to Cornell for the support and facilities provided and particularly to my colleagues in the Department of Geological Sciences.

Thanks are also due to my secretaries Tanya Scheible and Robyn Curtin, who, with their characteristic efficiency, converted illegible handwritten manuscripts into accurately-typed chapters.

Finally, a deep tribute to my wife Gun, and children Kristina and Peter, particularly for their cheerful and loyal acceptance of the need to sacrifice so many holidays "down the coast" while this book was being written.

Contents

PART I

COMPOSITION AND CONSTITUTION OF THE EARTH

CHAPTER 1

The Mantle–Crust System

1.1 Principal Subdivisions of the Earth's Interior

The principal subdivisions of the Earth's interior are based on the depth distribution of seismic wave velocities (Fig. 1.1). The *crust* is·defined as the region extending from the surface to the *Mohorovicic Discontinuity*, (Moho) which lies at a depth of 30–50 km beneath most continents and about 10–12 km beneath most oceans. Underlying the crust, and extending to a depth of about 400 km, lies the *upper mantle*. Below this depth, extending between 400 and 1000 km, lies the *transition zone*. This is characterized by high-velocity gradients on the average and by the presence of two or more seismic velocity discontinuities. In contrast, the *lower mantle*, extending from 1000 to 2900 km, is characterized by a moderate and relatively uniform increase of seismic velocities with depth. The boundary of the Earth's *core* is reached at a depth close to 2900 km. This is marked by a major first-order seismic discontinuity for *P* waves and by cessation of transmission of *S* waves, implying a liquid state. The core is divided into two regions, the liquid *outer core* and the *inner core*, as shown in Figure 1.2. The latter region transmits *S* waves as well as *P* waves and is therefore believed to be solid. The dimensions and masses of the individual regions are given in Table 1.1.

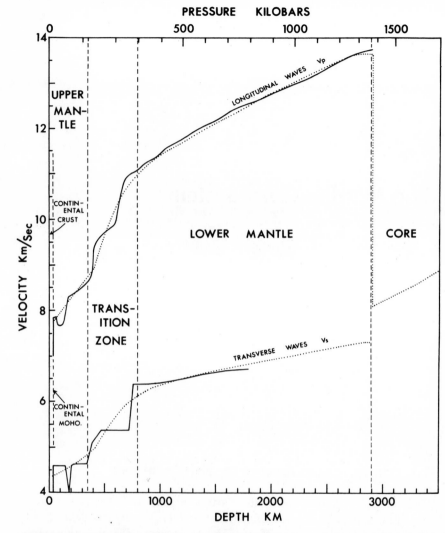

Figure 1.1 Variation of seismic *P* and *S* wave velocities throughout the mantle and outermost core. *P* waves—solid line: Johnson (1967, 1969); *S* waves—solid line: Nuttli (1969); broken lines: Jeffreys (1939). (From Ringwood, 1975a, with permission).

Table 1.1 Dimensions and masses of the internal layers of the Earth

Region	Depth to boundaries (km)	Mass (10^{25} g)	Fraction of total mass
Crust	0–Moho	2.4	0.004
Upper mantle	Moho–400	62	0.10
Transition zone	400–1000	100	0.17
Lower mantle	1000–2900	245	0.41
Outer core	2900–5154	177	0.30
Inner core	5154–6371	12	0.02

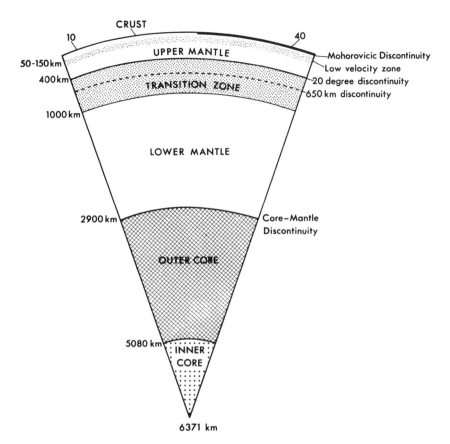

Figure 1.2 Principal internal subdivisions of the Earth. (From Ringwood 1975a, with permission.)

1.2 The Crust

Although its volume is of minor proportions, the *crust* nevertheless possesses considerable global geochemical significance, representing the end result of extremely complex and efficient differentiation processes that have concentrated a large portion of the Earth's total inventory of certain key incompatible elements[1] into this thin surficial region. A typical section through a region of stable continental crust is shown in Figure 1.3. Beneath the sedimentary veneer is an upper layer which, from its geochemical and geophysical properties, is inferred to possess an overall granodioritic composition (Table 1.2). Below this layer, extending to depths of 15–25 km,

---

[1] Incompatible elements are defined as those possessing ionic radii and/or ionic charges that do not permit their ions to substitute readily into the principal crystalline phases of the mantle. As a result, they are strongly partitioned into magmas during partial melting processes. Examples are Cs, Rb, K, Ba, Pb, La, Ce, U, Th, Ta, Nb, and P.

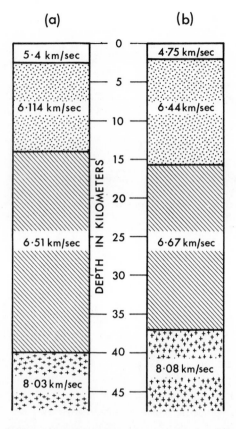

Figure 1.3 Crustal sections for two typical stable continental regions—the Precambrian shields in (a) central Wisconsin and (b) upper Michigan (from Steinhart and Meyer, 1961). (From Ringwood, 1975a, with permission.)

seismic *P* velocities increase to values mostly between 6.7 and 7.2 km/sec and are more widely variable on a regional scale. Interpretations based on geophysical, petrological, and geochemical data indicate that in most evolved continental regions, this lower layer, although highly heterogeneous, probably consists mainly of acidic-intermediate rocks in the garnet granulite metamorphic facies. This region may be residual in nature, arising from repeated metamorphic and partial melting processes that have affected the lower crust, resulting in strong depletions of volatiles and incompatible elements as compared to the upper crust. The lower boundary of the continental crust is set by the Moho, where the seismic *P*-wave velocity increases in a relatively short depth interval to values mostly in the range 8.2 ± 0.2 km/sec, characteristic of the upper mantle. The depth to the Moho beneath most stable continental regions is usually between 35 and 45 km, with an average close to 40 km. The mean chemical composition of the continental crust has been estimated by Taylor (1977) and is believed to be similar to that

Table 1.2 Average composition of (A) continental shield crystalline surface rocks (Poldervaart, 1955) and (B) the entire continental crust (Taylor, 1977)

Component	A wt.-%	B wt.-%
SiO_2	66.4	58.0
TiO_2	0.6	0.8
Al_2O_3	15.5	18.0
Fe_2O_3	1.8	—
FeO	2.8	7.5[a]
MnO	0.1	—
MgO	2.0	3.5
CaO	3.8	7.5
Na_2O	3.5	3.5
K_2O	3.3	1.5
P_2O_5	0.2	—

[a] All iron as FeO.

of an intermediate igneous rock, corresponding to the most abundant magma type (andesite) erupted in regions of lithosphere subduction (Table 1.2).

The structure of the oceanic crust is fundamentally different from that of the continental crust. Its average thickness in most regions is only about 7 km, at which depth the Moho is encountered. Beneath the oceanic sediment layer (0.45 km average thickness), the crust is believed to consist mainly of rocks of tholeiitic basaltic composition (Table 1.3) with smaller amounts of serpentinite, derived by hydration of peridotitic rocks ultimately derived from the mantle and diapirically intruded into the crust, particularly along fracture zones. Compared to the continental crust, large regions of which are over two billion years old, the oceanic crust is relatively youthful, being formed by differentiation of the mantle at mid-oceanic ridges and later subducted into the mantle beneath oceanic trenches on a timescale of about 10^8 years. We will return to the subject of the evolution of the crust in Section 3.2.

Table 1.3 P velocities and thicknesses of oceanic crust layers with standard deviations (from Hill, 1957; and Raitt, 1963)

Layer	Thickness (km)	Velocity (km/sec)
Sea water	4.5	1.5
1	0.45	2
2	1.71 ± 0.75	5.07 ± 0.63
3	4.86 ± 1.42	6.69 ± 0.28
4	—	8.13 ± 0.24

1.3 The Upper Mantle

The Peridotite Layer

The upper mantle extends from the Moho to a depth of about 250–400 km (Figures 1.1 and 1.2). It is characterized generally by rather low *P*- and *S*-wave velocity-depth gradients, except perhaps in the vicinity of the low-velocity zone (LVZ), and by the existence of pronounced lateral variations in velocity distributions. In most regions of the Earth, the *S*-wave velocity and possibly the *P*-wave velocity pass through a minimum between depths of about 70 and 150 km. This minimum may be caused by the occurrence of a small degree (~ 1 percent) of partial melting along grain boundaries in this region. It is of considerable importance as a zone of low viscosity and tectonic decoupling of the overlying rigid lithosphere from underlying more mobile regions of the mantle or asthenosphere.

The *P*-wave velocity of most regions of the uppermost mantle beneath both continents and oceans lies in the range 8.2 ± 0.2 km/sec. This property, together with certain broad petrological and chemical limitations, effectively restricts the mineralogical composition of the upper mantle to some combination of olivine, pyroxene(s), garnet, and, perhaps, in restricted regions, amphibole. The two principal rock types carrying these minerals[2] are *peridotite* (olivine–pyroxene) and *eclogite* (pyroxene–garnet). Both types may carry some amphibole. Complete mineralogical transitions between the two major rock types are rare. Petrological reasons for this dichotomy are discussed in Ringwood (1975a).

The densities of *fresh* eclogites which might be derived from the upper mantle average about 3.5 g/cm^3, whereas the densities of corresponding fresh peridotites average about 3.3 g/cm^3. Thus, in principle, an independent determination of density of the upper mantle could provide evidence on the relative abundances of these rock types. The density difference between the crust and uppermost mantle as obtained from isostatic considerations is close to 0.4 g/cm^3. In combination with independent estimates of the mean density of the crust, the density of the uppermost mantle is found most probably to be between 3.3 and 3.4 g/cm^3, implying that this region is dominantly composed of peridotite. This conclusion is supported by several observations of large-scale lateral anisotropy of seismic velocities in the upper mantle (e.g., Morris et al., 1969). As seen in Figure 1.4, peridotites frequently show large degrees of anisotropy of seismic velocities owing to the intrinsic elastic properties of their principal mineral, olivine, whereas eclogites show much smaller degrees of anisotropy.

---

[2] For convenience of non-Earth scientists, the basic chemical formulae of the principal minerals of the upper mantle are approximately as follows. Olivine $(Mg_{0.9}Fe_{0.1})_2SiO_4$; Ortho-pyroxene $(Mg_{0.9}Fe_{0.1})SiO_3$; Clinopyroxene $(Mg_{0.9}Fe_{0.1})CaSi_2O_6$; Spinel-chromite $MgAl_2O_4$–$FeCr_2O_4$; Pyrope garnet $(Mg_{0.7}Fe_{0.3})_3Al_2Si_3O_{12}$; Clinopyroxene commonly contains significant quantities of jadeite $NaAlSi_2O_6$ in solid solution, while pyrope garnet usually contains a substantial amount of grossularite $Ca_3Al_2Si_3O_{12}$ in solid solution.

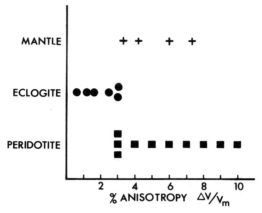

Figure 1.4 Percent anisotropies (expressed as difference between maximum and minimum observed velocities, ΔV, divided by the mean velocity V_m, for peridotites, eclogites and the upper mantle. (From Ringwood, 1975a, with permission.)

The geophysical inference that peridotite is the dominant rock type in the upper mantle is strongly supported by petrological observations. Kimberlite pipes containing diamonds are of frequent occurrence throughout one million square miles of southern Africa and a comparable region of Siberia. They are also known in India, Brazil, the United States, and Australia. These pipes are known to have been derived from depths as great as 150 km (necessary for the thermodynamic stability of diamonds) and to have ascended upwards into the crust very rapidly (owing to exsolution of volatiles). As a result, they have torn off samples of wall rocks from their conduits and transported them to the surface as *xenoliths*. It is of great significance that, in the African and Siberian kimberlites where sampling has been the most thorough, peridotite xenoliths are found to be much more common than eclogite xenoliths. If the sampling is representative, then the upper mantle must be composed mainly of peridotite with eclogite a minor but widely distributed constituent. The most common mineral assemblage among the peridotites is garnet lherzolite—average composition: olivine 64 percent, orthopyroxene 27 percent, clinopyroxene 3 percent, and pyrope-rich garnet 6 percent. Bulk chemical composition ranges are given in Table 1.4.

A second suite of volatile-rich magmas derived from shallower depths of the mantle than the kimberlites have also transported xenoliths of mantle rocks to the surface. Over 1000 such occurrences are known, both in oceanic and continental regions. These comprise the alkali basalt–basanite–nephelinite suite (Table 1.5). Again, the dominant type of xenolith is peridotite, containing spinel in place of garnet, in consequence of a lower pressure origin. Typical bulk compositions of such xenoliths are given in Table 1.4. They are generally similar to garnet lherzolite xenoliths from kimberlites.

A third line of petrological evidence on the composition of the upper mantle comes from the study of peridotite bodies occurring in the Earth's crust, both in continental and oceanic regions. Many of these are now believed to represent actual slices of the upper mantle which have been intruded into the crustal environment by tectonic mechanisms connected with plate tectonic processes. As shown years ago by Ross et al. (1954), and since then by many others, the bulk chemical and mineralogical compositions of these alpine peridotites are very similar to those of the peridotite xenoliths of the alkali basalt suite.

The overall conclusions arising from numerous petrological and geophysical studies is that beneath both continents and oceans, a layer exists in which the principal rock is a variety of peridotite possessing a very limited range of bulk compositions. It should be recognized that many other rock types also occur in this region (e.g., various kinds of eclogites and that a considerable degree of local petrological heterogeneity is present. However, these rock types appear to be much less significant volumetrically than peridotite. See Ringwood (1975a), Chapters 3–6, for a more detailed discussion of this topic.

Basalt Genesis and the Pyrolite Model

Magmas of the basaltic suite have been erupted in copious volumes throughout geological time at localities scattered all over the Earth's surface, both continental and oceanic. We now understand that basaltic magmas are formed by partial melting processes in the upper mantle, these liquids representing the low melting point components of the system, being analogous to the eutectic liquids formed during the melting of simple binary or ternary systems. Clearly, a fundamental property of much of the upper mantle is that it should be capable of yielding basaltic magmas when subjected to appropriate partial melting processes. However, the vast majority of peridotites derived from the upper mantle and sampled as xenoliths in kimberlites and alkali basalts, or as large intrusions into the crust, possess abundances of "incompatible elements[3]" such as K, U, Th, Ba, Rb, La, Ti, and P which are much too low to permit them to yield the observed compositions of basalt magmas when partially melted. Most peridotites also do not contain sufficient Na, Ca, and Al for them to serve as satisfactory sources for basaltic magmas. The general abundance patterns of most of these peridotites are those appropriate to refractory residues that might remain after basaltic magmas have been extracted.

We thus arrive at the interpretation that most naturally occurring peridotites possess a *complementary* rather than a parental relationship to basaltic magmas. It is inferred that there must exist a primitive source below the peridotite layer from which basalt magmas have not yet been extracted. This primitive material must be the source of modern basalts. For convenience, this primitive source material has been denoted by the term "pyrolite," implying a nonspecific pyroxene–olivine rock, capable of yielding a basalt magma on partial melting. These considerations lead us to the chemically zoned model for the upper mantle shown in Figure 1.5.

---

[3] Defined on page 5.

Table 1.4 Compositions of peridotite xenoliths from diamond-bearing kimberlite pipes (1,2) and alkali basalts (3,4)

	1[a]	2[b]	3[c]	4[d]
SiO_2	44.5 –47.9	46.5	44.4	45.0
TiO_2	0.02– 2.3	0.3	0.04	0.07
Al_2O_3	1.1 – 3.3	1.8	1.7	3.0
Cr_2O_3	0.2 – 0.5	0.4	0.5	0.4
FeO	5.9 – 8.4	6.7	8.9	8.0
MnO	0.1 – 0.2	0.1	0.1	0.1
NiO	0.25– 0.4	0.3	0.27	0.25
MgO	37.7 –45.7	42.0	42.3	39.7
CaO	0.9 – 3.5	1.5	1.6	3.2
Na_2O	0.06– 0.4	0.2	0.1	0.2
K_2O	0– 0.4	0.2	0.04	0.04
P_2O_5	0– 0.05	0.02	—	—

[a] 1. Composition range in 15 analyzed garnet-peridotites (Carswell and Dawson, 1970).

[b] 2. Mean of nine analyzed garnet peridotites (Carswell and Dawson, 1970).

[c] 3. Mean of 27 analyses of spinel peridotite xenoliths occurring in alkali basalt from Puy Beanite (Hutchinson et al., 1970).

[d] 4. Mean of 20 analyses of spinel peridotite xenoliths occurring in the alkali basalt of Rocher du Lion (Vilminot, 1965).

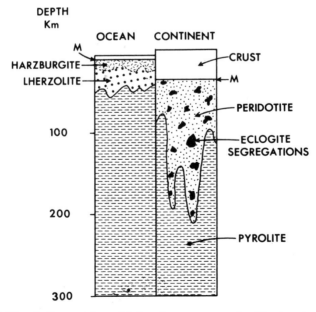

Figure 1.5 Chemically zoned model for the upper mantle. Harzburgite (olivine + orthopyroxene + chromite) and lherzolite (olivine + orthopyroxene + clinopyroxene + spinel) are varieties of peridotite. (From Ringwood, 1975a, with permission.)

Material of pyrolite composition is believed to be the dominant constituent of the upper mantle.

In a general sense, the composition of pyrolite must lie between those of basalt and peridotite. This composition may be obtained from the study of chemical equilibria among the various classes of basaltic magmas and their respective peridotitic residua using the techniques of experimental petrology at high pressures and temperatures (e.g., Green and Ringwood, 1967; Green, 1971; Ringwood, 1975a). This approach is combined with the use of trace element distributions between basaltic and peridotitic components in the light of experimentally determined crystal–liquid partition coefficients (e.g., Gast, 1968; Kay et al., 1970).

These studies have led to the recognition of a family of basaltic-type magmas (Table 1.5) formed by widely varying degrees of partial melting of mantle pyrolite. Thus, nephelinites and basanites may comprise 1–5 percent partial melts, alkali basalts 5–10 percent, tholeiites 15–25 percent, and picrites and komatiites 30–60 percent partial melts of pyrolite (Green and Ringwood, 1967; Gast, 1968; Kay et al., 1970; Green, 1970a, 1972; Brey and Green, 1976). Likewise, a series of complementary refractory peridotitic components representing the residual materials remaining after

Table 1.5 Compositions of typical members of the principal classes of basaltic magmas. Numbers 1–6 represent magmas formed by increasing degrees of partial melting of their source region

	Olivine nephelinite 1[a]	Olivine basanite 2[b]	Alkali olivine basalt 3[c]	Oceanic tholeiite 4[d]	Picrite 5[e]	Komatiite 6[f]
SiO_2	39.3	44.6	46.5	47.2	44.0	45.2
TiO_2	3.9	2.9	2.7	0.7	0.6	0.3
Al_2O_3	9.5	11.7	13.9	15.0	8.3	6.2
Fe_2O_3	5.1	3.0	2.6	3.4	2.2	—
FeO	10.7	9.4	9.5	6.6	8.8	11.5[g]
MnO	0.2	0.2	0.12	0.1	0.2	0.2
MgO	13.9	13.9	9.7	10.5	26.0	28.6
CaO	11.2	7.7	10.4	11.4	7.3	5.6
Na_2O	3.0	3.7	2.8	2.3	0.9	0.6
K_2O	1.5	2.0	0.7	0.1	0.06	0.04
P_2O_5	2.3	1.0	0.35	0.07	0.07	0.02
$\dfrac{MgO}{FeO^g + MgO}$ (Mol)	0.70	0.73	0.60	0.66	0.81	0.82

[a] 1. Scotsdale, Tasmania (Green, 1970b)

[b] 2. Mt. Leura, Victoria (Green, 1970b)

[c] 3. Hawaii (average) (Kuno et al., 1957)

[d] 4. DSDP leg 37 sample JSC 56 (Blanchard et al., 1976)

[e] 5. Baffin Bay (Clark, 1970)

[f] 6. Yilgarn block, Western Australia (Nesbitt and Sun, 1976)

[g] All iron as FeO.

these varying degrees of partial melting can be recognized. Garnet and spinel lherzolites—comprising the mineral assemblage olivine + orthopyroxene + clinopyroxene ± spinel ± pyrope garnet—represent the residuum remaining after small degrees of partial melting, and are complementary to nephelinites, basanites, and alkali basalts, while harzburgites (olivine + orthopyroxene) probably represent the residues remaining after extraction of olivine tholeiite magmas. Finally, the residuum remaining after the largest degrees of partial melting, resulting in formation of komatiite magmas, is dunite (Sun and Nesbitt, 1977). A review of the use of these methods to constrain the pyrolite composition is given in Ringwood (1975a).

Limiting compositions for pyrolite can also be obtained from additional complementary approaches. Studies of peridotites occurring as xenoliths in kimberlites and alkali basalts on the one hand, and as massive intrusions into the Earth's crust on the other, have revealed a continuum of rocks ranging between pure dunite, harzburgite, and lherzolite. This range corresponds to CaO and Al_2O_3 contents of 0–4 percent and Na_2O contents of 0–0.3 percent; then there is a sharp cutoff–lherzolites (of noncumulate origin) containing more than these quantities of CaO, Al_2O_3, and Na_2O are rare. It seems clear that those lherzolites containing about 3–4 percent of CaO and Al_2O_3 are residual from processes involving loss of magmas (e.g., nephelinites) produced by the smallest degrees of partial melting of pyrolite. Some types of high-temperature peridotites (e.g., the lherzolites of Tinaquillo, the Lizard, and Ronda) also seem to belong in this category. Nevertheless, these rocks display strong depletions and internal fractionations among incompatible elements (e.g., the light rare earths, K and U) and have clearly lost a low melting-point component, presumably a basanite or nephelinite.

Rare-earth element (REE) abundances exhibited by a typical high-temperature peridotite are shown in Figure 1.6. The extreme internal fractionation and depletion of light REE are to be remarked. Also shown are the REE abundances of a typical nephelinite. The complementary nature of these patterns is immediately obvious. Cosmochemical considerations (discussed later), combined with the observed near-chondritic REE patterns in oceanic tholeiites, strongly indicate that pyrolite originally possessed a near-chondritic REE relative abundance pattern. Production of this pattern from the complementary patterns of the peridotite and nephelinite (Fig. 1.6) requires the addition of about 1–2 percent nephelinite to the peridotite. This is consistent with the experimental evidence that nephelinites are produced by a very small degree of partial melting of pyrolite. Such considerations make it possible to calculate an ideal pyrolite composition by combining 99 percent of the peridotite with one percent of the nephelinite.

An alternative approach to seeking the least fractionated ultramafic residual material as an approximation to the pyrolite composition, is to search for natural ultramafic liquids that represent the greatest degrees of partial melting of pyrolite source material. It is hoped that these approaches may converge.

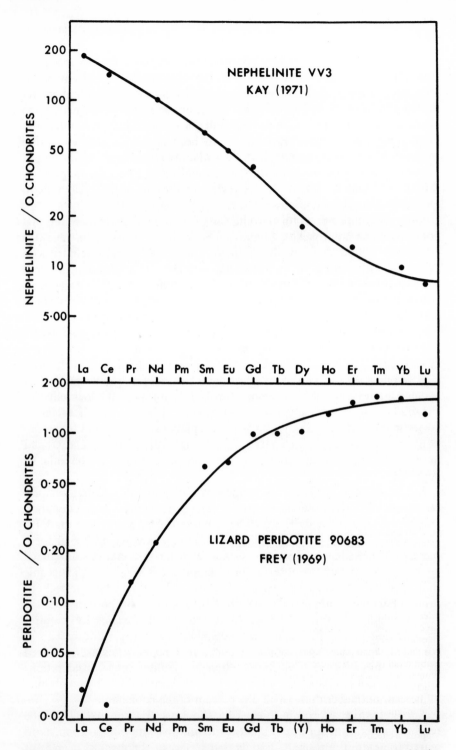

Figure 1.6 Rare earth elemental abundances in the Lizard high temperature peridotite and in a Cape Verde nephelinite. Rare earth abundances have been divided, element by element, by the corresponding abundances, in ordinary chondrites. (From Ringwood, 1975a, with permission.)

   Important evidence of the existence of natural ultramafic magmas approaching the pyrolite composition has been presented by Viljoen and Viljoen (1969a,b). These authors described the occurrence of a series of thin peridotitic lava flows from the Archaean Komati formation of the Barberton region of South Africa. Field and petrologic evidence demonstrated that these "peridotitic komatiites" had been erupted under subaqueous conditions and were completely molten when erupted. Convincing evidence establishing the latter characteristic is supplied by the widespread occurrence of "quench textures" (Fig. 1.7) within the flows. Experimental investigations (Green et al., 1975) have shown that these textures can be reproduced only by rapid quenching of ultramafic liquids representative of the compositions of komatiites. Similar rocks have been described from Archaean regions in Australia (Glikson, 1970; Nesbitt, 1972) and Canada (Naldrett and Mason, 1968; Arndt et al., 1977).

   Experimental studies (Green et al., 1975; Arndt et al., 1977) showed that olivine (Fo_{93}) was on the liquidus of a typical peridotitic komatiite composition to pressures beyond 40 kbar, and the overall phase relationships were consistent with derivation of this komatiite by very high degrees (≥ 60 percent*) of partial melting of pyrolite, leaving only olivine (Fo_{93}) as a residual phase.

   For the purposes of the present study, it is important to note that the composition of peridotitic komatiite formed by the largest degree of partial

---

   * *Note added in proof.* Arndt (1977) has argued that it is improbable that komatiite magmas were formed by as much as 60 percent of batch melting of mantle pyrolite. He argues that beyond a limit of about 50 percent (by weight) of partial melting, the rate of segregation of magma will be rapid compared to the rate of uprise of the mantle diapir which provides the heat of fusion for melting by adiabatic decompression. Arndt thus suggests that komatiites have formed by a process of "sequential melting" and the more ultrabasic varieties represent partial melts of mantle that has already been markedly depleted in low melting components at an earlier stage of magma generation. He appeals to the light rare earth depletion often observed in komatiites and to the high CaO/Al_2O_3 ratios in Barberton Land komatiites as evidence for this latter process. Arndt concludes that the compositions of komatiites therefore, do not provide information about the primary unfractionated composition of the mantle prior to magma generation.

   In the author's opinion, there is substantial merit in Arndt's arguments, but they are taken too far. The light rare earth depletions in most komatiites are generally smaller than those typically found in oceanic tholeiites (Sun and Nesbitt, 1977). The latter are believed to be caused by prior extraction of only a small proportion of liquid, which did not affect the bulk major element composition significantly, or indeed that of the heavy REE and other moderately in compatible elements such as Ti, Y, Zr, Nb, and Sc. Nesbitt and Sun (1976) and Sun and Nesbitt (1977) demonstrated that the relative abundances of all of these elements, and also Ca and Al, in the source regions of many komatiite magmas were essentially chondritic. Any substantial (e.g., >5 percent) degree of partial melting in an event prior to the extraction of komatiite magma would have caused marked internal fractionation of these elements, and gross fractionation of light REE.

   It seems, therefore, that any previous loss of magma from the source regions of many komatiite magmas has been decidedly minor, and that valid estimates of mantle composition can therefore be obtained. Nesbitt and Sun (1976) and Sun and Nesbitt (1977) have employed this approach effectively and concluded that the bulk composition of the source region of komatiites was very close to that of pyrolite.

   Once the degrees of melting exceeded about 50 percent by weight, Arndt's arguments become more compelling. It may well be that some komatiites possessing non-chondritic Ca/Al ratios and very low ($<3 \times$ chondrites) and strongly fractionated REE abundances were formed by the mechanism he suggests.

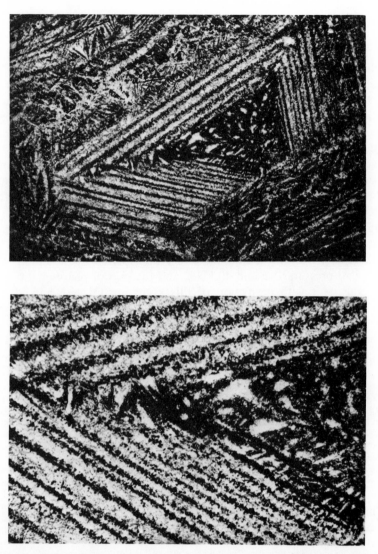

Figure 1.7 "Quench" komatiitic peridotite, which originally crystallized rapidly from the molten state. The original intergrowth of olivine and pyroxene has since been replaced by alternating layers of tremolite, magnetite–chlorite, and serpentine. Magnification × 20. (From Viljoen and Viljoen, 1969a, with permission.) These textures have since been reproduced experimentally by rapid quenching of melts of komatiite composition (Green et al., 1975).

Table 1.6 Pyrolite composition

	Least fractionated ultramafics 1[a]	Peridotitic komatiite liquids 2[b]	Peridotitic komatiite liquids 3[b]	Average pyrolite composition 4[c]
SiO_2	44.9	46.5	44.8	45.1
TiO_2	0.1	0.2	0.2	0.2
Al_2O_3	3.2	3.6	5.3	3.3
Cr_2O_3	0.5	0.4	—	0.4
Fe_2O_3	—	1.0	1.0	—
FeO	7.6	9.4	9.5	8.0
MnO	0.1	0.2	0.2	0.15
NiO	0.3	—	—	0.2
MgO	40.0	33.0	34.3	38.1
CaO	3.0	5.1	4.4	3.1
Na_2O	0.2	0.5	0.35	0.4
K_2O	0.0006	0.2	0.01	0.03
P_2O_5	—	0.01	0.015	0.02
$\dfrac{MgO}{MgO + FeO}$[d]	0.90	0.85	0.85	0.89

[a] 1. Representing lherzolite residuum subjected to small degree of partial melting (Tinaquillo peridotite—Green, 1963).

[b] 2, 3. Ultramafic liquids representing higher degrees of partial melting of pyrolite. Column 2 represents a sample of peridotitic komatiite from Barberton Land (Green et al., 1975), while Column 3 represents a corresponding sample from Mt. Ida in Western Australia (Nesbitt, 1972).

[c] 4. Obtained by several different methods (Ringwood, 1975a). As noted in the footnote of his Table 5.2, the methods used led to an unrealistically high value of Al_2O_3 (4.6 percent). In column 4, this has been adjusted downwards by normalizing to CaO using Al_2O_3/CaO ratios observed in West Australian and Canadian peridotitic komatiites. Minor elements are based on Sun and Nesbitt (1977).

[d] All iron as FeO.

melting of pyrolite closely approaches that of the ultramafic composition residual after extraction of magmas representing very small degrees of partial melting. These limiting compositions, given in Table 1.6, thus bracket the ideal pyrolite composition. Another estimate of the pyrolite composition based on these and other methods is also given in Table 1.6. It is emphasized that this is not to be taken as a *unique* composition. There is probably a significant compositional range for pyrolite in the upper mantle. Present indications are that this range is quite small for most major elements and for compatible trace elements,[4] but may be very large for some incompatible trace elements.

A detailed geochemical study of the compositions and petrogenesis of a range of Archaean komatiitic magmas from Australia, South Africa, and Canada has been carried out by Nesbitt and Sun (1976) and Sun and Nesbitt

[4] Defined on page 30 of chapter 2.

(1977). They demonstrated the existence of a complete continuum of liquid compositions ranging from tholeiitic basalt–picrite–basaltic komatiite–peridotitic komatiite. These studies led to an estimate of the probable composition range of the Archaean upper mantle. They were also of fundamental importance in defining the abundances of many minor elements in pyrolite, a topic to which we shall return shortly.

1.4 The Transition Zone

It can be seen from Figures 1.1 and 1.8 that, on the average, seismic velocities and density increase very rapidly with depth between about 350 and 900 km. Major seismic and density discontinuities occur at around 400 and 650 km, and there is also evidence of smaller discontinuities around 350, 500, and deeper than 650 km. Bullen (1936) first demonstrated that density increased with depth more rapidly throughout the transition zone than could be explained by self-compression of homogeneous material within the Earth's gravitational field. He concluded that chemical and/or phase changes must occur in this region. Birch (1952) inferred from his study of the elastic properties of the mantle that phase changes were primarily responsible for the inhomogeneity between depths of 350 and 900 km, although chemical changes were also possible. Birch also concluded that the rates of increase of seismic velocities and density between depths of 900 and 2700 km were

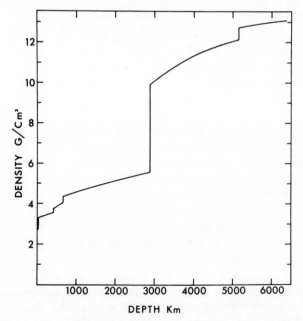

Figure 1.8 Distribution of density with depth throughout the Earth. The model is from Dziewonski et al. (1975) and is very similar to the classical Model A derived by Bullen (1940).

consistent with this region being homogeneous. No further phase changes or chemical changes were required to explain the observations. He proposed that the lower mantle consists of a series of closely packed high-pressure polymorphs of major upper mantle minerals, these new phases having elastic properties and densities similar to those of close-packed "oxide" phases such as corundum, periclase, rutile, and spinel.

Birch's hypothesis was finally confirmed by direct experimentation at very high pressures mainly by Ringwood and coworkers in Canberra during the period 1958–1972 and by Akimoto and coworkers in Tokyo from 1965 onwards. Comprehensive reviews of this work are given in Ringwood (1970a, 1975a) and Akimoto (1972). Further important contributions since 1972 have been made by others, particularly Liu (1975, 1977a). Shock wave investigations of Hugoniot equations of state of rocks and minerals under very high pressures have also provided key data necessary for the interpretation of the structure of the mantle (e.g., Al'tshuler et al., 1965; McQueen et al., 1967).

It is of interest to consider the series of phase transformations which are believed to occur with increasing depth in a mantle of pyrolite composition, based on the experimental data referred to above. These are summarized in Figure 1.9. At a depth of about 350 km,[5] pyroxene forms a complex solid solution with preexisting garnet. The solid solution involves the formation of a novel class of garnet end-members of the types $Mg_3^{VIII} (MgSi)^{VI}Si_3^{IV}O_{12}$ and $Ca_3^{VIII} (CaSi)^{VI}Si_3^{IV}O_{12}$, in which one-quarter of the silicon atoms are octahedrally coordinated. (The superscripts in the preceding formulae denote the coordination number of cations by oxygen anions.) The transformation involves an increase in the effective density of the pyroxene component of pyrolite of about 10 percent.

At a slightly greater depth, around 400 km, a major transformation occurs in olivine, the most abundant mineral in the upper mantle, and is probably responsible for the major seismic and density discontinuity occurring around this depth. The relevant phase relations are shown in Figure 1.10. Olivine transforms to the beta-Mg_2SiO_4 structure, which is about 8 percent denser than the olivine and is closely related to the spinel structure.

At greater depths, probably around 500–550 km, the calcium silicate $(CaSiO_3)$ component of the garnet transforms to the extremely dense perovskite structure (Ringwood and Major, 1971; Liu and Ringwood, 1975), while β-Mg_2SiO_4 transforms to the true spinel γ-Mg_2SiO_4 structure, accompanied by a density increase of about 2 percent (Ito et al., 1974).

Liu (1975) has shown that at a pressure of about 250 kbar Mg_2SiO_4 spinel disproportionates to an assemblage of $MgSiO_3$ possessing an orthorhombic perovskite structure plus MgO (rocksalt). This transformation is probably responsible for the major seismic discontinuity around 650 km. Another transformation in this region is of $MgSiO_3$–Al_2O_3 garnet, first to

---

[5] This depth is based on the results of Ringwood (1972). However, Akaogi and Akimoto (1977) concluded that the pyroxene–garnet transition may occur at a greater depth than the olivine to β-Mg_2SiO_4 transition. Further work is necessary to clarify the situation.

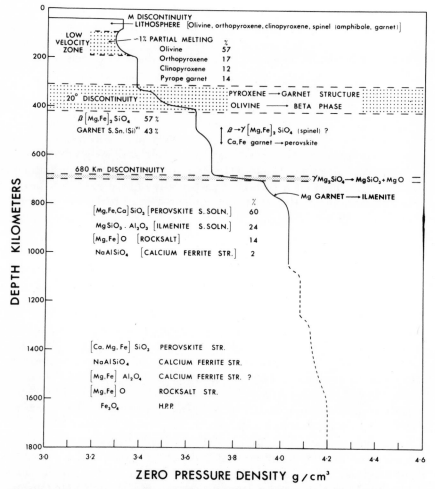

Figure 1.9 Mineral assemblages and corresponding zero-pressure densities for a model mantle of pyrolite composition based on the results of static high-pressure, high-temperature experimentation.

the ilmenite structure (Ito et al., 1972; Liu, 1976; Ringwood and Major, 1968), followed by transformation to the perovskite structure (Liu, 1974, 1975). Also in this region, sodium is expected to be present in a high-pressure modification of $NaAlSiO_4$, possessing the extremely dense calcium ferrite structure (Liu, 1977b; Reid et al., 1967).

Ringwood (1975a) has shown that the above series of phase transformations, discovered in the laboratory and occurring in material possessing the bulk chemical composition of pyrolite, provides a satisfactory explanation of the positions and magnitudes of the seismic discontinuities in the transition zone and also provides an adequate explanation of the density change through this region.

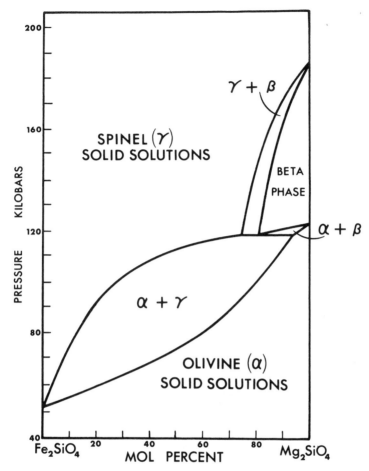

Figure 1.10 Phase relationships in the system Mg_2SiO_4–Fe_2SiO_4 at high pressures and at 1000°C. Based on Ringwood and Major (1966, 1970), Akimoto and Fujisawa (1968), Akimoto et al. (1976a), and Ito et al. (1974).

1.5 The Lower Mantle

Birch (1952) concluded that the variation of seismic velocities and density between depths of 900 and 2700 km could be explained by self compression of homogeneous material within the Earth's gravitational field, and that there was no evidence for the occurrence of *major* phase transformations or of substantial changes of chemical composition in this region. However, subsequent seismic studies (e.g., Johnson, 1969) have suggested the existence of minor seismic discontinuities, indicative of the existence of further small phase transitions (and/or chemical changes). It should be emphasized (Fig. 1.1) that these are very small when compared to the major velocity changes occurring in the transition zone.

Many previous discussions of the constitution of the lower mantle have been based on a comparison of the observed velocities and densities in this region with the corresponding properties of mixtures of oxides: MgO + FeO + SiO_2 (stishovite) as inferred from elasticity and shock-wave data (e.g., Anderson, 1967, 1968). This was justified, in part, by a widespread belief that the lower mantle indeed consisted of a physical mixture of oxide phases formed by the disproportionation of transition zone minerals under pressure (e.g., Birch, 1952), and also by the convenience of using simple oxide phases as a comparative yardstick.

Studies of this kind suggested that the density of the lower mantle was a few percent greater than would be expected for a mixture of oxides isochemical with pyrolite (Clark and Ringwood, 1964; Anderson, 1967, 1968, 1970; Ringwood, 1970a; Davies and Dziewonski, 1975). Although this discrepancy might appear rather minor, it has been the subject of a great deal of discussion and is important, both to the dynamics of the mantle and to cosmochemical considerations.

In principle, the density differential might be caused by an increased FeO/MgO ratio in the lower mantle, as strongly argued by Anderson (1967, 1968, 1970), or by additional phase transformations to minerals that are intrinsically denser (i.e., closer-packed) than the mixed-oxide assemblage, the bulk composition of the lower mantle being similar to that of the upper mantle (Clark and Ringwood, 1964; Ringwood, 1970a, 1975a). In the former case, a stable stratification of the mantle would be implied, preventing mass transport (e.g., via convection currents) between the upper and lower mantle. Moreover, Anderson's models required the overall iron abundance in the Earth (relative to Mg and Si) to be substantially higher than in Type 1 carbonaceous chondrites and in the sun, implying that large-scale fractionation of iron from silicates had occurred in the solar nebula prior to accretion of planets, as suggested by Urey (1952). On the other hand, if the second alternative were correct, mass transfer processes between the upper and lower mantle would be possible, whilst the total abundance of iron in the Earth (relative to lithophile elements) could be similar to that in Type 1 carbonaceous chondrites and in the sun.

One of the principal arguments used by Anderson (1967, 1968) to support an increased iron content in the lower mantle was an offset in plots of seismic velocities versus density within the transition zone (see, for example, Press, 1970). According to systematic empirical relationships between velocities, densities and mean atomic weights developed by Birch (1961), such plots had been believed to remain linear in chemically uniform material undergoing phase transformations. The observed offset was consistent with an increase in mean atomic weight caused by a higher iron content in the lower mantle. However, Liebermann and Ringwood (1973) and Liebermann (1974) showed that a similar offset occurred during those phase transformations in which the coordination number of cations by oxygen anions increased. Since the phase transformations in the lower part of the transition zone were of this category, an increase in iron content was not necessary in order to explain the offset.

In principle, the properties most useful in distinguishing between the alternative hypotheses are the seismic velocities V_P and V_S, and the elastic ratio K/ρ (K = bulk modulus, ρ = density) within the lower mantle. Phase transitions to denser states cause increases in velocities and elastic ratio, whereas increases in iron content cause decreases in these properties. Ringwood (1970a, 1975a) examined the data and tentatively inferred that the elastic ratio in the lower mantle was significantly higher than would be expected for a mixed oxide assemblage of pyrolite composition, implying that the mineral assemblage present in this region was intrinsically closer packed than mixed oxides. This inference was placed on a more concrete foundation by Davies and Dziewonski (1975) on the basis of a study of seismic velocities in the lower mantle. Anderson and Jordan (1970) and Anderson et al. (1971) responded by proposing a mixed oxide model in which the proportions of SiO_2 and FeO were considerably increased over the pyrolite composition, while MgO was correspondingly reduced. The larger amount of SiO_2 had the effect of increasing the overall elastic ratio and seismic velocities, while the increase of FeO raised the density.

Much of the earlier ambiguity from attempts to interpret the properties of the lower mantle in terms of mixed oxides arose from uncertainties in the elastic properties of stishovite and wüstite (Graham, 1973). These have since been determined more accurately by Liebermann et al. (1976), Olinger (1976), Sato (1977) and Jackson et al. (1977). Using these improved values, Davies and Dziewonski (1975) and Liebermann et al. (1976) showed that the properties of the lower mantle cannot readily be explained by oxide mixtures possessing stoichiometries ranging from those of olivine to pyroxene composition (Table 1.7). In particular, the properties of the $(Mg_{0.64}Fe_{0.36})$ O + SiO_2 mixture proposed by Anderson (1970) and Anderson et al. (1971) to represent the composition of the lower mantle are in serious disagreement with the corresponding properties of the lower mantle derived by Davies and Dziewonski (1975). The unsatisfactory nature of Anderson's composition is also demonstrated directly by shock-wave data on the densities of a pyroxenite [composition $(Mg_{0.86}Fe_{0.14})$ SiO_3] obtained by McQueen et al. (1967). This rock attains densities of 5.15–5.3 g/cm^3 at pressures of 950–1100 kbar, which are in close agreement with the densities actually obtained for the mantle in this pressure range (Dziewonski et al., 1975). The much more iron-rich pyroxenite composition $(Mg_{0.64}Fe_{0.36})$ SiO_3 advocated by Anderson would be about 0.3 g/cm^3 denser at the same pressures and well above the mantle density range.

Ringwood (1962, 1966a, 1970a) had suggested that the lower mantle is probably composed, not of mixed oxides, but of normal binary compounds like $MgSiO_3$ and Mg_2SiO_4, which occurred in crystal structures intrinsically denser than the isochemical mixed oxides. In particular, he suggested that $MgSiO_3$ might adopt the perovskite structure at very high pressures,

Table 1.7 Estimate of density (ρ_0), bulk modulus (K_0) and elastic ratio (K_0/ρ_0) of oxide mixtures and their components, compared to the corresponding values for the lower mantle (corrected to ambient conditions)

Material	ρ_0	K_0	K_0/ρ_0
MgO (periclase)	3.58	1.63	45.5
FeO (wüstite)[a]	5.91	1.82	30.8
SiO_2 (stishovite)[b]	4.29	2.8	65.3
$2(Mg_{0.88}Fe_{0.12})O + SiO_2$ olivine stoichiometry	4.04	2.0	49.5
$(Mg_{0.88}Fe_{0.12})O + SiO_2$ pyroxene stoichiometry	4.11	2.2	53.5
$(Mg_{0.64}Fe_{0.36})O + SiO_2$ pyroxene stoichiometry[c]	4.37	2.2	50.3
Lower mantle[d]	4.15	2.7	66

[a] Jackson et al. (1977).

[b] Liebermann et al. (1976); Sato (1977); Olinger (1976).

[c] Mantle composition according to D. L. Anderson (1970) and D. L. Anderson et al. (1971).

[d] Davies and Dziewonski (1975).

thereby achieving a density a few percent higher than that of isochemical periclase + stishovite. This hypothesis was supported by detailed studies of the crystal chemistry of perovskites, including the high-pressure synthesis of $ScAlO_3$ orthorhombic perovskite, which has a molar volume similar to that of the proposed $MgSiO_3$ perovskite (Reid and Ringwood, 1974, 1975).

Discovery of the predicted orthorhombic perovskite form of $MgSiO_3$ by Liu (1974), using diamond anvil techniques at a pressure of about 250 kbar, was a milestone. This was followed by his demonstration (Liu, 1975) that Mg_2SiO_4 spinel disproportionates at around 250 kbar directly into $MgSiO_3$ (perovskite) plus MgO (rocksalt) without an intervening episode of disproportionation into $MgO + SiO_2$ (stishovite), which had previously been believed to occur. The density of $MgSiO_3$ perovskite is 3.5 percent higher than the isochemical mixture of $MgO + SiO_2$ (stishovite). Moreover, other binary phases intrinsically denser than isochemical mixed oxides were synthesised. These included $Sc_2Si_2O_7$ pyrochlore (Reid and Ringwood, 1974), $NaAlSiO_4$ calcium ferrite (Liu, 1977b), and ultra dense modifications of $MgAl_2O_4$ and Fe_3O_4 formed under shock loading (McQueen and Marsh, 1966). Shock-wave data (McQueen and Marsh, 1966; Jackson et al., 1977) also show that pure Mg_2SiO_4 adopts a structure intrinsically denser than isochemical mixed oxides in the deep mantle.

In the light of this data, it now seems highly probable that the lower mantle consists of an assemblage of mineral phases a few percent denser than isochemical mixed oxides. By a depth of 1000 km, experimental data indicate that pyrolite would crystallize in the following mineral assemblage: $MgSiO_3$ perovskite, $CaSiO_3$ perovskite, $(Mg,Fe)O$ rocksalt, and $NaAlSiO_4$ (calcium ferrite).

Further transitions to slightly denser states appear quite likely to occur between 1000 and 2700 km, and may be responsible for the very small seismic discontinuities observed in the lower mantle by Johnson (1969) and others. The perovskite structure has a large number of polytypes and it is possible that various modifications of this basic structure may become stable below 1000 km.

The conclusion that the mineral assemblage present in the lower mantle is more closely packed than an isochemical mixture of component oxides effectively resolves the density discrepancy referred to earlier. There seems little doubt that material of pyrolite composition, occurring in a mineral assemblage a few percent denser than isochemical mixed oxides, is capable of explaining the density and seismic distributions observed throughout most of the lower mantle (Ringwood, 1970a, 1975a; C. Y. Wang, 1972; H. Wang and Simmons, 1972; Mao, 1974; Davies and Dziewonski, 1975).

We should note here that the uncertainties involved in extrapolating the distributions of density and elastic properties through the lower mantle to atmospheric pressure and low temperature are such that a small change in iron content in the lower mantle cannot be absolutely precluded. The position is that there are no valid grounds at present for positively maintaining that the FeO/MgO ratio of the lower mantle is different from that of the upper mantle. In the absence of such evidence, and in view of our conclusion (above) that the pyrolite composition is capable of explaining the properties of the lower mantle within their resolution limits, we are entitled to conclude that the *mean* composition of the lower mantle is probably similar to that of the upper mantle.

The seismic properties of the lowermost 200 km of the mantle are anomalous (Bullen, 1963; Cleary, 1974; Haddon and Cleary, 1974). There is evidence for a decrease in velocities and localized inhomogeneities which cause scattering of seismic waves. The structure of this region is further discussed in Section 7.3.

CHAPTER 2

Geochemistry of the Mantle

2.1 Chemical Homogeneity versus Heterogeneity

The degree to which the mantle can be considered chemically uniform depends partly on the scales at which phenomena are analyzed. Many key geophysical observations (e.g., seismic velocities) relate to scales of 10–100 km. A key result discussed in Chapter 1 is that at these scales, the overall pyrolite composition derived for the upper mantle is capable of explaining the essential features of the distribution of seismic velocities and density with depth throughout the transition zone and lower mantle to depths of about 2700 km, when the effects of known phase transformations are taken into account. In particular, there is no convincing evidence requiring the existence of any major radial chemical zoning in the mantle,[1] e.g., a substantial increase of the FeO/MgO ratio in the lower mantle. This suggests that the pyrolite upper mantle composition might be applicable throughout the entire mantle. At the very least, this would be a reasonable assumption in attempting to estimate the bulk composition of the mantle.

---

[1] Except for the depleted peridotite layer in the uppermost mantle (Fig. 1.5) and for the lowermost 200 km overlying the core–mantle boundary.

At smaller scales, the mantle displays various degrees of heterogeneity, arising in part from differentiation processes associated with the generation and subduction of lithospheric plates (Fig. 2.1). One example of this is differentiation of pyrolite beneath mid-oceanic ridges to form lithospheric plates consisting of a mafic crust overlying a layer of refractory peridotite. Subsequently, the plates are subducted into the mantle beneath oceanic trenches where a second stage of partial melting of the oceanic crust is believed to occur, leading, via a complex series of petrochemical processes, to the formation of calc-alkaline (andesitic) magmas. These magmas form rocks that are ultimately incorporated into continents which grow with time. The mantle is thus being subjected to a process of irreversible chemical and petrological differentiation, one product of which is the continental crust, while the other is the refractory residium of mafic crust (later transformed to eclogite and other high-pressure derivatives) and peridotite, which sinks into the mantle. The petrological and geochemical characteristics of this refractory peridotite and eclogite are such that they are unable to provide basalt of observed compositions by a further stage of partial melting (Ringwood, 1975a). An important effect of the deep subsidence of these refractory plates, shown in Figure 2.1, is the displacement of fertile pyrolite into the low-velocity zone, where it provides a continuing source of further basaltic volcanism by partial melting beneath mid-oceanic ridges.

Palaeomagnetic observations imply that large-scale plate motions have occurred more or less continuously since the early Precambrian (McElhinny, 1973). This suggests that a considerable proportion of the mantle—perhaps 30–50 percent of its volume—may have been irreversibly differentiated in this manner (see, for example, Dickinson and Luth, 1971; Ringwood, 1975a). Thus, the pyrolite now present in the low-velocity zone, which is believed to represent the source of modern oceanic basalts, may ultimately have been derived from considerable depths in the mantle. The remarkable uniformity in (major element) compositions of the least differentiated oceanic tholeiites all over the world and throughout geological time thus takes on added significance, suggesting a corresponding uniformity in source regions extending to considerable depths in the mantle.

It will be important in the future to establish whether subducted plates sink below the limit of the deep-focus earthquake zones, around depths of 700 km, into the lower mantle. Studies of the thermal and seismic structures of sinking plates (Oxburgh and Turcotte, 1970; Fitch, 1975, 1977) imply that the plates are still substantially cooler and therefore denser than the surrounding mantle to depths exceeding 650 km. Moreover, the phase transformations occurring in this region do not constitute a barrier to convection (Ringwood, 1975a). The absence of deep-focus earthquake foci below 700 km cannot be construed as evidence that plates do not sink deeper than this level. Recent seismic studies have indicated the presence of velocity anomalies in the lower mantle beneath subduction zones, which have been interpreted to imply that plates have indeed sunk into the deep mantle (Jordan, 1975, 1977). If it can finally be demonstrated that convective

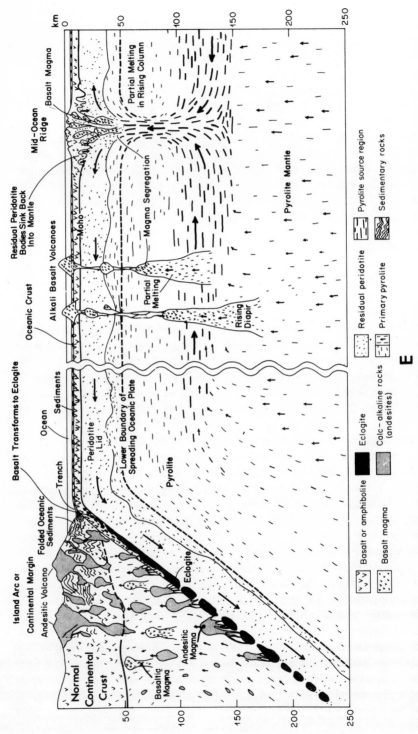

Figure 2.1 Petrologic model of plate generation and consumption. (From Ringwood, 1969, with permission of the American Geophysical Union.)

interchange of material occurs on a mantle-wide scale, then our previous inferences concerning the relevance of the pyrolite composition to that of the bulk composition of the entire mantle will have been further strengthened.

Other Geochemical Heterogeneities

The irreversible differentiation of pyrolite into basalt (eclogite) + peridotite (Fig. 2.1) provides one important source of chemical heterogeneity in the mantle. Because of the small dimensions of eclogite blocks (probably 5–10 km) and the similarity of elastic properties between peridotite and pyrolite, it is difficult to detect and analyze the extent of this heterogeneity by geophysical methods. We note, however, that this is believed to be an evolutionary process, and that early in the Earth's history the degree of heterogeneity caused by this process of irreversible differentiation was relatively small.

The mantle has also been subjected to very complex chemical differentiation processes involving the incompatible elements (Gast, 1968). Typical examples are the processes which have caused depletions and enrichments of the light rare earths, and of fractionations of Rb from Sr and Pb from U in the source regions of basalts. A great deal of attention is currently being given to these processes in connection with the hypothesis of mantle plumes (see, for example, Schilling, 1973). They clearly testify to very complex heterogeneities in the mantle relating to the distribution of these elements and the occurrence of complex fractionation processes that are poorly understood. Nevertheless, these affect elements which, in the main, are present only in small concentrations, and appear to be decoupled from the major element chemistry. The results of Sun and Nesbitt (1977) imply that the degree of such minor element heterogeneity was smaller in the Archaean. Apparently, the fractionation processes have been evolutionary in nature.

Other classes of regional chemical heterogeneities have been recognized and doubtless many remain to be discovered. One example is provided by the high CaO/Al_2O_3 ratio of the komatiites from Barberton Land (Viljoen and Viljoen, 1969a,b), as compared to the more normal values (approximately chondritic) for those from Western Australia (Nesbitt and Sun, 1976) and Canada (Arndt et al., 1977). Moreover, the olivines from Archaean komatiites in general seem to be richer in nickel than most of the olivines of Phanerozoic ultramafic rocks.

It is dangerous to make general statements about the degree of chemical heterogeneity or homogeneity of the mantle without numerous qualifications, and the topic has been explored above all too briefly. Nevertheless, the author does not believe that the observed and inferred degrees of chemical heterogeneity in the mantle are likely to be of such a magnitude as to seriously invalidate estimates of the *bulk* mantle composition based on the inference that this approximates the mean composition of the upper mantle pyrolite (Table 1.6).

2.2 Abundance Patterns of Compatible[2] Elements and Their Significance

In contrast to the wide dispersions of abundances of incompatible elements in upper mantle rocks, the dispersions of abundances of many compatible elements appear, on the basis of currently available data, to be remarkably low. For example, in fresh ultramafic rocks of mantle origin, the vast majority of Ni, Co, and Mn abundances are within $\pm$ 50 percent of the mean values (Goles, 1967; Steuber and Goles, 1967). Although the data are more limited, a similar pattern seems to apply to Ga and Ge in ocean floor basalts and, presumably, also in their source regions (Argollo, 1974; Wedepohl, 1974; Frey, et al., 1974). The dispersion of zinc abundances in ultramafic rocks and in basalts also appears to be low (Rader et al., 1963; Baedecker et al., 1971; Wedepohl, 1974; Gurney and Ahrens, 1973; Archbald, 1978).

One of the most striking characteristics of mantle-derived alpine ultra-mafics and lherzolite xenoliths in alkali basalts is their uniformity of $MgO/(FeO + MgO)$ ratios, which lie mostly within the range 0.89–0.93 (molecular). These displays of uniformity on a worldwide basis imply that a corresponding degree of uniformity extends to large proportions of the upper mantle, on both a regional and local scale. Although abundances of incompatible elements are highly variable, the abundance ratios of certain of these (e.g., Pb and U) are relatively constant, as shown by the small range of μ values ($^{238}U/^{204}Pb$) characteristic of basalts and upper mantle–crust system (Tatsumoto, 1978; Cumming and Richards, 1975).

The present homogeneity of distribution of compatible elements may result from (a) solid-state diffusion, (b) diffusion in the liquid state during partial melting and magma generation, (c) convective mixing, or (d) it may be an original characteristic of the mantle, established during accretion. Hofman and Magaritz (1977) have evaluated the first two factors, based on their experimental measurements of diffusion coefficients in silicate systems. They show that the first process is capable of causing homogeneity only over scales of about 10 cm since the Earth's formation. On the other hand, liquid-state diffusion is much more rapid, and large volumes of the mantle can be presumed to have been present in an incipiently melted state in the low-velocity zone (approx. 70–150 km) for periods on the order of 10^9 years. Hofman and Magaritz (1977) have demonstrated that in this situation, where homogenization occurs via diffusion through an interstitial liquid which wets the grain boundaries, volumes with dimensions on the order of 1 km would be homogenized over a timescale of 10^9 years. It might also be expected that local homogenization could occur during the uprise of a diapir, leading to larger degrees of partial melting and the separation of magmas. However, in this case also, on the basis of Hofman and Magaritz's

---

[2] Compatible elements are defined as those possessing ionic radii and charges which permit them to substitute readily in the principal minerals of the mantle.

results, it seems difficult to explain homogenization occurring over scales exceeding about 1 km. Most probably, the effect would be much smaller.

The possible role of convection in causing homogeneity was also considered by Hofman and Magaritz. They point out that the scales of convective mixing and diffusion–homogenization are totally different. It would be extremely difficult for solid-state convection processes to cause efficient homogenization and mixing throughout the mantle at scales of 10–100 km. Moreover, if the mantle were initially highly chemically inhomogeneous (major elements) at scales exceeding 100 km, mantle dynamics would be dominated by density differences associated with these compositional differences. Regions of compositionally "light" material would rise vertically as diapirs, while regions of compositionally "heavy" material would sink. The end result would be a stable chemical stratification within the Earth's gravitational field, representing a larger-scale degree of chemical heterogeneity than was originally present.

We conclude, therefore, that the homogeneity of distribution of compatible elements at the kilometer scale was probably an original characteristic of the Earth's mantle, presumably established during accretion. This conclusion has important cosmochemical implications. The compatible elements possess vastly different behavior under the cosmochemical conditions that might have prevailed during the formation of the Earth. For example, they vary greatly in volatility (Mg, Fe, Mn, Zn, Ge, Ga), siderophile nature (Mg, Fe, Ni, Co, Si, Ge), and as indicators of the redox state (FeO/MgO). The fact that many of these elements seem to be relatively well mixed at the kilometer scale in large regions of the Earth's mantle, places some important constraints on the processes which governed the accretion of the Earth.

Limitation on Melting of Mantle

Our previous inferences concerning the applicability of the upper mantle pyrolite composition to the entire mantle and the low dispersions of some compatible minor elements within mantle rocks have an obvious bearing on the questions as to whether the mantle was totally or extensively melted during the formation of the Earth.

When large bodies of magma (e.g., the Bushveldt and Stillwater complexes) cool within the Earth's crust, they display strong fractionations caused by crystallization differentiation. In general, the efficiency of this process is expected to increase with the dimensions of the body of magma. If the mantle had been totally or extensively melted during formation of the Earth, the inevitable operation of crystallization differentiation should have resulted in tremendous petrological and chemical diversity on a very large scale. This is not observed. The evidence previously discussed, implying a large degree of chemical homogeneity throughout the mantle, more especially at an early stage of its evolution before the smaller-scale differentiation processes connected with plate tectonic and dynamical processes had become prevalent, provides strong evidence that the mantle did not experience a large degree of melting during the formation of the Earth.

This conclusion is further supported by studies of the abundances of many moderately incompatible elements in pyrolite, as discussed in Section 2.3. Several of these elements, including Ti, Y, Zr, Nb, Sc, and the heavy REE are present in pyrolite approximately in their primordial ratios, when compared to major elements such as Mg, Si, Ca, and Al. If the mantle had ever been totally melted, intense fractionation of these elements would have occurred. At present, it is only the highly incompatible elements, e.g., La, Ce, U, Th, Rb, and K that display marked fractionations in the source regions of basaltic magmas. As discussed subsequently, this behavior is to be attributed to numerous episodes involving small degrees of partial melting within the mantle, which have occurred over geological time, rather than to an episode of complete melting and differentiation.

The conclusion reached above, that the mantle was not totally or extensively melted during the formation of the Earth, is far from new. It was clearly stated by Urey (1952, 1954, 1962a), Vinogradov (1959, 1961, 1967), Rubey (1951, 1955), and others. One of the arguments cited by these workers was the small size of the Earth's crust, compared with expectations for differentiation of a completely melted mantle. A vast amount of geochemical and geophysical data on the mantle has accumulated since Urey, Vinogradov, and Rubey reached their conclusions. As reviewed in the present chapter, these data point firmly in the same direction.

2.3 Major Elements and Involatile Lithophile Elements in Pyrolite as Compared to Their Primordial Abundances

Cosmochemical Background

It is now generally agreed that except for highly volatile elements, such as H, He, C, N, and O, and the inert gases, abundances in Type 1 carbonaceous (Cl) chondrites closely approach the primordial abundances of the parental solar nebula in which the Earth was formed. This conclusion is based on several lines of evidence: nucleosynthesis theory and systematics; studies of the chemical and genetic relationships between the different classes of chondritic meteorites; abundances of elements in cosmic rays; and, finally, comparison with solar abundances. The abundances of elements in Type 1 carbonaceous chondrites are plotted against the solar photosphere abundances (J. E. Ross and Aller, 1976) in Figure 2.2. Agreement in most cases is remarkably good and within experimental error. The discrepancies in boron and lithium are caused by thermonuclear reactions in the sun. The elemental abundances in Type 1 carbonaceous chondrites thus provide an

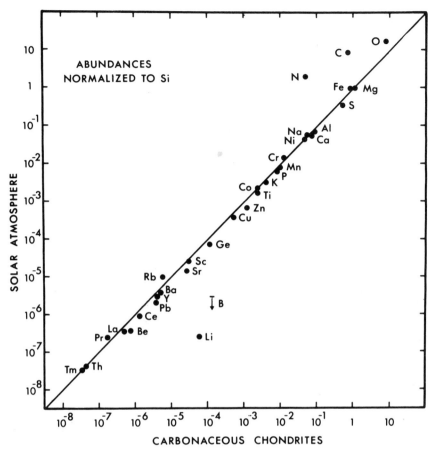

Figure 2.2 Comparison of elemental abundances in Type 1 carbonaceous chondrites with those in the solar photosphere. All abundances are normalized on the basis of Si = 10^6. (From J. E. Ross and Aller, 1976.)

appropriate baseline from which to explore and interpret chemical fractionation processes that have affected planets and other classes of meteorites.

As discussed further in Part II, virtually all current theories of formation of the planets hold or imply that processes involving differential volatility or condensation played an important role in establishing the present compositions of planets. Likewise, similar processes were largely responsible for the chemical fractionations displayed by most classes of meteorites (see, for example, Anders, 1968). Therefore, we will proceed in later parts of this section to compare the bulk compositions of pyrolite with the primordial C1 abundances, in order to elucidate the roles of volatilization and other fractionation processes during the formation of the Earth.

The condensation temperatures of many elements and compounds from a gas phase of solar composition at 10^{-4} atmospheres are given in Table 6.2. This provides a useful indicator of the relative volatilities of many elements in relevant cosmochemical environments. As seen from Table 6.2, a particular group of elements condenses at higher temperatures than the major components, iron and Mg_2SiO_4. This group of "high-temperature condensates" includes the oxides and silicates of Ca, Al, Sc, the rare earths, Y. Ba, Sr, Ti, Zr, Hf, U, Th, Ta, and Nb. Larimer and Anders (1970) demonstrated that the high-temperature condensates have been substantially fractionated as a group among the principal classes of chondritic meteorites. However, elemental fractionations within this group have been small, except in isolated cases. Accordingly, it is of interest to see whether this highly involatile group of elements have been fractionated with respect to the more volatile major elements Fe, Mg, and Si in the Earth, as compared to Cl chondrites.

Comparison with Mantle

A model mantle composition has been derived from the primordial C1 composition by removing siderophile elements (assigning them to the notional "core"), but leaving sufficient FeO in the composition to yield an MgO/MgO + FeO) ratio of 0.88, which corresponds to the pyrolite ratio. This "primordial mantle" composition is compared with pyrolite in Table 2.1.

Table 2.1 Primordial model mantle composition derived from Type 1 carbonaceous chondrites[a] compared with pyrolite composition

	Primordial model mantle wt.-%	Pyrolite wt.-%
SiO_2	48.2	45.1
TiO_2	0.15	0.2
Al_2O_3	3.5	3.3
Cr_2O_3	0.7	0.4
MgO	34.0	38.1
FeO	8.1	8.0
MnO	0.5	0.15
CaO	3.3	3.1
Na_2O	1.6	0.4
K_2O	0.15	0.03
$\dfrac{CaO}{(MgO + SiO_2)}$	0.040	0.037
$\dfrac{Al_2O_3}{(MgO + SiO_2)}$	0.043	0.040

[a] Based on average of chemical analyses of Orgueil and Ivuna (Wiik, 1956) on a volatile and sulphur-free basis. Siderophile components (NiO and most of the FeO) have been removed and assigned to the model "core" following the procedure of Ringwood (1966a), leaving sufficient FeO in mantle to produce an MgO/(MgO + FeO) ratio of 0.88.

The overall agreement in abundances of most major elements is very close. In particular, the abundances of the key components of the high-temperature group (CaO and Al_2O_3) relative to the major components (MgO + SiO_2) is very close, implying that the high-temperature condensates, as a group, were not substantially fractionated from magnesium silicates during the formation of the Earth.

The principal differences between pyrolite and the primordial model mantle composition are that silica is modestly depleted in pyrolite (by 17 percent relative to MgO), while Cr_2O_3 (43 percent), MnO (70 percent), Na_2O (75 percent), and K_2O (80 percent) are depleted by increasingly large factors. As seen in Table 6.2 this is also the order of increasing volatility of these elements. We will return to this topic in the next section, after considering the abundances of several other involatile lithophile elements of the "high-temperature condensate" group in pyrolite.

Studies of the abundances of heavy rare earths in peridotites of mantle origin indicated that the parental mantle contains about twice the *ordinary chondritic*[3] bulk abundances of the heavy rare earths on a weight-for-weight basis (Frey and Green, 1974; Loubet et al., 1975). This implies that the abundances of heavy rare earths relative to (MgO + SiO_2) in the mantle are similar to the corresponding relative abundances in C1 chondrites. Moreover, the *relative* abundances of the heavy and intermediate rare earths in oceanic tholeiites (see, for example, Kay et al., 1970) show them to be present in the same proportion as in chondrites.

Sun and Nesbitt (1977) have recently obtained an estimate of the abundances of several elements of the high-temperature condensate group from a study of the compositions of a suite of komatiite magmas from Australia, Canada, and South Africa. They demonstrated (see also Nesbitt and Sun, 1976) that Al, Ca, Ti, Y, heavy REE, Zr, Nb, and Sc were present in the source region at levels amounting to twice their (wt.-%) abundances in ordinary chondrites and that this factor probably applied also to the light REE, Ba, and Sr.

---

[3] In many trace element studies, particularly those dealing with the rare earths, the abundances in terrestrial materials are compared with corresponding bulk abundances in *ordinary chondrites* on a weight-for-weight basis. This convention was adopted initially because analytical data for many elements in ordinary chondrites were of higher precision and more abundant than in Type 1 carbonaceous chondrites.

At first sight, there might appear to be a contradiction between the above statement that the abundances of heavy rare earths (relative to MgO + SiO_2) in the Earth's mantle are similar to those in Type 1 carbonaceous chondrites and the preceding statement that the absolute abundances of these elements are present at about twice the *ordinary chondritic* levels. However, this is not the case. When we derive the model mantle composition from the primordial C1 composition (Table 2.1) by removing iron into the "core" and volatilizing 17 percent of the total SiO_2 (Section 2.4), the abundances of all lithophile elements are increased by a factor of 1.6 in the model mantle composition. Moreover, on a weight-for-weight basis, C1 chondrites are intrinsically enriched in these "refractory" lithophile elements (Ca, Al, Ti, Zr, Sc, REE, U, Th, etc.) by an average factor of 1.4 compared to ordinary chondrites (Larimer and Anders, 1970; Larimer, 1971). Thus a model mantle composition derived from the C1 primordial composition would contain about 2.2 times the ordinary chondritic abundances of these elements, on a weight-for-weight basis, in agreement with the estimates for the Earth's mantle cited above.

The above studies, in conjunction with previous discussions, provide strong grounds for concluding that this entire group of high-temperature condensates possess abundances in the Earth's mantle (relative to the major components, $MgO + SiO_2$) that are generally similar to their corresponding relative abundances in Type 1 carbonaceous chondrites (Ringwood, 1975a; Sun and Nesbitt, 1977).

Some workers have attempted to estimate the abundances of U and Th in the mantle on the basis of a thermal equilibrium model (Birch, 1958), which assumes that the total heat lost per unit time over the Earth's surface is equal to the total radiogenic heat generated within the Earth by U, Th and ^{40}K. Using an estimate of 1.95 μcal cm^{-2}sec^{-1} for the mean global heat flow (Williams and Von Herzen, 1974), this would imply a mean uranium content of 0.042 ppm within the mantle, which amounts to about 3.2 times the ordinary chondritic abundance. A similar enrichment is found for thorium. Since U and Th are members of the high-temperature condensate class, it is apparent that a significant discrepancy exists between the abundances of high-temperature condensates in the mantle obtained from heat flow considerations on the one hand, and the geochemical–petrological arguments discussed previously.

In the author's view, estimates of U and Th abundances in the mantle obtained from geothermal arguments may possess substantial uncertainties. Firstly, it is doubtful whether the Earth is in thermal equilibrium. The episodic nature of large-scale plate motions, manifested for example by the break-up of Gondwanaland in the Mesozoic, may have been accompanied by a corresponding episodic transfer of previously accumulated radiogenic heat to the surface by convection, and this heat may still be contributing to the present surface heat flow. It is also possible that the time constant for heat loss from the Earth is sufficiently long so that a significant portion of the present heat flux is derived from "original heat" caused by the exothermic core formation process (Chaper 3).

A second source of uncertainty concerns the chondritic abundances of uranium. The data of Morgan (1971) show that this varies over a three-fold range in ordinary chondrites (observed falls) and over a two-fold range among Type 1 carbonaceous chondrites (observed falls). In view of these uncertainties, it is preferable to estimate the abundances of involatile lithophile elements in the mantle from the geochemical and petrological evidence discussed earlier.

2.4 Abundance Patterns of Volatile and Siderophile Elements

Numerous studies of the different classes of meteorites (see, for example, Anders, 1968, 1971) have demonstrated that their compositions have been strongly influenced by processes involving selective volatility or condensation. Therefore, it is appropriate to explore the influence of these processes on the composition of the mantle.

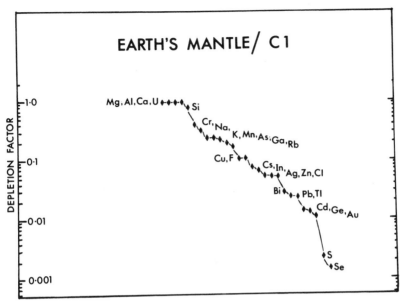

Figure 2.3 Abundances of elements in the Earth's mantle compared to primordial abundances as given by C1 chondrites. (From Ringwood and Kesson, 1977, with permission.)

Ringwood and Kesson (1977) have compared the abundances of many elements in pyrolite with their abundances in Type 1 carbonaceous chondrites, normalized to magnesium in both these cases (Fig. 2.3). As noted earlier, representative lithophile elements that are less volatile than silicon are present in primordial proportions. However, a wide range of other elements that are more volatile than magnesium (Fig. 2.3) are seen to be depleted by variable factors. The least volatile element to be significantly depleted in the Earth (by 17 percent) is silicon. If we consider only the non-siderophile elements, the degrees of depletion generally increase with increasing volatility as Si, Cr, Mn, Na, K, As, Ga, Rb, and F become progressively more depleted, until a plateau is reached consisting of Cs, In, Zn, and Cl, possessing depletion factors in the range 0.05–0.08. There can thus be little doubt that the terrestrial depletions are primarily a consequence of selective condensation/volatilization processes. In some cases, e.g., Cu, Ag, Bi, Pb, Tl, Cd, Ge, Au, S, and Se, the observed depletions in the mantle have been enhanced by their siderophile tendencies, which have caused variable degrees of entry into the Earth's core (Ringwood and Kesson, 1977). However, the observed depletions of these elements are much too high to have been caused solely by siderophile behavior. Elements that are much less volatile and much more strongly siderophile (e.g., Ni and Co) have not been depleted in the mantle to the same extent as Bi, Pb, Cd, Tl, and Ge.

Although the depletion of silicon in pyrolite is relatively small (~ 17 percent) when compared to the primordial abundances as given by Cl chondrites, it seems to be well established. Ringwood (1958, 1959) suggested that this depletion might be caused by entry into the Earth's core. However, the geochemical conditions required to reduce silica to silicon, thereby permitting it to enter the metal phase during the formation of the Earth, are rather extreme, and silicon is not now considered to be an important component of the core (Ringwood, 1977b). As seen in Figure 6.8, silicon is substantially more volatile than magnesium in a system of solar composition. In view of the quite strong depletions of Cr, Mn, and Na in the mantle, which may reasonably be attributed to volatility (Table 6.2), it seems quite likely that the small depletion of silicon is also caused by this factor.

Siderophile Elements

Siderophile elements are those that become preferentially partitioned in the metal phase when metallic iron (or nickel–iron) and silicate systems are equilibrated. Such elements are spoken of as being more "noble" than iron. Some, including Ni, Co, Ir, Pt, and Re, are less volatile than silicon under appropriate cosmochemical conditions (Table 6.2), either as metals or oxides, so that they are likely to be present in the Earth in their primordial abundances. Others, such as Cu, Ag, Ge, and As, are more volatile than silicon, so that they are likely to be depleted in the Earth compared to the primordial abundances.

In Table 2.2, column 1, we have compared the abundances of some siderophile elements in pyrolite with their estimated abundances in the core. The former data are obtained from geochemical observations (Ringwood and Kesson, 1977). The latter are based on the assumption that the abundances of these siderophile elements in the bulk Earth correspond to the Cl abundances, combined with knowledge of the mass ratios of core to mantle. It is seen that the siderophile elements are depleted in the upper mantle (pyrolite), and it may confidently be assumed that they have become dominantly segregated into the core. The partition coefficients for many of these elements between iron and appropriate silicate phases have been measured or estimated (mostly from meteoritic data) and are also given in Table 2.2 (column 2). The remarkable feature (Ringwood, 1966a,b) is that the elements for which partition coefficients are available are more abundant in the upper mantle (by factors varying from 10 to over 100) than would be expected if these elements had been partitioned into the core under (low-pressure) equilibrium conditions. From what is known of the general thermodynamic properties of Ir, Os, Pt, Pd, and Ag compounds, the same conclusion probably applies to them. The discrepancies in the cases of Cu, Au, Pd, Ag, Ge, and As are probably greater than indicated in Table 2.2, since these elements are moderately volatile and may well have been intrinsically depleted in the bulk Earth due to this property by factors of 3–10 (Fig. 2.3), so that ratios in column 1 should be multiplied by factors in this range.

Table 2.2 Distribution of some siderophile elements between mantle and core, as compared to experimental Fe metal–silicate coefficients.

Element	Observed $K\left(\dfrac{\text{Mantle}}{\text{Core}}\right)$	Experimental $K\left(\dfrac{\text{Silicate}}{\text{Fe metal}}\right)$
Involatile		
Co	0.1	0.005
Ni	0.08	0.0006
Re	0.001	5×10^{-4}–10^{-6}
Os	0.004	
Ir	0.002	Probably similar to Au and Re
Pt	0.01	
Volatile		
Au	0.01	$< 3 \times 10^{-5}$
Cu	0.14	0.02–0.003
Ge	0.013	0.001
As	0.4	< 0.01

Likewise, sulphur and selenium are moderately volatile so that appropriate correction factors should also be introduced.

It is clear that these relationships must have a vital bearing on the conditions under which core formation occurred in the Earth. Ringwood (1959) and Brett (1971) suggested that a large increase in the distribution coefficient of nickel between silicates and metal caused by high pressures was responsible for the entry of nickel into the silicates of the deep mantle. This increase was believed to be caused by the $P.\Delta v$ term in the free energy of the metal–silicate exchange reaction. However, Ringwood (1966a, 1971) abandoned this explanation when it became clear that several other siderophile elements were similarly overabundant in the upper mantle, since it would have been highly coincidental for the $P.\Delta v$ terms to have the same sign for all these equilibria. Accordingly, he proposed a model of core formation under disequilibrium conditions, in which material of primordial composition incorporated within the Earth at an early stage of accretion, was rapidly mixed into the mantle under conditions whereby equilibration with metallic iron did not occur. An analogous model, in which the primordial component was added to the Earth at a late stage of accretion after the core had segregated, and thereby became incorporated into the upper mantle only, was advocated by Turekian and Clark (1969).

Another significant feature of the geochemistry of the siderophile elements concerns their dispersions in mantle rocks. Compatible siderophile elements, such as Ni, Co, Ge, Ga, and perhaps Cu, display relatively small dispersions (previously discussed), while the more noble elements Au, Ir,

Re, Os, and Pt seem to display much larger dispersions (see, for example, Wedepohl, 1974; Crocket and coworkers, 1971, 1972, 1977). It is not clear whether these dispersions are caused by the incompatible geochemical behavior of these latter elements during crystal/liquid fractionation processes that have occurred in the mantle throughout geological time, as seems to be the case for the dispersions of La, Cs, Rb, K, U, and Pb, or whether these dispersions reflect initial inhomogeneous distributions within the Earth established soon after accretion. Moreover, hydrothermal and metamorphic processes may have caused a substantial degree of redistribution of these elements in mafic and ultramafic rocks. Although data are sparse, the author suspects that these factors may be insufficient to explain the dispersions of Au in basalts and Ir in ultramafics, and that primary distributional inhomogeneities may be indicated. Corresponding large variations of distributions of primordial inert gases in the mantle are also indicated by preliminary data (Graig and Lupton, 1976).

The oxidation state of iron in the mantle raises problems related to those discussed in connection with the "overabundances" of siderophile elements in the mantle. In order to explain the Fe^{3+}/Fe^{2+} ratios of fresh oceanic basaltic glasses and of unaltered primary mantle minerals (from xenoliths and peridotites), an Fe^{3+}/Fe^{2+} ratio of 0.05–0.1 is required for primary pyrolite. On the other hand, the Fe^{3+}/Fe^{2+} ratios of basalts, pyroxenes, and spinels that have equilibrated with metallic iron at high temperatures are much lower, probably by an order of magnitude (Ringwood, 1971).

The difference in oxidation states is reflected in the nature of the volatiles developed by degassing of the mantle. It is well known that the volatiles degassed from the Earth are dominantly composed of H_2O and CO_2, rather than H_2 and CO. Thus, the gas phase in equilibrium with an average Hawaiian tholeiite at 1200°C would have $H_2O/H_2 \sim 120$ and $CO_2/CO \sim 35$ (Holland, 1963). These ratios agree closely with Rubey's (1951) estimates for the composition of volatiles degassed from the Earth. The observed composition of the gas phase is thus compatible with the oxidation state of the pyrolite mantle, which indeed constitutes the ultimate redox state buffer. On the other hand, if the mantle had equilibrated with metallic iron, the gas phase would be dominated by CO and H_2, rather than CO_2 and H_2O (at 1200°C, $CO_2/CO \sim 0.05$, $H_2O/H_2 \sim 0.1$).

CHAPTER 3

Composition and Formation of the Core

3.1 The Role of Metallic Iron

At a depth of 2900 km, a major seismic discontinuity occurs, marking the boundary of the Earth's core (Fig. 1.1). Below this depth, S waves are no longer transmitted, implying a liquid state, while density increases by about 80 percent. The traditional interpretation, based partly on these physical properties and partly on analogy with meteorites, has been that the outer core consists of a liquid iron–nickel solution, while the inner core (which transmits S waves) consists of an iron–nickel alloy solidified under the very high pressures (3.3 Mbars) attained at a depth of 5154 km (Fig. 1.2). However, an alternative hypothesis advanced by Lodochnikov (1939) and Ramsey (1948, 1949) proposed that the core was composed of silicates that had been transformed to the metallic state by high pressures. Thus, the core–mantle boundary was interpreted to represent an isochemical phase transition.

This hypothesis has been widely discussed (see, for example, Bullen, 1963; Levin, 1970) and has been supported by some workers, largely on the grounds that it may permit a simpler interpretation of density differences between planets than the iron core hypothesis. However, Birch (1963) has used shock-wave data on the equations of state of many metals and silicates to pressures exceeding a megabar to demonstrate that the phase change hypothesis of the core is extremely improbable. The key data are shown in Figure 3.1, in which hydrodynamical sound velocities of many metals and other materials are plotted against density.

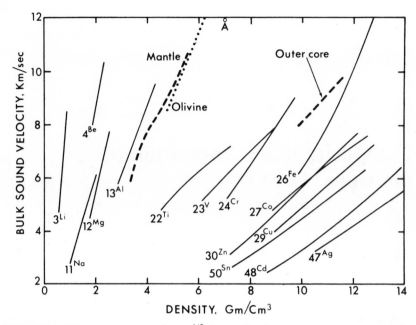

Figure 3.1 Bulk sound velocity, $(\partial P/\partial \rho)_s^{1/2}$, versus density along Hugoniot compression curves for metals and for a pure olivine rock [Twin Sisters dunite, $MgO/(MgO + FeO)$ = 0.90]. Figures attached to curves denote atomic numbers. The corresponding curves for the Earth's mantle and outer core are also indicated. (After Birch, 1963.)

Figure 3.1 shows that the relationships are highly systematic. For any given material, velocities and density are approximately linear over a very broad pressure range; moreover, the locations of the velocity/density lines are strongly determined by (mean) atomic weight. As expected, the observed velocity/density relationship for the mantle falls in the light element region. However, for the present purposes it is important to note that the corresponding relationship for the core falls in the region between V, Cr, Fe, and Co, providing convincing evidence that the mean atomic weight of the core is similar to those of these elements. Cosmochemical and nucleosynthesis considerations show that iron is the only element in this region of the periodic table which is sufficiently abundant to permit it to be a major component of the core. As emphasized by Birch (1963, 1965), it seems inescapable, on the basis of the relationships shown in Figure 3.1, that iron is the principal component of the core.

This conclusion is complemented by additional shock-wave data. Silicates such as Mg_2SiO_4 and SiO_2 have been shocked to pressures up to 5 Mbar without transforming to the hypothetical dense metallic state (Al'tschuler and Kormer, 1961; Al'tschuler et al., 1965; Trunin et al., 1965). Levin (1970) and others have suggested that these negative results may not be significant because large overpressures are generally needed to achieve phase transformations during the microsecond intervals for which materials are subjected to peak shock pressures. This objection, however, is hardly

convincing. The shock-wave overpressures needed to achieve phase transformations in most silicates and oxides exceed the equilibrium pressures
by only 100–300 kbar (McQueen et al., 1967; Ahrens et al., 1969), whereas
the hypothetical transition to a dense metallic state is not achieved at a
pressure over 3 Mbar higher than that at the core–mantle boundary. Because
of the rapid increase of temperature along the Hugoniot, the overpressure
needed to produce phase transformations should decrease with increase
of total shock pressure.[1]

A third line of evidence relating to this issue is provided by empirical studies
of elastic properties of materials during compression and during phase
transformations. It is well established, on both theoretical and experimental
grounds, that the bulk modulus of materials increases sharply as volume
decreases, as a result of the repulsion terms in lattice energy expressions.
During the homogeneous compression of most oxides and silicates, the
relationship between bulk modulus (K) and density (ρ) is approximately
$K \propto \rho^n$, where $3 < n < 4$ (O. L. Anderson, 1966). Where the increase of
density is caused by phase transformations, the exponent n may be as small as
2 (Liebermann and Ringwood, 1973; Liebermann, 1974). However, the
general empirical and theoretical result[2] is that phase transformations
accompanied by density changes greater than about 5 percent are expected
to result in correspondingly large increases in bulk modulus, K.

Bullen (1940) demonstrated that the bulk moduli of the lower mantle and
core were almost identical, despite the much higher density of the core. If the
core–mantle boundary was caused by an isochemical phase transformation,
an increase in bulk modulus by at least a factor of three would be expected.
It does not appear possible to explain the similarity in bulk moduli of both
regions, combined with large density differences, on the basis of an isochemical phase transition.

3.2 Light Elements in the Outer Core

Shock-wave data (Al'tschuler et al., 1958, 1962, 1965, 1968; McQueen and
Marsh, 1960, 1966) demonstrate that the density of the outer core is substantially smaller than that of iron–nickel under similar pressure and temperature (P, T) conditions.[3] According to Figure 3.2, the discrepancy is

---

[1] Vereshchagin et al. (1975) and Kawai and Nishiyama (1974a,b) claim to have transformed
MgO and SiO_2 to phases exhibiting metallic electrical conduction under very high static pressures. The densities and elastic properties of these phases are not known. Interpretation of the
experimental results is not free from ambiguity because of the possibility of electrical insulation
loss during the experiments.

[2] A possible exception to this generalization may occur where the low-pressure phase has an
exceptionally high intrinsic bulk modulus, as the result of a relatively open structure in which
the atoms are linked by extremely strong, directed covalent bonds, as in diamond. This situation
does not prevail in the case of closely packed lower mantle silicates.

[3] However, the density and seismic velocities of the *inner* core below a depth of 5154 km are
consistent with the interpretation that this region consists of essentially pure nickel–iron
(Anderson, 1977).

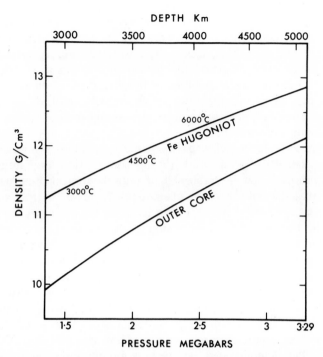

Figure 3.2 Variation of density with pressure in the outer core (Dziewonski et al., 1975) and along the Hugoniot for pure iron (McQueen and Marsh, 1966). Temperatures on the Hugoniot are also indicated. (From Ringwood, 1977b; with permission.)

about 11.5 percent near a pressure of 1.4 Mbar. However, it seems likely that this comparison refers to solid iron (in the shock experiments) versus the liquid state in the core. If an allowance is made for volume change on melting, the deficit is about 8 percent (McQueen and Marsh, 1966). Moreover, the hydrodynamical sound velocity of the outer core is substantially higher than expected for pure iron (see, for example, D. L. Anderson et al., 1971; Davies and Dziewonski, 1975). These properties have been generally interpreted to imply that the outer core contains about 5–15 percent of an element or elements of low atomic weight, e.g., H, C, N, O, Mg, Si, and S (Birch, 1952; Brett, 1976). The identification, both of these light elements and the manner in which they became incorporated in the core, has important implications for the origin of the Earth. The situation has been reviewed by Brett (1976) and Ringwood (1977b). It does not now appear likely that the principal light elements in the core are comprised of H, C, N, Mg, or Si, though small amounts of these elements may be present.

In recent years, considerable support has developed in favor of the view that sulphur is most likely to be the principal light element in the outer core (see, for example, Goldschmidt, 1922; Mason, 1966; Murthy and Hall, 1970; D. L. Anderson et al., 1971; and many others). Ahrens (1976) has examined the composition of the outer core in the light of shock wave data on the densities of FeS, Fe, and Ni under very high pressures. Assuming that the additional light element in the core is sulphur, he finds that about 14 weight-percent of this element would be necessary to account for the lower density of the outer core as compared to that of pure iron–nickel alloy of appropriate composition. This is equivalent to 4.5 percent of sulphur in the bulk Earth, and implies that the Earth must have accreted about 44 percent of the primordial abundance of sulphur, assuming the latter to be given by Type 1 (Cl) carbonaceous chondrites.*

In the parental solar nebula, sulphur would be condensed as FeS only at temperatures below 650°K (Table 6.2). Nearly all metals are fully condensed at temperatures higher than this (exceptions are Pb, Bi, In, and Tl). Sulphur is thus an extremely volatile element in a nebula of solar composition.

Studies of the composition of the Earth's mantle, as discussed in Section 2.4, demonstrate that it is depleted by factors ranging from 1.3 to over 100, in a wide range of elements compared to the primordial abundances as given by Cl chondrites. The general sequence of increasing depletion factors (Fig. 2.3) corresponds quite well to the sequence of condensation of elements and phases from a cooling gas of solar composition (Table 6.2), strongly implying that the terrestrial depletions are primarily consequences of selective condensation/volatilization processes. In the case of some elements (e.g., Cu, As, Au, Ge, S, and Se), depletions in the mantle are also caused in part by their siderophile nature, which has resulted in preferential entry into the Earth's core (Section 2.4).

A formidable problem confronting the hypothesis that sulphur is the principal light element in the Earth's core lies in explaining why such a highly volatile element is not as depleted in the bulk Earth as several non-siderophile elements that are much less volatile than sulphur in the solar nebula. For example, Cr, Mn, Na, K, Rb, F, Cs, Zn, and Cl are all less volatile than S (Table 6.2) and yet appear to be depleted in the Earth by much larger factors. Depletion factors of these elements compared to the primordial Cl and solar abundances range from 0.3 to 0.03 as compared to the *implied* terrestrial depletion factor for sulphur of 0.44.*

The problem of retaining sulphur while losing less volatile elements is highlighted by the compositions of meteorites. In the chemical processing by which the ordinary chondrites were formed from primordial material, about 80 percent of the sulphur was lost, but Na, K, Rb, Mn, and Cr remained essentially intact. The highly oxidized, metal-free assemblage of the Karoonda C3 equilibrated chondrite indicates that it condensed at lower temperatures than ordinary chondrites, yet its sulphur content is smaller. Moreover, iron meteorites contain, on the average, only about 1 percent of sulphur, yet many of them evidently formed in an environment (prior to metal segregation) where Na, K, and Rb were undepleted, as indicated by the compositions of silicate inclusions.

We must, therefore, acknowledge the very severe difficulties encountered by models which propose that the Earth, accreting in a nebula of solar composition, condensed over 40 percent of the primordial complement

* See footnote on p. 59.

of sulphur, but nevertheless lost much larger proportions of several elements[4] that are much less volatile than sulphur. It does not seem possible to devise possible conditions of condensation and accretion within a nebula of *solar composition* by which sulphur is retained, and these other elements lost.

The thermodynamic conditions under which the Earth could nevertheless have accreted large quantities of sulphur while losing Cr, Mn, Na, K, etc. by volatilization were investigated by Ringwood (1975a, 1977b). The condensation of sulphur is governed by the equilibrium:

$$Fe + H_2S \; \rightleftharpoons \; FeS + H_2, \qquad K = \frac{H_2}{H_2S}$$

In order to condense sulphur as FeS at temperatures at which the above elements would be volatile, it would be necessary to reduce the hydrogen/sulphur ratio in the bulk system by factors varying from 100 to 300. A fractionation of this kind could hardly be accomplished in the gas phase. It would require an initially cold ($\ll 0°C$) nebula in which the dust particles consisted of an oxidized, volatile-rich condensate similar to Type 1 carbonaceous chondrites in which sulphur was fully condensed. The dust particles then became strongly concentrated relative to gas by sedimentation, possibly combined with coagulation. The degree of enrichment of primordial dust relative to hydrogen is such that solid material of Cl composition constituted between 40 and 80 percent of the total mass of the system. The Earth is thus required to form largely from material of Cl chondrite composition, as, for example, in the models of Ringwood (1960, 1966a,b, 1975a).

These models are discussed in Section 7.2 and found to be untenable in their present form. Therefore, it is desirable to enquire into the possibility that elements other than sulphur might be present in substantial quantities in the outer core.[5]

3.3 Iron Oxide as a Major Component of the Outer Core

Dubrovskiy and Pan'kov (1972) pointed out that at high pressure, ferrous oxide would be expected to transform to the denser metallic state. They suggested that the density of the outer core could be explained if it consisted

---

[4] Lewis (1971) and Hall and Murthy (1971) suggested that the bulk Earth is not depleted in K and Rb. Most of the Earth's complement of these elements is believed to reside in the core. This explanation is not applicable, however, to non-siderophile and non-chalcophile elements (at terrestrial redox states) such as Cr, Mn, Na, F, Zn, and Cl, which are clearly depleted. A detailed critique of the proposals of Lewis, and Hall and Murphy is given by Ringwood (1977b).

[5] Other components which have been suggested by various authors to be present in the Earth's outer core include carbon, silicon, and magnesium oxide. The cases for these components have been discussed by Ringwood (1977b) and found to be much weaker than for sulphur or oxygen. Stevenson (1977) has recently suggested an important role for hydrogen. The main problem with hydrogen lies in devising a method by which the appropriate amount can be trapped in the Earth during accretion. This problem has been investigated by Ringwood (1977b, 1978). It appears that a small amount of H may well be present, but not nearly in sufficient quantities to cause the observed density deficit.

of a mixture of about 60 percent liquid metallic FeO, plus 40 percent nickel–iron. A related suggestion was put forward by Bullen (1973a,b) who proposed that the outer core consists of iron suboxide Fe_2O. These suggestions derived from efforts to interpret the density distributions within the Earth and terrestrial planets, and chemical evidence bearing these hypotheses was largely ignored.

Because of the technological importance of the Fe–O system, a considerable amount of chemical information relevant to this hypothesis exists, particularly in the metallurgical literature. Ringwood (1977b) investigated the feasibility of the hypothesis in the light of this data. A key issue relates to the solubility of FeO in molten iron. This is very small, in the vicinity of 0.6 percent, near the liquidus at atmospheric pressure (Fig. 3.4). However, the solubility of molten FeO in molten Fe increases rapidly with increasing temperatures in the range 1500–2000°C. Distin et al. (1971) showed that solubility of oxygen in molten iron in this temperature interval is accurately described by the following relationship

$$\log_{10}(\text{wt.-}\%\ 0) = \frac{6380}{T(°K)} + 2.765.$$

The high quality of their experimental results is shown in Fig. 3.3. They imply that the solubility of oxygen in molten iron at 2000°C reaches 3.1 atomic percent, equivalent to 6.2 mol-percent of $Fe_{1/2}O_{1/2}$. The above relationship obtained by Distin et al. (1971) has been used to calculate the solubility of FeO in iron at higher temperatures, outside the experimental

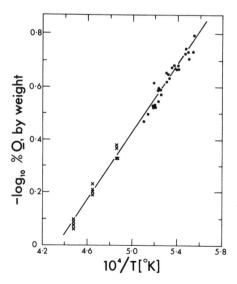

Figure 3.3 Solubility (in $\log_{10}$ weight-percent) of oxygen in liquid iron versus reciprocal of absolute temperature in range 1800–2230°K. Dots refer to data of Distin et al (1971) and crosses to C. R. Taylor and Chipman (1943). (From Distin et al. 1971.)

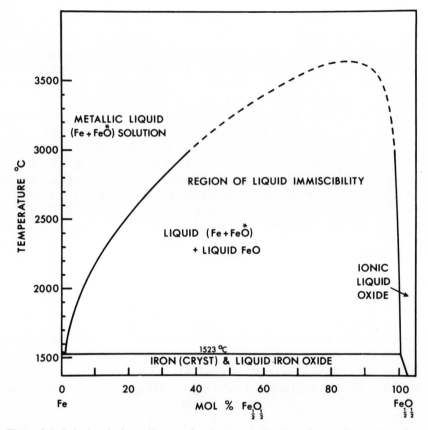

Figure 3.4 Calculated phase diagram for the system Fe–Fe$_{1/2}$O$_{1/2}$ at 1 atm., based on experimental measurement of solubility relationships between 1500 and 2000°C. (From Ringwood, 1977b, with permission.) Asterisk and FeO refer to a metallic solution.

range. Data from the same workers were also used to calculate the solubilities of Fe in FeO—results are shown in Figure 3.4. It is seen that approximately 40 mol-percent of Fe$_{1/2}$O$_{1/2}$ would dissolve in liquid iron at 3000°C, while at 3500°C the proportion of dissolved Fe$_{1/2}$O$_{1/2}$ is expected to be in the vicinity of 70 mol-percent. The temperature at the base of the mantle probably exceeds 3000°C (Higgins and Kennedy, 1971; Jacobs, 1975), so that extensive solubility of FeO in iron should be possible in the core.[6]

The effect of high pressures on this solubility was also considered by Ringwood (1977b). The partial molar volume of FeO when in solution in molten Fe was found to be about 18 percent smaller than the molar volume of pure, near-stoichiometric FeO. Therefore, high pressures will tend to

---

[6] Several other transition metals are known to display extensive or complete miscibility with their corresponding oxides at liquidus temperatures. This behavior is displayed by the systems Ti–TiO, V–VO, Ni–NiO, Nb–NbO, Zr–ZrO$_2$, and Y–Y$_2$O$_3$.

Table 3.1 Calculated solubilities of $Fe_{1/2}O_{1/2}$ in molten iron in equilibrium with $Mg_{0.88}Fe_{0.12}O$ at 2500°C as a function of pressure P (after Ringwood, 1977b)[a]

Pressure (kbar)	Solubility Mol. % $Fe_{1/2}O_{1/2}$
0	2
100	4
200	7
300	11
400	18
500	31
600	52
700	87

[a] The pressures calculated in Table 3 of Ringwood (1977b) are too low by a factor of two. The error arose because the molar volume of ferrous oxide used in the calculations was that of FeO rather than $Fe_{1/2}O_{1/2}$. The pressure calculated for the nonmetal-to-metal transition in FeO at 2500°C in that paper should accordingly be increased from 158 to 316 kbar.

increase the solubility of FeO in molten iron. Semiquantitative calculations of this effect, including the effect of the $P.\Delta v$ term on the free energy of the solution equilibria, show that high pressure will depress and reduce the extent of the liquid immiscibility field between the (Fe–FeO) metal solution and the oxide solution. Such calculations indicate that above about 500 kbar, complete miscibility should occur between metallic iron and iron oxide above the solidus. The resultant phase diagram would be qualitatively similar to that displayed by the analogous system Fe–FeS (Hansen and Anderko, 1958).

Iron oxide is an important component of the Earth's mantle; however, its thermodynamic activity may be reduced by compound formation and solid solution in other minerals. Ringwood (1977b) made a semiquantitative estimate of the significance of this effect by calculating the solubilities of FeO in molten iron in equilibrium with a possible oxide component of the lower mantle, $(Mg_{0.88}Fe_{0.12})O$, at a temperature of 2500°C as a function of pressure. Results are given in Table 3.1. We see that at this temperature, the metal phase in equilibrium with the possible lower mantle mineral assemblage should contain large amounts of FeO at pressures in the vicinity of 600 kbar and that solubility increases very rapidly with pressure in the interval 500–700 kbar.

Although these calculations were necassarily oversimplified, their "margin of safety" is rather broad. It is predicted that the liquid metal phase in equilibrium with the minerals of the Earth's lower mantle would contain large amounts of FeO at pressures around 600 kbar and temperatures as low as 2500°C. The pressure at the base of the mantle is 1.4 Mbar, and at the center of the Earth, 3.6 Mbar. Temperatures within the core almost certainly exceed 2500°C. Since FeO is already *known* to be an important component of the Earth's mantle, it seems very likely, in the light of this evidence, that oxygen (in the form of the component FeO) is also an important constituent of the core. The amount of FeO in the core necessary to account for the observed density is difficult to estimate. A calculation on the basis of the density of the covalent-metallic form of FeO implies that the core could contain as much as 65 weight-percent FeO (Ringwood, 1977b). Alternative estimates by Dubrovskiy and Pan'kov (1972) yield FeO contents in the vicinity of 57–68 percent. Another estimate is obtained from a comparison of the observed core density at 1.4 Mbars (Fig. 3.2) with the densities of pure Fe and of a mixture of (Fe_3O_4 + Fe) as obtained from shock-wave data at 1.4 Mbar. This method yields a value of 28 weight-percent FeO in the core.[7] In view of the uncertainties, we take a value of 44 ± 16 weight-percent FeO, which is equivalent to 10 ± 4 weight-percent of oxygen.

Incorporation of Iron Oxide into the Earth

At first sight, the incorporation of large amounts of FeO into the Earth at temperatures sufficiently high to cause the observed depletions of many volatile elements (e.g., Cr, Mn, Na, K, Cl, and Zn) poses problems analogous to those encountered in the case of sulphur. Equilibrium calculations (Ringwood, 1977b) show that hydrogen must be depleted relative to H_2O by factors in the range 100–300, as compared to the solar H_2/H_2O ratio, in order to condense FeO at sufficiently high temperatures to maintain the above elements in the gas phase. This requires an initially cold primordial solar nebula in which a low-temperature, highly oxidized, water-rich primordial condensate becomes strongly concentrated relative to the gas phase (mainly H_2), most probably by sedimentation into the ecliptic plane. The degree of concentration required of the solid condensate is such that it must comprise at least 70 percent by weight of the total system. Thus, the Earth is required to form by accretion from planetesimals dominantly of C1 carbonaceous chondrite composition, as in the models of Ringwood (1960, 1966a, 1975a). These models, however, encounter certain difficulties which will be discussed in Section 7.2.

A more attractive means of introducing large amounts of FeO into the Earth can be envisaged in terms of a process by which the Earth accreted from a physical mixture of components formed under different P,T conditions in the parental nebula, and which therefore possessed contrasting compositions. The simplest example of this is a two-component model whereby the Earth's composition is interpreted in terms of a mixture of highly reduced, metal-rich, devolatilized material formed at high tempera-

---

[7] Incorrectly given as 44 weight-percent FeO on page 123 of Ringwood (1977b).

tures, with a smaller amount of primordial, oxidized, volatile-rich material similar to Cl chondrites and formed at low temperatures (Table 8.1).

The abundances of volatile (non-siderophile) elements in the mantle (Fig. 2.3) suggest that the bulk Earth incorporated about 10 percent of the low-temperature component (Table 8.1, column A) during accretion (Ringwood, 1977b). When this component is mixed with the devolatilized component (B) during accretion of the Earth, it introduces directly only a minor amount of FeO. Most of the FeO required is produced by a chemical reaction between excess metallic iron (in component B) and oxidized species (principally H_2O) in component A:

$$Fe_{(B)} + H_2O_{(A)} = FeO + H_2.$$

(These reactions are discussed at greater length in Chapter 8.) In order to provide the required amount of FeO in the mantle and core, it would be necessary to assume that the Earth had accreted about 15 percent of component A (Ringwood, 1977b). This is somewhat higher than the 10 percent of component A necessary to explain the abundances of volatile elements in the mantle. This discrepancy, if real, is not serious, and plausible explanations can be advanced (Section 8.5; Ringwood, 1977b).

Possibility of Disproportionation of FeO in the Lower Mantle

When wüstite is cooled at atmospheric pressure, it disproportionates at 500°C into magnetite plus metallic iron:

$$4\,FeO = Fe_3O_4 + \alpha Fe, \qquad \Delta v = +3.0\ cm^3$$

Because the disproportionation is accompanied by an increase in volume, the application of pressure favors the stability of wüstite. However, under shock loading at very high pressure ($\geq 600\ kbar$), magnetite ($\rho_0 = 5.20\ g/cm^3$) transforms to a very dense phase possessing an estimated zero pressure density of $6.40\ g/cm^3$ (Davies and Gaffney, 1973). Moreover, αFe transforms to the denser ε polymorph at about 130 kbar (15°C). Thus, at pressures corresponding to those in the deep mantle, the reaction.

$$4\,FeO = Fe_3O_4\ (\text{h.p.p.}) + \varepsilon Fe, \qquad \Delta v = -5.8\ cm^3$$

is accompanied by a large decrease in volume, implying that the disproportionation reaction is strongly favored by pressure. The free energy of this reaction will be dominated by the $P.\Delta v$ term at mantle pressures. In this situation, the high-pressure polymorph of Fe_3O_4 (h.p.p.) may be formed as a separate phase in the lower mantle according to the simplified equilibrium: $MgFeSi_2O_6$ (perovskite) $+ MgO = 2\,MgSiO_3$ (perovskite) $+ \frac{1}{4}Fe_3O_4$ (h.p.p.) $+ \frac{1}{4}Fe(\varepsilon)$.

If this disproportionation reaction occurs as suggested above, the effect of pressure upon the solubility of h.p.p. Fe_3O_4 in molten iron may not be large, because of the intrinsically high density of Fe_3O_4 (h.p.p.). However, under these circumstances, Fe_3O_4 (h.p.p) would be present as a separate phase possessing unit thermodynamic activity. The effect of temperature upon the solubility of Fe_3O_4 (h.p.p.) in molten Fe will be similar to that of FeO (Figs. 3.3 and 3.4), since the stoichiometry of the discrete iron oxide phase is not an important factor. In both cases, we are considering the solubility of *oxygen* in molten iron. We conclude that if much of the FeO in the deep mantle (initially present in rocksalt and perovskite-type phases) ultimately disproportionates to form Fe_3O_4 (h.p.p.) plus εFe under sub-solidus conditions, then the iron oxide phase will dissolve extensively in molten iron when temperatures exceed 2500–3000°C (Fig. 3.4) so that "FeO" is still expected to be an important component of the core. These phase relationships have an important influence on the process of core formation, discussed in the next section.

3.4 Formation of the Core

Historically speaking, our ideas concerning the early thermal evolution of the Earth have been strongly influenced by concepts relating to its origin. A popular hypothesis (Jeans, 1917, 1919) during the first half of this century envisaged the solar system as being formed by condensation from hot gases torn out of the sun by tidal interaction with a passing star. Derivative hypotheses on the development of the Earth accordingly proposed a "hot" origin, in which the Earth was believed to have formed from the totally molten state (see, for example, Jeffreys, 1929). With the resurgence of hypotheses that proposed that the Earth had formed by accretion from initially cold or cool solid particles (planetesimals), as in the models of Schmidt (1944, 1958) and von Weiszäcker (1944), opinions swung around to favoring a corresponding cool origin for the Earth. The vast majority of papers on this subject written during the 1950s and mid-1960s assumed that the Earth accreted as a cool (mostly <1000°C) mixture of silicates and metallic iron, and was well below the solidus at all depths. Heating by long-lived radioactivity elements then caused internal temperatures to rise above the melting point of the metal phase, leading to differentiation and formation of the core, long after the accretion of the Earth. Among the numerous variants of this theme, the papers of Lubimova (1958), Safronov (1959), Urey (1952, 1962a), MacDonald (1959), Vinogradov (1961, 1967), Elsasser (1963), and Birch (1965) may be cited.

A new dimension was added with the geochemical arguments introduced by Urey (1952, 1954, 1957a, 1962a). He concluded that the Earth had retained the primordial abundances of many volatile elements, which, moreover, were not strongly concentrated near the surface, as would be expected if the Earth had crystallized from the molten state. According to Urey, these observations implied that the Earth had accreted under cool conditions and had never subsequently been extensively melted. Urey also pointed to the small volume of the crust and the evidence for only limited differentiation of the mantle as being incompatible with expectations based on a totally melted mantle which would be expected to undergo extreme fractionation by crystallization differentiation. Parallel arguments were introduced by Vinogradov (1959, 1961, 1967).

These arguments carried considerable weight and were widely cited by those who supported a cold origin for the Earth. However, the subsequent accumulation of much more accurate and voluminous data on abundances of many elements in meteorites and in materials from the Earth's mantle has shown that Urey was incorrect in one major aspect. There is little doubt that the Earth has been strongly depleted in a wide range of volatile elements (Fig. 2.3), an observation which requires extensive high-temperature processing, either in the nebula prior to accretion or during accretion of the Earth. However, contrary to a view earlier expressed by Ringwood (1966a), Urey's other argument that the degree of differentiation of the mantle is much smaller than would be expected during complete melting, remains firmly based, and is further supported by the evidence on mantle homogeneity discussed in Section 2.2.

Urey (1952) was also the first to realize the significance of the very large amount of gravitational potential energy liberated during the segregation of the core from its original state, as metal dispersed throughout the Earth. Most of this energy would be converted into thermal energy during core formation. Urey estimated the energy production as about 700 cal/gm,[8] enough to heat the entire Earth by over 2000°C. He suggested that segregation of much of the core occurred gradually, throughout geological time, thus driving mantle convection and providing a tectonic engine, continuing to the present day. This model was explored and developed further by Runcorn (1962, 1965) and Munk and Davies (1964). However, Ringwood (1960) noted that the core formation process as postulated by Urey was likely to be intrinsically unstable, and once started would proceed to completion in a very short time. The evolution of heat resulting from metal segregation would cause the effective viscosity of the mantle to fall exponentially with temperature, thereby increasing the rate of core formation, leading to a runaway process. A similar conclusion was reached by Elsasser (1963) and Birch (1965). However, these latter authors continued to maintain that core formation did not occur until a large time interval ($> 5 \times 10^8$ years) had elapsed, during which substantial internal heating by long-lived radioactive elements occurred, until the melting point of iron was exceeded.

Ringwood (1960) pointed out that segregation of the core would cause a significant fractionation of lead from uranium, since the molten iron would be expected to carry down substantial quantities of lead, but not uranium. Since the Earth's "age" of 4.55×10^9 years as measured by the lead isotope method refers to the time when, to a first approximation (and neglecting second-order differentiation effects), the present overall Pb/U

---

[8] A more accurate estimate by Flasar and Birch (1973) gives 640 cal/gm.

ratio of the upper mantle/crust system was established, the "age of the Earth" as determined in this manner actually dates the time of formation of the core.

This argument was subsequently quantified by Oversby and Ringwood (1971), who measured the partition coefficient for lead between a series of relevant iron alloys and a basaltic melt, and obtained a relationship between the partition coefficient and the time which had elapsed between accretion of the Earth and segregation of the core. They concluded that the time interval between accretion and core formation was probably smaller than 10^8 years and almost certainly smaller than 5×10^8 years. It was pointed out by Ringwood (1960) and Oversby and Ringwood (1971) that the early period of core formation represents a basic boundary condition for all hypotheses relating to the origin and early evolution of the Earth.

The time interval between accretion and core formation was reconsidered by Vollmer (1977) on the basis of small subsequent revisions to the uranium decay time constants and isotopic composition of primordial lead. Vollmer concluded that core formation was not completed within 10^8 years after accretion, but more probably occurred over a period of about 5×10^8 years after accretion. However, Vollmer assumed that deviations in isotopic compositions of oceanic basaltic leads from the lead growth curve of the terrestrial crust were caused not by minor fractionations of lead from uranium in the source regions of basalts (Gast, 1968), but by the core formation process itself. A wide range of geochemical evidence, including neodymium–samarium isotope systematics cast doubt upon the first part of this assumption.

The problem was also examined by Gancarz and Wasserburg (1977) in relation to the lead isotopic composition of the 3.6-billion-year-old Amitsoq Gneiss of West Greenland. They found that providing the Earth's age was close to 4.55×10^9 years, and that most of the lead entered the metal phase during core formation as indicated by the experimental partition coefficient data cited earlier, the most probable time of core formation was around 4.4×10^9 years, i.e., about 1.5×10^8 years after accretion. For a wide range of initial assumptions, core formation occurred within 4×10^8 years after accretion.[*]

Because of the extremely exothermic and catastrophic nature of the core formation process, it must have been completed well before the formation of the oldest crustal rocks (Hanks and Anderson, 1969). With the discovery of 3.8-billion-year-old crustal rocks in Greenland (Black et al., 1971), this geologic constraints pushes the probable time of core formation back to over 4×10^9 years, in general support of the estimate mentioned earlier.

---

[*] *Note added in proof*: Pidgeon (1978) has been able to constrain the time of core formation still more tightly, based on studies of isotopic composition of the lead in an exceptionally well-dated ($3453 \pm 16 \times 10^6$ years) conformable ore deposit, "Big Stubby," in Western Australia. Pidgeon concluded that core formation occurred within 10^8 years of formation of the Earth (taken as 4.55×10^9 years).

An alternative hypothesis of "heterogeneous accretion" maintains that the iron core accreted before the surrounding silicate mantle (Orowan, 1969; Turekian and Clark, 1969). The time of formation of the core thus corresponded to the age of the Earth. However, this model does not result in heating from the energy of core segregation, as in models where the iron and silicates were originally mixed during accretion. A critical discussion of this version of the heterogeneous accretion model is given in Section 7.3. The evidence against it appears to be compelling.

Thermal Regime During Core Formation

Assuming the Earth to have accreted initially as a mixture of silicate and metal particles, formation of the core soon after accretion requires that the melting point of the iron alloy had been exceeded throughout substantial volumes of the interior, thereby permitting the metal to segregate into bodies sufficiently large (e.g., ~ 100 km dimension) to sink through the solid but plastic mantle into the core. Thus, a large source of "initial heat" is required. This is most plausibly supplied by partial conservation of the gravitational potential energy present in the parental dust and its transformation into thermal energy during accretion. Heating by the long-lived radioactive elements K, U and Th is not a significant factor. During the first 500 million years after accretion, these elements (assumed to be present in abundances estimated in Section 2.3) would heat the Earth by only about 300°C. The role of the short-lived radioactive nuclide ^{26}Al (7×10^5 years) may yet warrant consideration as a possible heat source. However, to be effective, the Earth would need to have accumulated in a short time ($< 10^6$ years), which would also lead to strong heating from accretional energy (Chapter 8).

The temperature distributions within the Earth after accretion are given in Figure 3.5, for two very different accretional models. In the first (Ringwood, 1960, 1975a), the Earth is assumed to accrete on a short time scale, so that surface temperatures reached during most of the accretion period are sufficiently high to evaporate the volatile elements inferred to be depleted in the mantle. The model, which is described further in Section 7.2, requires an accretion time scale of less than 10^6 years. A further contribution to the initial temperature distribution within the Earth's interior arising from adiabatic heating during self-compression (including phase changes) is also incorporated in the net post-accretional temperature distributions of Figure 3.5.

The second temperature distribution is based on Safronov's (1964, 1972a) model of accretion of the Earth over a period of 10^8 years from a hierarchy of planetesimals possessing sizes of up to several hundred kilometers. Safronov showed that where a substantial proportion of the mass is accreted in the form of large planetesimals (dimension > 100 km), a much greater proportion of thermal energy is trapped during accretion, as compared to the case where small objects only are accreted. A corresponding allowance for adiabatic heating during self-compression has also been added to Safronov's post-accretional temperature distribution in Figure 3.5.

Although based on radically different assumptions, the two estimates in Figure 3.5 are not grossly dissimilar, the maximum temperature difference between the curves being in the vicinity of 1000°C. Wetherill (1976a) has reconsidered the post-accretional temperature distribution within the Earth on the basis of Safronov's model and has shown that substantially higher temperatures could be attained by a reasonable variation of Safronov's assumptions concerning the size distribution of planetesimals. Thus, the differences in internal temperature distribution for accretional time scales of 10^6–10^8 years need not be very large.[9]

The estimated melting point of iron with depth (Higgins and Kennedy, 1971) is shown in Figure 3.5, in relation to estimates of the mantle liquidus and solidus by Kennedy and Higgins (1973), to a depth of 2000 km. Although the uncertainties in these estimates should not be ignored, it appears likely that the melting point of pure iron with depth lies between the solidus and liquidus of the silicate mantle to a depth of at least 1000 km, and probably 2000 km. This corresponds to a region containing about half the mass of the Earth. We note, moreover, that the melting curve for iron and the mantle liquidus are much higher than either of the two post-accretional temperature profiles given in Figure 3.5. This is also true of the mantle solidus, except within the uppermost 100 km.

We concluded earlier that segregation of the core occurred close to 4.55×10^9 years ago, and that most of the Earth's mantle remained unmelted during the process. In order to begin the segregation of metal from silicate, it is necessary to exceed the melting point of the metal phase so as to permit it to segregate into large bodies. However, the similarities of iron and mantle melting intervals down to great depths would seem to make it difficult to achieve this without melting the silicate phase. Moreover, the fact that the melting curves of iron and mantle materials are much higher than the expected post-accretional temperature distribution increases the difficulty of early core formation.

The most plausible solution to these problems is the proposal by Murthy and Hall (1970) that the melting point of iron (but not of silicates) was lowered considerably by another component. The presence of large amounts of a light element in the core is firmly established (Fig. 3.2). Murthy and Hall (1970) proposed that this additional component consists of sulphur

---

[9] This difference may be further reduced if the effects of chemical energy trapped during accretion (Florenskii, 1965) are considered. For the case of "slow" accretion (10^7–10^8 years), material arriving at the Earth's surface is evaporated and rapidly recondensed, resulting in the formation of highly metastable, disequilibrium mineral assemblages, including much glass. These metastable assemblages become buried under a highly porous surficial layer characterized by low thermal conductivity. When temperatures rise to about 600°C, the metastable mineral assemblages recrystallize, achieving equilibrium, and considerable heat is evolved. This may easily amount to 100 cal/gm, which would cause a temperature increase of about 400°C. Much of this additional chemical heat may have been retained within the interior of the Earth during accretion.

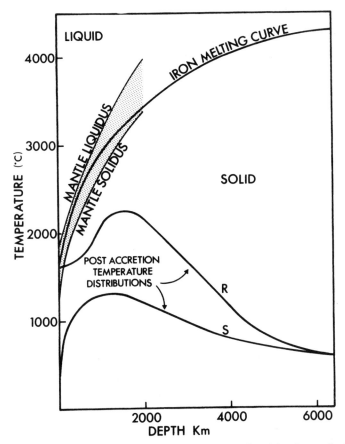

Figure 3.5 Estimates of melting temperatures of iron and melting interval of mantle (Higgins and Kennedy, 1971; Kennedy and Higgins, 1973) together with estimates of post-accretional temperature distributions within the Earth and prior to core formation. S = Safronov (1972a), R = Ringwood (1975a).

(as FeS). The viewpoint favored in that paper maintains that the additional component is composed mainly of FeO, though significant quantities of sulphur could also be present. This is expected to have a comparable effect in reducing the melting point of iron. Thus, 1 percent of FeO in solution is known to decrease the melting point of iron by 11°C, whereas 1 percent of FeS reduces the melting point by only 6°C (Hansen and Anderko, 1958). It has been suggested (Section 3.3) that under high pressures, complete miscibility may occur between Fe and FeO (as in the Fe–FeS system) and we would therefore expect FeO to greatly reduce the melting point of iron at pressures sufficiently high for miscibility to occur. These are probably attained at depths of about 1500 km.

If the principal light element in the core were sulphur, segregation of metal into large "drops" would be expected to occur relatively close to the surface, as in the model of Elsasser (1963). The drops would then sink into the core, heating the mantle at greater depths by viscous dissipation, so that the loci of "drop" segregation moved downwards with time. However, of oxygen were the principal light element in the core, melting and segregation of metal would be likely to occur first at depths below 1500 km. After sinking of the resultant drops, a region of less dense silicate would have remained, which would be gravitationally unstable relative to the overlying unmelted silicate–iron mixture. The metal-free light material would rise convectively to be replaced by a batch of silicate–iron mixture from above, and so the segregation process would continue, accompanied by strong convection in the outermost 2000 km or so of the Earth.

It has sometimes been assumed that once the temperatures in the Earth's interior had risen sufficiently to permit metal segregation, the additional 600–700 cal/gm liberated during rapid formation of the core would cause extensive or complete melting of the mantle, e.g., Ringwood (1960). This, however, is not correct. The process of core formation would necessarily result in strong convection throughout the Earth leading rapidly to the establishment of an adiabatic gradient. Upper mantle temperatures (required for metal segregation) are probably in the vicinity of 1500°C, and the adiabatic temperature distribution would prevail at greater depths. Because of the drastic effect of pressure upon the melting point of silicates, which causes the melting point gradient to be much greater than the adiabatic gradient, the mantle below depths of 200 km or more would have been solid (Figs. 3.5 and 3.6).

As core-forming energy was pumped into the Earth, the mean temperature of the convecting mantle depended on the rate at which energy could be dissipated at the Earth's surface. If the core-forming process took more than a million years, most of this energy would be radiated away by the 100–200-km-deep layer of magma near the surface, and the net temperature distribution within the solid Earth at greater depths would remain unchanged. However, even if core formation were a catastrophic process (see, for example, Ringwood, 1960, 1975a) occurring, say, in a period on the order of 10^3–10^4 years, the temperature at the Earth's surface would be unlikely to exceed about 2000°C. Above this temperature, silicates would be evaporated from the surface of the magma ocean, a process which provides an effective thermostat. In this case, the molten layer would not extend below about 400 km. The internal temperature distribution would then be as in Figure 3.6. The upper mantle would be largely or completely molten. However, because of the rapid increase of melting point with pressure, the pyrolite solidus would intersect the adiabat at a depth of 400 km. Below this depth, and extending to the core, the mantle would be solid, since the adiabatic gradient remained much smaller than the melting point gradient.

Fractional crystallization of the primitive upper-mantle magma ocean has been discussed by Ringwood (1975a). Because of the nature of the phase relationships involved, most of this magma ocean would crystallize to form olivine + orthopyroxene + clinopyroxene + garnet cumulates. Be-

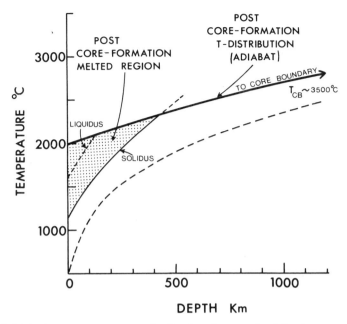

Figure 3.6 Maximum temperature distribution in mantle following catastrophic core formation, in relation to pyrolite solidus and liquidus. The mantle temperature distribution followes an adiabat from 2000°C, based on data of Birch (1952) and Verhoogen (1956). Note extensive molten and partly molten zone extending to a depth of 400 km. Also shown is an estimate of present mantle temperature distribution (broken line). (From Ringwood, 1975a, with permission.)

cause of the very small maximum pressure at which residual liquids in the system can differentiate to form silica-oversaturated liquids, only a very small thickness (~ 3 km average) of silica-rich granophyre would be formed. This material, which would be strongly enriched in incompatible elements, may have formed the nucleus of the continental crust. The underlying cumulate layers would have cooled to form a dense lithospheric plate which would presumably have been subducted into the mantle at an early stage of evolution of the Earth (Ringwood, 1975a).

Note added in proof: On the basis of additional shock wave data, Ahrens (*J. Geophys. Res.* **84**, 985–1008, 1979) now estimates the possible sulphur content of the core to be in the range 9–12 wt.-% S. This would imply that the Earth captured 28–38 percent of the primordial sulphur abundance (as given by C1 chondrites). Although these values are smaller than the estimate of 44 percent retained sulphur used on page 45 of this chapter, the geochemical arguments based on this latter estimate remain valid when the revised sulphur values are employed.

CHAPTER 4

Boundary Conditions for the Origin of the Earth

In the previous chapters we have reviewed much evidence relating to the composition and internal constitution of the Earth. We will now attempt to utilize this evidence in formulating a set of boundary conditions for theories of origin.

(a) Abundance of Iron in the Earth

The major element compositions of the mantle and core (Chapters 2 and 3) may be combined in their known proportions to obtain the total abundances of the principal metallic elements Mg, Si, Fe, Al, and Ca in the Earth. These are compared with the primordial abundances of these elements in Table 4.1. Apart from minor depletion of Si in the Earth (discussed in Section 2.4) the relative abundances of these major elements are identical within their uncertainty limits. In particular, there is no evidence for substantial fractionation of iron from silicates, a basic postulate in the cosmogonic hypotheses of Urey (1952), Anders (1968, 1971), and many others.

(b) Abundance of "High-Temperature Condensates" in the Earth

From Table 4.1, we see that the terrestrial abundances of the principal members of this group, Ca and Al, are similar (relative to Fe, Mg, and Si)

Table 4.1 Comparison of major element bulk composition of Earth (as obtained in Chapters 1 and 2) with corresponding compositions for Type 1 carbonaceous chondrites[a]

Atom fractions (Mg = 1.0)	Cl Chondrites	Earth[b]
Si	0.95	0.80
Mg	1.00	1.00
Fe[c]	0.90	0.92
Al	0.08	0.07
Ca	0.07	0.06

[a] Average of Orgueil and Ivuna (Wiik, 1956).

[b] Core assumed to contain 12 weight-percent of light elements.

[c] Ni has been included with Fe.

to their primordial abundances. As discussed in Section 2.3, this characteristic applies also to many less abundant elements of this group, e.g., Ti, Zr, Nb, Sc, Y, REE, and probably Ba, Sr, U, and Th.

It appears that the bulk composition of the Earth is related very simply to the primordial composition given by Cl chondrites and the sun. There is no need to employ the complex seven-component fractionation models employed by Anders (1968, 1971) and coworkers to explain the composition of the chondritic meteorites, in order to explain the composition of the Earth. Our planet yields no evidence that these processes played a significant role in its genesis. This suggests that the formation of the Earth may have been much more simple, in a cosmochemical sense, than envisaged in the theories of Anders and coworkers.

(c) Depletion of Volatile Elements in the Earth

The Earth has been depleted in a wide range of elements that are more volatile than silicon under appropriate cosmochemical conditions. For non-siderophile elements, there is a strong correlation between the degree of depletion and the degree of volatility. The overall composition of the Earth can be matched by a mixture of about 10 percent of a low-temperature nebula condensate (similar to Cl chondrites) that has retained the primordial abundance of volatile metals, with 90 percent of a high-temperature condensate that has been subjected to varying degrees of loss of volatile metals, in amounts which correlated with their volatilities.

Boundary conditions (a)–(c) above imply that the present composition of the Earth can be derived very simply from the primordial composition (as given, for example, by Cl chondrites), simply by reduction of most of the oxidized iron and nickel in Cl chondrites to the metallic phase, accompanied by varying degrees of loss of volatile components (Ringwood, 1962, 1966a,b). This relationship has often been *assumed* in the past, as in the "chondritic earth model." It is only comparatively recently, however, that this assumption has become independently justified by our increased understanding of the structure and composition of the Earth, as reviewed in Chapters 1–3.

(d) Chemical Homogeneity of the Primordial Mantle

Differentiation processes connected with the dynamical behavior of the mantle, as manifested ultimately in the formation and subduction of lithospheric plates, have caused a substantial degree of local heterogeneity within the mantle throughout geological time. However, source regions of basaltic magmas that have not been subjected to these processes appear to be relatively homogeneous in major element composition and particularly so in their distributions of compatible elements.[1] This degree of homogeneity appears to have been more prevalent early in the Earth's history, before the operation of these local differentiation processes. The compatible elements seem to have been well mixed at the kilometer (or smaller) scale. They include elements exhibiting highly variable degrees of volatility (Mg, Fe, Mn, Zn), (Al, Ga), (Si, Ge); differing siderophile nature (Mg, Ni, Co, Mn), (Si, Ge), (Al, Ga); and also components that are indicators of redox conditions (FeO/MgO).

The high degree of homogeneity in the mantle of groups of elements possessing extremely different cosmochemical behavior in the solar nebula provides strong constraints on the nature of the accretion process, as discussed in Chapter 8.

In contrast to the low dispersions of compatible siderophile elements (Ni, Co, Ge, Fe) in the mantle, another group of siderophiles including the platinum metals and gold seem to exhibit wide dispersions in their abundances. There are also indications of wide dispersions in concentrations of primordial inert gases within the mantle.

(e) Thermal State of Mantle after Formation of the Earth

The degree of homogeneity of major element and compatible minor element distributions throughout the mantle, as inferred to be present early in the Earth's history, implies that the mantle was not completely or extensively melted during the formation of the Earth.

---

[1] However, as discussed in Section 2.1, important fractionations of *incompatible* elements have occurred in the source regions of basaltic magmas throughout geological time.

(f) Siderophile Elements in the Mantle

Many siderophile elements (e.g., Ni, Co, Cu, Au, and Re) are present in the upper mantle at levels from 10 to over 100 times higher than would be expected if this region had equilibrated with metallic iron, which had subsequently settled into the core. Moreover, the upper mantle is too oxidized, as manifested by Fe^{3+}/Fe^{2+} ratios and by the CO_2/CO and H_2O/H_2 ratios of gases evolved, to have been in equilibrium with metallic iron.

(g) Composition of the Earth's Core

Although the core consists mainly of iron, its density requires that it contain 5–15 percent of an element of relatively low atomic weight. The principal light element component is believed to be oxygen, although a significant amount of sulphur may also be present.

(h) Time of Formation of the Core

The core formed very soon after the accretion of the Earth had been essentially completed. This required a source of heat sufficient to cause temperatures in large regions of the Earth's interior to exceed the melting point of the metal phase at this early stage. The most likely source of this energy is partial retention of gravitational energy of accretion.

(i) Early Thermal History

The core must be formed rapidly, via an episode of intense convection which established an adiabatic gradient throughout the Earth. The highly exothermic nature of the core forming process would have led to extensive melting in the uppermost mantle. However, the strong increase of melting point with depth in the mantle prevented melting to depths greater than a few hundred kilometers. This melted and differentiated region was probably subducted into the mantle at a later stage, so that little or no direct evidence of its occurrence is now accessible.

PART II
ORIGIN OF THE EARTH

CHAPTER 5

Protostars, Disks, and Planets

5.1 Cosmogonic Background

For some centuries, attempts to understand the origin of the solar system were preoccupied with explaining its remarkable dynamical characteristics. It is only during the last few decades that the comparable importance of physicochemical properties as clues to origin has become generally appreciated.

Some key physical properties are summarized below (see also Figure 5.1 and Table 5.1).

(a) The orbits of the planets are coplanar, so that the solar system has the configuration of a very flat disk.

(b) Planets move in nearly circular orbits around the sun, which itself rotates in almost the same plane as the planets.

(c) Planetary revolutions are all direct, as is the solar rotation. Moreover, most of the planets themselves rotate in the direct sense.

(d) Spacings of planets are described by a simple relationship, Bodes' Law, wherein the mean distance of the nth planet from the sun, r_n, is given by the relationship.

$$r_n = r_0 \, \beta^n$$

where $\beta \approx 1.73$.

(e) The planets fall into two groups: an inner group of *terrestrial* planets possessing high densities, and an outer group of *major* planets possessing low densities (Table 5.1).

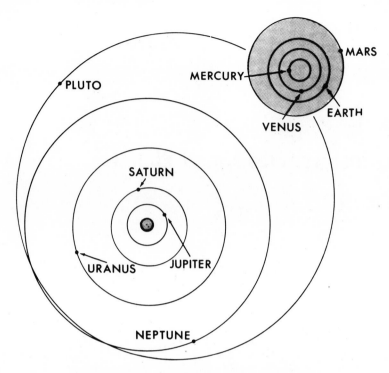

Figure 5.1 Orbits of the planets in the solar system.

(f) The regular satellite systems surrounding Jupiter, Saturn, and Uranus are closely analogous to the solar system itself, displaying relationships similar to a, b, c, d and e above.

These remarkable regularities would not be expected for an independently formed, self-gravitating system of celestial bodies. Such a system would normally be characterized by highly elliptical orbits deviating grossly from coplanarity, as seen in multiple star systems.

Kant (1755) and Laplace (1796) proposed that the dynamical properties of the solar system might be explained if the planets had formed within a largely gaseous, flat disc rotating around the sun. The discoidal configuration was derived from an originally gaseous *nebula* which flattened into a disk as a result of rotation and self gravitation perpendicular to the axis of the disk. In Laplace's version, as the disk contracted, circular rings were shed periodically at the periphery, due to rotational instability. These rings aggregated to form planets, necessarily revolving in circular, coplanar orbits. In later developments of this nebula hypothesis, the roles of dust particles and planetesimals within the disk have been emphasized. Gravitational forces, combined with viscous dissipation between dust grains and gas, necessarily cause the solids to collect into the equatorial plane and describe circular orbits.

Table 5.1 Properties of the planets

	Relative mass	Density (g/cm^3)	Mean distance from sun (AU^a)	Revolution period (years)
Sun	332,488	1.4	—	—
Mercury	0.055	5.44	0.39	0.24
Venus	0.815	5.24	0.72	0.62
Earth	1.0	5.52	1.0	1.0
Mars	0.108	3.94	1.52	1.88
Jupiter	317.9	1.33	5.20	11.86
Saturn	95.2	0.70	9.54	29.46
Uranus	14.6	1.3	19.18	84.01
Neptune	17.2	1.67	30.06	164.8
Pluto	0.1?	2?	39.4	247.7

^{a} Au = "astronomical unit," the mean distance of the Earth from the sun (approx. 150 × 10^6 km, or about 93 × 10^6 miles).

Today, this general class of hypotheses is seen to possess a basic elegance and appeal to elementary "common sense" and has become firmly entrenched. Nevertheless, it was later abandoned by most astronomers, particularly during the first half of this century, in the face of some severe difficulties that were encountered. The most serious of these is that the sun possesses nearly 99.9 percent of the mass of the solar system, but only 2 percent of the angular momentum. This does not appear to be a likely development from a Laplacian disk-nebula in which the sun should be revolving approximately at its rotational instability limit, 200 times faster than presently observed. Moreover, if all the angular momentum of the planets were transferred to the sun, the centrifugal force at the radius of Neptune would be less than 1 percent of the gravitational force towards the sun, so that rotational instability at that orbit appeared quite impossible. Finally, the minute proportion of the mass of the system now present in the planets seemed most difficult to explain on the basis of a contracting, rotating nebula.

These seemingly insuperable problems led to the formulation of a host of alternative hypotheses, most of which invoked ad hoc assumptions of low intrinsic probability. The most widely discussed alternatives involved tidal interaction during a close approach of two stars, as proposed, for example, by Jeans (1917) and Jeffreys (1916). In addition to their contrived nature, these hypotheses were found to be subject to a wide range of specific physical objections, equally as compelling as those which had led to the eclipse of the nebula hypothesis. The problem seemed to be reaching an impasse.

However, almost unnoticed, developments in astrophysics during the 1920s and 1930s began to undermine the most severe difficulties formerly encountered by the nebula hypothesis. With more accurate determinations of the relative abundances of elements in the sun, the enormous preponderance of hydrogen and helium became recognized. It followed that the planets represented only a very small proportion of the matter which had once existed in this region of the parental solar nebula. Vast amounts of hydrogen and helium must have been lost during the formation of planets. The mass of that part of the nebula parental to the planets was perhaps 10–100 times as great as the present total mass of planets. A correspondingly large amount of angular momentum must also have been lost from the system.

Moreover during the same period, it became recognized that the present slow rotation of the sun is but one example of a general rule that main-sequence stars of solar type have very slow rotations, whereas on numerous theoretical grounds rapid rotations would be generally expected. It follows that some mechanism must exist for the loss of angular momentum from stars.

At least a partial answer to this mystery was soon provided by Alfvén (1942, 1954) who showed that the magnetic field of the sun could couple it to a surrounding ionized envelope of dispersed matter, causing a transfer of angular momentum to the envelope and consequent braking of the sun's rotation. This proposal was further developed by Hoyle (1960), who suggested that the sun had transferred most of its angular momentum to the primordial solar nebula in this manner. A related process was suggested by Schatzman (1962), who showed that hydromagnetic coupling of the sun to ejected ionized material in the form of solar flares or the solar wind would effectively remove angular momentum from the sun.

Further means of solar braking involving coupling the contracting protosun to the surrounding nebula by means of turbulent viscosity (sometimes combined with magnetic fields) have been proposed by Schatzman (1967) and Prentice (1974). Although it is premature to regard the angular momentum problem as having been finally solved, it no longer constitutes the formidable barrier to a nebula-type hypothesis that it once seemed.

With the demise of these classical objections, the path has been reopened for further development of nebula hypotheses, early variants of which were proposed by Schmidt (1944, 1958), von Weiszäcker (1944), Ter Haar (1950), Kuiper (1951), Levin (1949), and Safronov (1954, 1972a). These hypotheses recognized the key role played by the separation of solid particles and planetesimals from the gas component of the nebula, followed by the accretion of solids to form planets. The chemistry of planets was strongly controlled by the conditions under which solids separated from gases, and has been extensively investigated by Urey (1952). We will return to this topic in later chapters. Useful historical summaries of the development of theories of the origin of the solar system have been presented by Kuiper (1955), Ter Haar and Cameron (1963), and Hartmann (1972).

It is worth remarking that the situation where a "common sense" hypothesis which seems to explain a wide range of observations, yet is apparently confronted by one or two seemingly insurmountable physical (or chemical) objections, is not uncommon in interdisciplinary fields such as astronomy and earth science. The history of the nebula theory shows that it is sometimes premature to be overawed by a single "insurmountable" objection to a hypothesis which otherwise provides a satisfactory explanation for most of the relevant data and observations. Such "insurmountable" objections not infrequently disappear with the passage of time. A classic case was the controversy between the physicists and geologists concerning the age of the earth, around the turn of this century. The geologists had marshalled an imposing array of evidence from many fields, all implying an ancient origin for the earth which, however, appeared to be contradicted by a single, simple but formidable physical (geothermal) argument. With the subsequent discovery of natural radioactive heat sources, the physical objection to an ancient earth collapsed.

Less dramatic examples of such situations have continued to arise in the planetary sciences. Some are to be found in this book.

5.2 Development of Protostars

Theoretical Studies

Stars are born within interstellar clouds of gas and dust. In the earliest stages, owing to initial perturbations not yet fully understood, a gas–dust cloud becomes gravitationally unstable and collapses on the free-fall time scale of 10^5–10^6 years. During this stage, the cloud is transparent to its own radiation and little heating occurs. As the density rises, the system becomes more opaque to radiation, so that temperature increases. This, along with the density increase, results in a build-up of pressure which eventually becomes strong enough to support the central regions against gravity, so that collapse is temporarily halted. The central region continues to be compressed and heated as material from the outer regions is added to the equilibrium core. When the central temperature reaches 1800°K, endothermic dissociation of hydrogen molecules provides a further heat sink which permits collapse to proceed still further, leading to full ionization and hydrostatic equilibrium of the bulk of the material. At this stage, according to the model developed by Hayashi (1966), the star may become fully convective because of high opacity, which results in highly efficient transport of energy. A sudden and drastic increase in luminosity occurs as the star evolves along the so-called "Hayashi" track. The maximum luminosity reached may exceed, for a few years, a thousand times the present solar luminosity. The luminosity then falls sharply with further slower gravitational contraction over a timescale of 10^7 years, while internal temperatures continue to increase. When the central temperature reaches about 10^7 °K, nuclear reactions commence and spread outward, replacing the gravitational energy source. This marks the arrival of the star on the main sequence of the Hertsprung–Russell (HR) diagram.

The contraction of a spherical protostar has also been investigated by Larson (1969, 1972a, 1973), who obtained results significantly different from those of Hayashi. Larson found that collapse is highly nonhomologous, leading to a dense hydrostatic central core surrounded by an extended cloud of in-falling material, which is accreted to the nucleus. The luminosity during this stage is governed mainly by conversion of kinetic energy of the in-falling material into thermal energy and ultimately into radiation at the shock front at the boundary of the core. The maximum luminosity in this model is only about 30 times that of the present sun. The differences between Larson's model and the fully convective Hayashi model are partly to be attributed to initial conditions, particularly the initial density of the parental interstellar cloud, and have yet to be fully resolved (Bodenheimer, 1972). The time scale for the formation of the sun according to both models is on the order of 10^7 years.

The models discussed above do not allow for the effects of rotation and magnetic fields. The nonhomologous nature of the collapse accords well with observations of " cocoon " protostars—central hot nuclei surrounded by cool shells of dust and gas radiating in the infra-red (see, for example, Hartman, 1972). In the presence of rotation, Larson's (1972b) results indicate that this residual material would form an unstable circumstellar ring, rather than a disk. On the other hand, theoretical studies (see, for example, Ostriker, 1972) show that a disk configuration could be stable providing that a massive central condensation is present.

The theoretical problem of forming a circumstellar disk during the collapse of an initially homogeneous parental gas–dust cloud to form a protostar is not yet solved. It is possible that the influence of magnetic fields within the collapsing cloud may play an important role. Nevertheless, empirical observations, mentioned in the next section, provide strong evidence that circumstellar disks have indeed formed around some protostars.

A rather different model of stellar formation has been proposed by Cameron (1973a) and Cameron and Pine (1973). Although this particular model has since been abandoned (Cameron and Pollack, 1976; Cameron, 1977), it is discussed here because of the very great (and continuing) influence it has had on interpretations of chemical fractionations within the solar nebula.

Cameron assumed that the parental solar system cloud contained *two* solar masses of material. He concluded that the initial rotating spherical cloud possessed sufficient angular momentum to cause it first to collapse perpendicular to its axis, thereby forming a disk (stellisk) not possessing any pronounced central condensation of matter. Strong convection currents and other dissipative mechanisms within the disk then led to outward transport of angular momentum, accompanied by central condensation of mass to form the sun (Fig. 5.2). High temperatures ($\sim 2000°K$) were developed in the region now occupied by the terrestrial planets. The material in this region was totally vaporized, so that planetary accretion and formation was strongly influenced by selective condensation from the gas phase

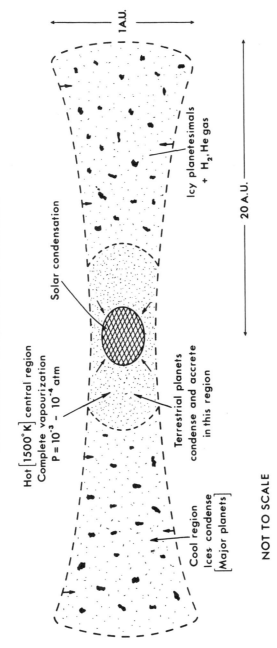

STELLISK : 2 SOLAR MASSES

Hot [1500° K] central region
Complete vapourization
$P = 10^{-3} - 10^{-4}$ atm

Solar condensation

1 A.U.

Icy planetesimals
+ H_2, He gas

20 A.U.

Terrestrial planets
condense and accrete
in this region

Cool region
Ices condense
[Major planets]

NOT TO SCALE

Figure 5.2 Primordial solar nebula (stellisk) according to the model of Cameron and Pine (1973). (From Ringwood 1975a, with permission.)

73

(Chapters 6 and 7). A notable feature of the hypothesis is that nearly one solar mass must have been removed from the system in the form of an intense solar wind (T-Tauri phase) combined with other dissipative mechanisms.

Cameron (1975a, 1977), and Cameron and Pollack (1976) have since recognized some difficulties with this hypothesis. These include the problem of disposing of the excess mass, the dynamical instability of the stellisk, and the short time scale which dissipative processes impose on the formation of planets. Cameron (1977) is currently developing an alternative hypothesis representing something of a compromise between the above stellisk model and the collapse model of Larson.

Observational Studies of Protostars

The nonhomologous collapse models of (Larson 1969, 1972a, b) and Bodenheimer (1972) predict the occurrence of cool clouds of dust and gas surrounding hot, central condensations. This phase of development may have been observed in some infrared stars consisting of a circumstellar shell of silicate (as shown by emission spectra) dust particles radiating in the infrared at temperatures of 300–1000°K, and surrounding a young, luminous star (Bodenheimer, 1972; Neugebauer et al., 1971; Strom et al., 1975b; Werner et al., 1977). There is a problem of interpretation, however, since the circumstellar infrared radiation may sometimes be derived from condensed matter ejected from evolved older stars (Neugebauer et al., 1971). Nevertheless, it is believed that in some cases, particularly where they are found to be associated with young star clusters embedded in dense clouds of dust and gas, the infrared stars probably represent protostars in the process of contraction. One of the more notable features is the rapid variability of luminosity in some of these objects. The brightness of FU Orionus increased by a factor of 250 over a period of 120 days. Herbig (1966) argued that this flareup might be identified with the collapse of a protostar, whilst Narita et al. (1970) proposed that the flareup was caused by rapid flattening of a previously absorbing circumstellar dust shell into a rotating disk. Many other young stars displaying rapidly fluctuating luminosities have been discovered—especially among the T-Tauri class (see below). It appears that major degrees of segregation and collapse of dust shells surrounding young stars can occur on very short time scales.

Further observations have suggested that some stars are surrounded by circumstellar *disks* of gas and dust. VY Canis Majoris is a variable star with infrared excess and is a source of OH, H_2O, and SiO maser radiation. It has been interpreted by Blerkom and Auer (1976) to be surrounded by a rotating equatorial disk, supporting Herbig's (1970) inference that this star is a young object, surrounded by a "solar-type" nebula. A different kind of disk, surrounding the highly evolved giant infrared star CIT6, has been described by Mufson and Liszt (1975). In this case, the disk is believed to have been ejected from the star, partly by rotational instability. Evidently, circumstellar disks may be formed by a multiplicity of mechanisms.

Thompson et al. (1977) have recently carried out a detailed investigation of the young O-type star, MWC 349. They conclude that it is surrounded by an opaque disk of dust and gas, seen approximately face-on and surrounded by a dusty HII region. The minimum mass is estimated to be about 1.5 percent that of the central star, the radius of the disk about 10^{13} cm and the thickness about $\frac{1}{20}$ of the radius. It is possible that the real mass is much larger than obtained, since the observations relate only to the portion of the disk radiating at accessible wavelengths. A large mass of cooler material may be present at greater distances. The luminosity of the disk has varied substantially with time. Densities and temperatures within the disk are compatible with accretionary processes and formation of solid planetesimals. Thompson et al. (1977) indeed suggest that planets may be forming, or have already formed in the disk surrounding this newly formed star. Clearly, these observations and inferences are of great interest and significance.

T-Tauri Stars

These are formed in regions of the Milky Way possessing high concentrations of dust and gas, such as the Orion Nebula, and are associated with young massive stars. Their brightness varies rapidly and they are characterized by rapid rotation, strong magnetic fields, and circumstellar shells of dust and gas which produce infrared radiation excesses. These shells are frequently found to be expanding rapidly outwards from the central star, implying substantial degrees of mass loss, via intense stellar winds. Kuhi (1964) inferred the occurrence of mass-loss rates exceeding 10^{-7} solar masses per year. His inference that a star of solar mass could lose 40 percent of its mass in this manner has been widely quoted. However, Kuhi implies that a preferable depletion estimate for a star of one solar mass may be about 0.2×10^{-7} solar masses per year. Moreover, he points out that in the view of the assumptions used, this figure probably overestimates the mass losses. Bodenheimer (1972) has discussed the situation and maintains that a substantial downward revision is necessary. Nevertheless, the phenomenon of mass loss at this stage is of great importance during the early development of the solar nebula.

T-Tauri stars are interpreted as approaching the main sequence, with a range of ages between 10^5–10^6 years (Strom et al., 1975b). These authors believe that loss of mass via stellar winds occurs at ages greater than 10^5 years. It has frequently been suggested that the sun passed through a T-Tauri phase, which was responsible for dissipation of that large part of the solar nebula which had failed to accrete into planets or planetesimals. This interpretation has not yet been finally confirmed. However, it is difficult to avoid the conclusion that the sun must have evolved through an extremely active early phase as its originally high rotational energy density was transferred to a greatly enhanced solar wind. Regardless of the detailed mechanisms involved, this phase of evolution must have been responsible both for braking the early rapid rotation of the sun and for dissipating the solar nebula.

5.3 The Formation of Disks

Although theoretical studies have not yet elucidated the mechanisms by which circumstellar disks might be formed around protostars, we discussed previously empirical evidence supporting the existence of such disks. The configuration of a discoidal structure orbiting around a massive central condensation can be recognized over a wide range of scales—galactic, stellar, and planetary. The Sombrero Hat galaxy shown in Figure 5.3, is indeed a spectacular example of this structure. The observations of disks surrounding some protostars support the general view that the planetary system evolved from a circumsolar disk of dust and gas. The close dynamical resemblances between the regular satellite systems of Jupiter, Saturn, and Uranus and the solar system itself likewise imply that these structures have evolved from circumplanetary disks. A relevant example is provided by the rings of Saturn, which might more properly be regarded as a disk containing gaps caused by resonances and perturbations. It seems likely that the

Figure 5.3 Disk structure surrounding massive central condensation—the Sombrero Hat galaxy (NGC 4594). Mt. Wilson and Palomar Galaxy Catalogue. Reproduced by courtesy of the Director of the Hale Observatories.

processes by which the solar system was formed share many aspects with those responsible for the formation of regular satellite systems. It will be argued later that the Earth's moon formed by the coagulation of such a disk.

Disks may be formed by several mechanisms. Schmidt (1944, 1958) hypothesized that the parental solar disk-nebula evolved from material captured when the sun passed through an interstellar cloud. This hypothesis does not explain two important dynamical properties of the solar system— the direct rotation of the sun in the same sense and nearly in the same plane as the planets, and the close analogies between the solar system itself and the regular satellite systems of the major planets. Schmidt's model also encounters other difficulties (Safronov, 1972a) and has not, therefore, found widespread support.

It is generally believed, following Laplace, that formation of the disk was an integral part of the formation of the star. Two possibilities based on observational interpretations of circumstellar disks have been considered. According to one of these, the disk might form out of material ejected from the central star. A model of this type was proposed for the solar system by Hoyle (1960). Rotational instability was believed to have set in when the contracting protosun reached the orbit of Mercury. Matter was then shed in the equatorial plane (cf., Laplace, 1796) and was hydromagnetically coupled to the sun. As the sun continued to contract, angular momentum was transferred to the circumsolar ring which was driven outwards to form a disk, ultimately parental to the planets. Although the basic process described by Hoyle probably has an important role in cosmogony, this specific model has been criticized (see, for example, Safronov, 1972a; Mestel, 1972) as requiring an improbable set of initial conditions; it also encounters a number of specific physical difficulties. Moreover, the recent interpretation of anomalies (Clayton and Mayeda, 1975) in the isotopic composition of oxygen in meteorites and in the earth, implying the existence of chemical inhomogeneities within the solar nebula, does not accord well with a model whereby the nebula was derived by condensation from a gas derived from well-mixed outer solar layers.

According to the second possibility, the circumstellar disk was derived from a small residue of the same cloud from which the sun was derived, but which possessed too much angular momentum to fall into the star. Nonhomologous collapse of the parental cloud led to the formation of a massive central condensation surrounded by a small remnant of the cloud in a spherical configuration. This remnant then contracted perpendicular to its rotation axis to form an orbiting disk, stabilized (temporarily) by the massive central condensation. It remains possible that transfer of angular momentum from the central nucleus into the disk by means of magnetic fields and/or turbulent viscosity played a significant role. There is widespread consensus that the sun's parental planetary disk formed in some such manner (see, for example, Huang, 1969), even though the details are not understood.

5.4 The Outer Planets and Their Regular Satellite Systems

We have already noted the remarkably close dynamical resemblances between the solar system on the one hand, and the regular satellite systems of Jupiter, Saturn, and Uranus on the other. It is possible that study of these satellite systems might provide insight into some aspects of the formation of planets.

Before proceeding, we need to refer to data on the composition of the sun as discussed in Chapter 6, Table 6.1. These show that the solar composition can be expressed in terms of three components that are useful in discussions of planetary chemistry. These are *gases* (hydrogen + helium) which make up 98 weight-percent of the solar mass, *ices* (H_2O, NH_3, CH_4) which amount to 1.5 percent by weight, and *rock* composed of the oxides of common metals and contributing 0.5 percent of the primordial mix.

Some interpretations of the constitution of Jupiter (Hubbard and Smoluchowski, 1973) marginally permit this planet to be of solar composition, providing some degree of segregation of helium towards the interior were possible. Such a segregation process would require some rather specialized and uncertain mechanisms. This "solar abundance" model for Jupiter would suggest that the planet had formed by a process of gravitational instability in the solar nebula, analogous to formation of the sun from an interstellar cloud. Calculations of the early evolution of Jupiter on this basis have been carried out by Bodenheimer (1974, 1977) and others. Many aspects of Jovian evolution, including the existence of an early high-luminosity phase and an envelope which might have been parental to the satellite system, are closely analogous with the results for early solar evolution.

Podolak and Cameron (1974), Stevenson and Salpeter (1976), and others have concluded, however, that providing Jupiter has a solar H/He ratio throughout the planet, the total ratio of (rock + ices) to gases ($H_2 + He$) in Jupiter is probably substantially higher than the solar ratio. If Jupiter were of solar composition, it would contain about 2 percent of rock-plus-ice, or about six earth masses of these components. By comparison, the outer planets Uranus and Neptune are dominantly composed of the latter components, each containing about twelve earth masses of rock-plus-ice.

Podolak and Cameron (1974) and Cameron and Pollack (1976) suggest that the growth of the giant planets (Jupiter and Saturn) began analogously to Uranus and Neptune, by the accretion of rock-plus-ice planetesimals to form a nucleus amounting to perhaps 20–30 earth masses, somewhat larger than the present sizes of Uranus and Neptune. These authors conclude that, as the mass of the nuclei grew by accretion of rock-plus-ices, their gravitational fields caused a strong concentration of hydrogen and helium towards them, forming a large atmosphere of gas. Finally, when the mass of the hydrogen + helium atmosphere became comparable with the mass of the nucleus, a hydrodynamic instability set in causing a large volume of the surrounding solar nebula to collapse onto the nucleus, increasing its size to about 60 earth masses. The final masses of Jupiter and Saturn were greatly augmented by further gravitational accretion of hydrogen and helium plus accompanying dust particles of ice and rock from the solar nebula onto these very large nuclei. Apart from the initial nucleus of rock plus ice, the material subsequently accreting on Jupiter and Saturn would be of solar composition.

The conditions required for gas-dynamic instability within the solar nebula appear to be rather critical, the process being favored by increasing nebula density and mass of nucleus, and opposed by increasing nebula temperature. Cameron suggests that in the cases of Uranus and Neptune the critical planetary nuclear masses were not achieved, so that the phase of hydrodynamic collapse of hydrogen and helium from the nebula did not occur.

Both classes of models for the origin of the giant planets, those of Boden-heimer on the one hand and Cameron and coworkers on the other, share essential elements in common with current models of formation of the sun. This correspondence tempts us to explore for analogies between the origins of regular satellite systems and of the planetary system itself.

The four large satellites of Jupiter—Io, Europa, Ganymede, and Callisto —rotate with coplanar circular orbits in the equatorial plane of the planet. The orbital radii increase regularly outwards according to a form of Bode's Law. The mean densities of Io, Europa, Ganymede, and Callisto are 3.50, 3.42, 1.95, and 1.65 g/cm^3, respectively (Cameron and Pollack, 1976). Thus, the densities decrease with increasing distance from the planet. The inner two of these satellites must be predominantly composed of *rock*, whereas the outer two must possess increasing amounts of *ices*.

These orbital characteristics strongly suggest formation from a parental disk or nebula which once rotated about Jupiter. Such a disk may well have formed during the dynamical collapse phase, because of excess angular momentum in the system. Formation in this manner requires that Jupiter was once rotating more rapidly, so that there was a continuum between Jupiter and its surrounding nebula. It would follow that some mechanism existed for extracting a large amount of angular momentum from proto-Jupiter.

Alfvén (1964a) had previously pointed out that the discoidal nebulae which were parental to Jupiter and Saturn probably possessed excess angular momentum which had to be removed before condensation into central bodies was possible. He suggested, furthermore, that the excess angular momentum had been transferred via hydromagnetic coupling to ionized light gases which were lost, a mechanism related to that previously proposed for braking the sun's rotation.

This suggestion has been followed up by Cameron and Pollack (1976), who conclude that much of the original angular momentum of Jupiter may have been transferred via hydromagnetic coupling to the parental satellite disk. This would have required a strong primordial Jovian magnetic field combined with an acceptable level of ionization within the disk. The net effect would have been to push the disk further away from Jupiter, so as to facilitate the dissipation of the disk and to retard the rotation of Jupiter.

The regular density increase of the satellites outward from Jupiter might be explained by accretion within the thermal field established by liberation of gravitational energy during the accretion of Jupiter (Ringwood, 1966b, p.98). This interpretation has been strongly supported by detailed calculations of Jupiter's early evolution (Graboske et al., 1975; Pollack and Reynolds, 1974).

The interpretation applied above to Jupiter and its satellites might also be applied to Saturn and its regular satellites (Pollack et al., 1977b). However, the Uranus system appears to have evolved differently. Uranus has five satellites possessing highly circular orbits lying in the equatorial plane of the planet. These satellites revolve in their orbits in the same sense as the rotation of the planet and form a more compact system than the regular satellite systems of Jupiter and Saturn. The unique character of the Uranus system is the inclination of its axis with respect to the normal to the plane of the ecliptic by 98 degrees, giving Uranus a retrograde motion.

Safronov (1972a) has explained the obliquities of planetary rotation axes to the ecliptic plane by the impacts of very large planetesimals during the final stages of planetary accretion. In the case of Uranus, he calculates that this planetesimal (actually of terrestrial planet dimensions) amounted to 7 percent of the mass of Uranus.

If the Uranus satellite system had formed by hydrodynamic collapse of portions of the solar nebula (as is believed have occurred in the cases of Jupiter and Saturn), the parental gaseous disk would have formed approximately in the plane of the ecliptic and resultant satellites could not have achieved orbits in the equatorial plane of Uranus (Ruskol, 1973). Cameron (1975b) has accordingly suggested that the disk was derived from material evaporated and spun off from proto-Uranus during the collision. Hydromagnetic coupling between ionized matter in the disk and Uranus via an assumed planetary magnetic field is hypothesized to have caused a transfer of angular momentum from planet to disk, thereby driving the disk outwards from Uranus, prior to coagulation to form satellites. In this case, we have a partial analogy with one of the modes of origin of circumstellar disks (Section 5.2), whereby the disk formed from material *ejected* from the star. We will examine, in Chapter 12, the possibility that the Earth's moon was formed by an analogous process.

5.5 Angular Momentum Relationships within the Solar System

An important relationship between angular momentum density and mass was noticed by Fish (1967) and Hartmann and Larson (1967). When angular momentum density (angular momentum per unit mass) is plotted against mass for the larger asteroids, the combined Earth–Moon system, the major planets, and the combined solar system, they are found to closely follow a simple relationship (Fig. 5.4),

$$A = k' M^{2/3}, \tag{1}$$

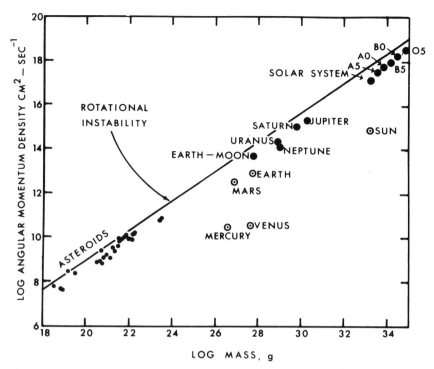

Figure 5.4 Angular momentum density versus mass for the larger asteroids, planets, combined Earth–Moon and solar systems, and rapidly rotating stars. (Based on Fish, 1967; Hartmann and Larson, 1967; and Kaula, 1968, p. 428.) Bodies represented by solid symbols are aligned along a line (not shown) given by the relation $A = k'M^{2/3}$, which is displaced downwards by a small factor from the line (shown) representing the approximate rotational instability criterion, $A = k''M^{2/3}$ where $k'' \approx 3k'$. Bodies represented by open symbols constitute exceptions to the above relationship. (From Ringwood, 1972, with permission.)

where $k' = 1.6 \times 10^{-5}$, A is the angular momentum density, and M is the mass, all in CGS units. Moreover, we see from Figure 5.4 that young, massive, rapidly rotating B and O stars fall upon the same line.

Since $A = \alpha R^2 \omega$, where α is the gyration constant, R is the radius, and ω is the rotational velocity of the planet, and assuming constant density, then M is proportional to R^3 and it follows that A is proportional to $\omega M^{2/3}$. Thus, the empirical relationship (1) is equivalent to stating that the rotational periods of the bodies and systems considered are independent of mass, and that this relationship holds over 14 orders of magnitude. This is a remarkable relationship, the significance of which has been emphasized by Alfvén (1964b).

The relationship suggests the operation of a fundamental physical process. The above authors point out that rotational instability leads approximately to a mass-independent constant period. Equating centrifugal force at the equator to gravity, one has

$$\omega^2 = 2\pi G \rho \phi, \tag{2}$$

where ρ is density, G is the gravitational constant, and ϕ is a constant varying between 0.19 and 0.36, depending on the model chosen (Alfvén, 1964a). Neglecting the variations in ρ and ϕ, (2) implies a constant period and a relationship of the form $A = k'' M^{2/3}$, which is plotted in Figure 5.4 as a straight line. However, k'' in the relationship is about three times larger than the empirical k' of Eq. (1). Thus the points for the bodies and systems indicated are displaced systematically below the theoretical rotational instability limit. In terms of periods, rotational instability of the present planets would require periods in the range 1.4–3.4 hours, whereas the most common periods in the solar system are 8–10 hours (Alfvén 1946b). The Earth–Moon system, if combined, would have a period of 4–5 hours, depending on density distribution.

The rotational instability line of Figure 5.4, of course, represents an upper limit to angular momentum density. The clouds from which stars were born almost certainly possessed much greater amounts of angular momentum, but much of this was removed during contraction by hydromagnetic coupling to ejected material and perhaps to surrounding preplanetary disks. Thus, it appears likely that the young, massive, rapidly rotating O and B stars have evolved from points much closer to the rotational instability line via loss of mass and angular momentum. Likewise, the solar system must have evolved downwards from the line as light gases were lost, carrying angular momentum.

Alfvén (1964a) and Cameron (1973b) have pointed out that the parental dust clouds from which the major planets were derived also possessed excesses of angular momentum which had to be dissipated before condensation into massive central bodies was possible, and suggested that hydromagnetic processes were responsible. This is particularly clear in the cases of Jupiter and Saturn. Following previous discussion, their satellites (now composed of rock-plus-ice) are believed to have been derived from parental disks of solar composition. The composition of these disks can be obtained by restoring the complementary ices-plus-gases of Table 6.1, which have been lost. These disks amount to about 2 percent of the mass of each giant planet. If the angular momentum represented by the disks were incorporated back into the giant planets, their present rotation periods of about 10 hours would be reduced to about 5 hours, placing them much closer to the rotational instability line.

The periods of the large asteroids have probably not been greatly modified by collisions (Hartmann, 1969) and these are plotted in Figure 5.4. It appears possible that these bodies were also formed closer to the rotational instability line of Figure 5.4, but have evolved downwards via loss of mass and angular momentum. Soon after their formation, the asteroids may have possessed mantles of ices which were subsequently evaporated when the sun passed through a T-Tauri or Hayashi phase. Evaporation would have occurred at the subsolar points and the evaporated material would have contained more angular momentum per unit mass than the parent bodies. Loss of angular momentum in this manner would cause a corresponding increase in rotation periods.

Finally, some comments are now made concerning the bodies which constitute exceptions in the sense that they deviate seriously from the empirical $A = k'M^{2/3}$ relationship (Fig. 5.4). The slow rotation of the sun is believed due to a hydromagnetic braking process as previously discussed. The rotation of Mercury is believed to have been retarded by solar tides, an explanation which might also apply to Venus (Goldreich and Soter, 1966). The Earth is known to have transferred most of its angular momentum to the moon via tidal interaction. The only clear exception is provided by Mars. This anomaly is discussed by Hartmann et al. (1975) and several possible mechanisms for removing angular momentum are considered. None of these are very convincing and the problem is regarded as unsolved. The most promising possibility invokes the scattering of high-eccentricity planetesimals by Jupiter and their impact on Mars. It is widely believed that Jupiter's perturbing influence prevented the accretion of a planet in the asteroid belt and the small mass of Mars may also result from his factor (Weidenschilling, 1975).

The preceding discussion does not imply that planets necessarily formed at the rotational instability limits, but it does suggest that most, if not all, planets originated in local systems possessing much greater angular momentum densities than are displayed by the present planets, and that mechanisms for removing this excess angular momentum have operated.

An alternative explanation of the proportionality between angular momentum density and (mass)$^{2/3}$ has been provided by Giuli (1968a,b). Based on a study of the accretion of planets from a system of direct eccentric orbits, and assuming that impact velocities were proportional to the escape velocity $(2GM/R)^{1/2}$ and that impact offsets were proportional to the radius R, Giuli found that planets would achieve direct rotations and would follow a relationship of the type $A = kM^{2/3}$ (Fig. 5.4). Kaula and Harris (1975) have also argued in favor of this mechanism.

5.6 Possible Relevance for Lunar Origin

In this chapter, we have described some very general characteristics of protostars, disks, and planetary systems which may possibly provide some perspectives on the dynamical aspects of lunar origin. We noted that disk structures can form at a wide range of scales—galactic, stellar, and planetary—and by at least two mechanisms. One kind of disk can be formed from material possessing excess angular momentum, left behind in orbit by a collapsing star or giant planet. Another may form from material ejected from a star or planet, by one of several mechanisms, including fission and major collisions (Uranus?).

In both cases of disk formation, transfer of angular momentum may occur between the central star or planet and the disk, caused by either hydromagnetic torques or turbulent viscosity. Satellites may subsequently form from both kinds of planetary disks, and planets from stellar disks.

A consideration of initial angular momentum densities among young stars, the outer planets, and large asteroids, allowing for transfer of angular momentum and loss of light gases, suggests that the minimum initial rotation period of most planets may have been about 5 hours or even less. If the Moon were fused with the Earth, the combined system would rotate with a period of 4–5 hours. The fact that the angular momentum density–mass relationship of the combined Earth–Moon system agrees closely with those systems previously discussed, particularly if the latter are shifted significantly upwards (Fig. 5.4) to allow for evolutionary losses of angular momentum, is a strong argument in favor of a common origin for the Earth–Moon system (Fish, 1967; Hartmann and Larson, 1967).

The mechanisms discussed above may well be relevant to the origin of the Moon, which may have formed by coagulation of a primitive circumterrestrial disk (Öpik, 1955, 1961, 1967). As discussed in Chapter 12, this disk could have been formed in at least three ways—fission from Earth, the disruption of a large planetesimal passing within Roche's limit, or by the impact of a large planetesimal on the Earth's surface. Processes of transfer of angular momentum between Earth and disk and loss of material, either as gases or solids, would have been involved in all three mechanisms. The point is that the fundamental processes which can reasonably be called upon to explain the origin of the Moon were probably not unique. It seems likely that they have operated in other stellar and planetary systems. Even the mass ratio of Moon to Earth is not necessarily anomalous when compared to the original mass ratios in other systems, prior to the loss of light gases (Ringwood, 1972). These considerations may, perhaps, place the origin of the Earth–Moon system in a somewhat different perspective from that which is often adopted.

CHAPTER 6

Aspects of Planet Formation in the Primordial Solar Nebula[1]

6.1 Mass of the Nebula

The initial bulk composition of the primordial solar nebula was essentially similar to that of the present sun (Fig. 2.2). At low temperatures, the nebula would consist of dust particles associated with a *gas* phase composed of hydrogen and helium. The dust particles would be composed of two principal classes of material: (a) *ices*, representing the frozen hydrides of O, C, N, Cl, plus Ne and Ar, and (b) *rock*, representing the relatively involatile metallic oxide components. The weight proportions of *gas*, *ice*, and *rock* in the nebula are given in Table 6.1. The overwhelming preponderance of hydrogen and helium is to be noted.

The terrestrial planets are composed mainly of rock, whereas the densities of the outer planets Uranus and Neptune (Table 5.1) show them to be composed mainly of ices. In contrast, the other outer planets, Jupiter and Saturn, are composed dominantly of hydrogen and helium. It is evident that very large fractionations of volatile components occurred during the formation of planets, the present masses of which represent only a small remnant of the original nebula. The terrestrial planets comprise only 0.5 percent of the material originally present in this region.

We can attempt to estimate the original mass of the nebula by adding in the complementary ices and gases that are believed to have been lost.

---

[1] We are referring to the discoidal portion of the nebula which was parental to the planets.

Table 6.1 Relative proportions (by weight) of *gases*, *ices*, and *rock* in the primordial solar nebula[a]

		wt.-%
Group I *gases*	H, He	98.0
Group II *ices*	C, N, O,[b] Ne, S, Ar, Cl (as hydrides, except Ne, Ar)	1.5
Group III *rock*	Na, Mg, Al, Si, Ca, Fe, Ni (as oxides)	0.5

[a] The data are based on solar photosphere abundances given by J. E. Ross and Aller (1976).

[b] The oxygen abundance in Group II is adjusted for oxygen combined with Group III elements.

This procedure is rendered uncertain principally by lack of agreement concerning the bulk composition of the giant planets, especially Jupiter. If Jupiter is assumed to be of solar composition (an interpretation which appears only marginally consistent with existing data), then the mass of the nebula would amount to about 1 percent of that of the sun. This might be regarded as the *minimum* mass of the nebula, as in the model of Hoyle (1960). Recent studies (see, for example, Podolak and Cameron, 1974; Stevenson, 1978) tend to favor models of Jupiter and Saturn which are enriched in rock and ice components relative to hydrogen and helium by factors of 5–20, compared to the primordial abundances. This would imply the former presence of much more hydrogen and helium in the outer planet region than in the "minimum mass" nebula. To this must be added the very uncertain masses of the cometary cloud and of material which was gravitationally ejected and lost from the solar system during accretion of the outer planets.

From considerations of this nature, Safronov (1972a) estimated that the nebula comprised about 5 percent of the solar mass. Most authors who have considered this problem arrive at an estimate for the mass of the nebula within a factor of two of Safronov's value. In the region of the terrestrial planets, the volume density for a nebula in this mass range would be from 10^{-8} to 10^{-9} gm/cm^3, while the pressure at 1 AU would be about 10^{-5} atm.

The radial mass distribution within the solar nebula has been discussed by Weidenschilling (1977), who constructed a model nebula by adding the solar complement of light elements to each planet using recent models of planetary compositions (Figure 6.1). It was assumed that the nebula was continuous and that the "solar equivalent" mass for each planet was spread throughout an annular zone extending approximately halfway to the orbits of the planets on either side.

The computed surface density from Venus to Neptune was found to vary approximately as $R^{-3/2}$. However, the zones of Mercury, Mars, and the asteroids appeared to be strongly depleted in mass (Fig. 6.1). The fit of Venus and Earth with the trend set by the outer planets suggests a common

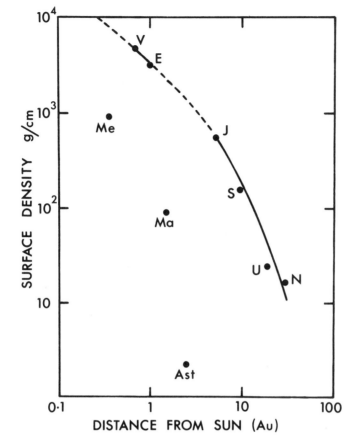

Figure 6.1 Surface densities in the primordial solar nebula obtained by restoring the planets to solar composition and spreading the resulting masses through contiguous zones surrounding their orbits. Anomalies at the orbits of Mercury, Mars, and the asteroids are believed due to specialized processes of mass loss. (After Weidenschilling, 1977.)

origin from a nebula with an initially monotonic variation of surface density. It seems very likely that the depletion of mass in the zones of Mars and the asteroids was caused by the strongly perturbing effect of Jupiter upon planetesimals in its region. These would be driven into eccentric orbits which would pass through the Mars–asteroids region, colliding with and disrupting the planetesimals that had formed there (Safronov, 1972a; Weidenschilling, 1975).

However, the low surface density in the region of Mercury is not so readily explained. Weidenschilling (1977, 1978) suggests from Figure 6.1 that the mass of Mercury should have been similar to that of Earth and Venus. However, a large amount of material, about 90 percent, was lost from this region by aerodynamic drag, which caused the material to spiral into the sun. This process is further discussed in Section 9.4.

In contrast to the above class of models, Cameron (1973a) has advocated a very large parental nebula, amounting to two solar masses (Fig. 5.2). However, as noted in Sections 5.2 and 7.3, this model has encountered some serious difficulties and Cameron (1977) is currently developing nebula models characterized by smaller masses.

6.2 Accretion of the Planets

Following the discussion in Chapter 5, we assume that a rotating circumstellar shell of dust and gas was left behind during contraction of the protosun. The effect of self gravitation parallel to the rotation axis caused the shell to flatten towards a discoidal configuration.

The processses by which dust particles began to adhere to one another and form a hierarchy of planetesimals of varying sizes are poorly understood. It is clear, however, that for accretion to commence, the relative velocities of dust grains had to be minimized so that they approached each other very slowly, thereby allowing surface chemical forces to cause them to adhere during gentle impacts. The presence of gas at this stage played a crucial role in damping random chaotic motions by viscous dissipation, thereby causing dust grains to rotate in nearly coplanar circular orbits with small relative velocities (Schmidt, 1944, 1958; Levin, 1949, 1972a; Urey, 1952; Safronov, 1954, 1972a; Hartmann, 1972; Cameron, 1973a). Dust grains began to adhere, building up a first generation of small aggregations or planetesimals, perhaps with dimensions up to a few centimeters. This stage of accretion is represented in the accompanying sketches (Fig. 6.2). These small planetesimals fell quite rapidly through the gas, reaching the equatorial plane in a period which could have been as short as 1–1000 years (Safronov, 1972a; Huang, 1973). Thus a thin, relatively dense disk of dust and small planetesimals collected quickly in the equatorial plane.

Edgeworth (1949), Safronov (1972a), and Goldreich and Ward (1973) demonstrated that such a disk would be characterized by an acute gravitational instability that would cause it to collapse rapidly to form a series of self-gravitating "cloudlets" of small planetesimals and dust. These cloudlets contracted to form a second generation of planetesimals approximately of asteroidal sizes (Fig. 6.2). The time interval for this process was on the order of 10^4 years.

The accumulation of planets from this generation of planetesimals according to the theory of Schmidt (1958), Safronov (1972a), and G. W. Wetherill (1976a) is depicted in Figure 6.3. Planetesimals were initially moving in circular, coplanar orbits, so`that approach velocities were low and a significant proportion of collisions led to effective fusion of the colliding bodies. A hierarchy of planetesimals with different sizes developed. Beyond a certain critical radius, on the order of a few hundred kilometers, the bodies developed gravity fields sufficiently large to perturb the orbits of approaching planetesimals, increasing their eccentricities and inclinations.

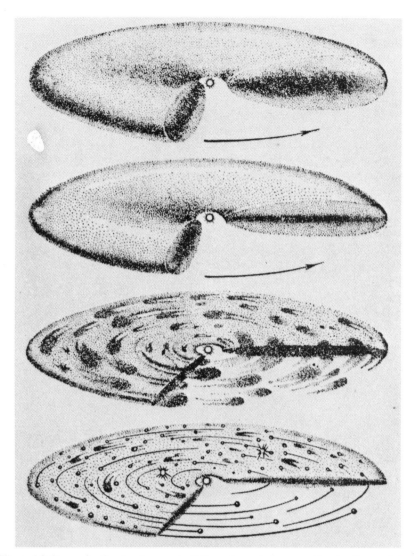

Figure 6.2 Stages in the development of the solar nebula according to the model of Schmidt and Safronov. Sedimentation of dust and small planetesimals to the ecliptic plane is followed by gravitational instability of the dust disk and formation of a population of asteroidal-sized planetesimals. (From Levin, 1972a, with permission.)

This represented a key stage in planetary accumulation, because at this critical radius the large planetesimals began to sweep up surrounding dust and small planetesimals not with their *geometric cross sections*, which are proportional to R^2, but with their *gravitational cross sections*, which are proportional to R^4 (Hartmann, 1969).

Thus, in a given zone in the solar system, if one planetesimal exceeded the critical radius, it would have grown very rapidly, sweeping up mass at the expense of surrounding planetesimals. In such a zone, this particular

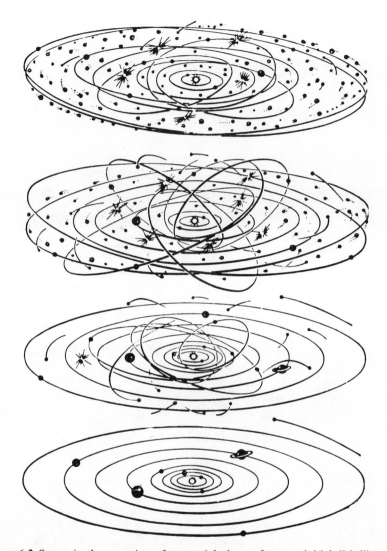

Figure 6.3 Stages in the accretion of terrestrial planets from an initial disk-like configuration of asteroidal-sized planetesimals, according to the Schmidt–Safronov model. The initially flat system of intermediate bodies first thickens due to their mutual gravitational perturbations, but returns to the flat configuration at the termination of planet formation owing to the averaging effects of large numbers of random planetesimal orbits (From Levin, 1972a, with permission.)

planetesimal would have grown much more rapidly than all others and would become a *planetary nucleus*. This nucleus would rapidly develop a large gravitational field and would dominate the subsequent accretional processes in its "feeding zone." Other planetesimals may have exceeded the critical radius for gravitational growth, but, in general, would have been very much smaller than the planetary nucleus. Thus, Safronov estimates that for most planets, the mass of the largest impacting planetesimals was from 10^{-2} to 10^{-3} of the mass of the planet.[2] We see from Figure 6.3 that although accretion commenced in a highly flattened disk of planetesimals rotating in coplanar circular orbits, the perturbations by growing planetary nuclei caused large increases of orbital eccentricities and inclinations of planetesimals, so that the effective thickness of the disk increased considerably during the accretion process. The present nearly circular, coplanar orbits of the planets result from an effective averaging of the chaotic motions of very large numbers of individual planetesimals. Safronov explains the present inclinations of the axes of rotation of planets as being caused by a random component of angular momentum contributed by impact of the "next largest planetesimal" upon the planetary nucleus.

The Schmidt–Safronov theory of accretion of planets is remarkably comprehensive and, within the framework of its assumptions, is relatively rigorous. Nevertheless, there is one unsatisfactory aspect which indicates that it is incomplete. This concerns the time scales calculated for the accretion of Mars, Uranus, and Neptune by these purely gravitational processes.

In general terms, the theory predicts that the rate of growth of mass of a planetary nucleus increases rapidly during the early stages (owing to increasing gravitational cross section of the nucleus), passes through a maximum, and then decreases because of exhaustion of material in the "feeding zone" of the planet.

The rate of growth (Levin, 1972a) is given by

$$\frac{dM}{dt} = \pi R_E^2 \, \rho V \propto R_E^2 \, \sigma, \tag{1}$$

where ρ and σ are, respectively, the volume density and surface density of accretable material; V is the mean velocity of planetesimals relative to the planetary nucleus; and R_E is the "effective radius" of the latter. R_E increases with decreasing V according to the formula

$$R_E^2 = R^2 \left(1 + \frac{2GM}{RV^2}\right) = R^2 \left(1 + \frac{V_e^2}{V^2}\right), \tag{2}$$

where R and V_e are the geometric radius of the planetary nucleus and the escape velocity at its surface, respectively.

The theory above leads to the conclusion that accretion of the earth (98 percent complete) occurred over a period of about 10^8 years, a figure which has been widely quoted. However, the theory also leads to a time scale for accretion of Uranus and Neptune on the order of 10^{11} years (Safronov, 1972a), i.e., more than 10 times the age of the solar system. The expedients adopted for avoiding this fundamental difficulty (see, for example, Levin,

---

[2] Uranus, with its rotational obliquity of $98°$, may have been hit by a "planetesimal" amounting to 7 percent of its mass (Safronov, 1972a).

1972b) are unconvincing. Further difficulties appear in the case of Mars, for which an accretional period of 2.6×10^9 years is calculated (Weidenschilling, 1976). Although there is a large uncertainty, crater counts support a much older surface than this. More seriously, such a relatively young age for completion of the accretion of Mars would imply a heavy bombardment of the Earth and Moon until 2×10^9 years ago. However, lunar studies imply that this period of rapid bombardment ceased at least about 3.9×10^9 years ago.

It seems, therefore, that there may be some factor which is overlooked in the Schmidt–Safronov treatment. Referring to Eq. (1) and (2), we see that the rate of growth of planets can be enhanced by increasing the surface density σ of accretable material, and/or by decreasing the relative (chaotic) velocities V of planetesimals, thus increasing the cross section for the sweeping-up process. These factors are not independent, since an increase in σ is likely to lead to a larger degree of gas–solid dissipation and, accordingly, to smaller V.

It is possible that a flaw in the Schmidt–Safronov treatment lies in their assumption of axial symmetry within the discoidal nebula, as shown in Figures 6.2 and 6.3. The effects of turbulence and magnetic fields in the parental cloud may well have produced an inhomogeneous "clumpy" structure at an early stage, and this may have been retained during collapse to the disk, so that the surface density σ of large regions of the nebula varied strongly with angular position as well as radius. A distant analogue of this postulated inhomogeneous distribution of matter within a disk may be provided by spiral galaxies. Studies of T-Tauri stars (see, for example, Gahm et al., 1975) have also revealed evidence that they are surrounded at planetary distances by discrete orbiting clouds of dust particles, radiating in the infrared.

McCrea (1960, 1972) and McCrea and Williams (1965) have developed theories of solar system origin invoking these properties. The sun is believed to have been surrounded by about 1000 small cloudlets (floccules) in random, chaotic motion as a result of earlier turbulence in the interstellar medium. This system of floccules, each of solar composition and at temperatures of about $100°K$, settled into the ecliptic plane. Collisions between floccules led to the development of a small number of large condensations which are analogous in some respects to the protoplanets of Kuiper (1951).

The significant aspect of these models is their highly inhomogeneous distribution of matter within the invariant plane as compared to Safronov's primordial nebula. The surface densities in the "patches of fog" (McCrea, 1972) or "protoplanets" within which planets ultimately accumulated are far higher than in Safronov's model, so that the accumulation of planets by sedimentation of planetesimals to the center of these "protoplanets" can occur much more rapidly—on a time scale of 10^4–10^5 years.

According to this hypothesis, the protoplanets (initially of solar composition), which were parental to the terrestrial planets, possessed only a transient existence because of marginal gravitational stability. After rapid sedimentation of solid planetesimals to their centers, they were dispersed by solar tidal action (since, at this stage, their densities became smaller than the corresponding Roche densities). The dissipation was also facilitated by heating caused by sedimentation of solid planetesimals towards the center.

McCrea's hypothesis would lead to relatively short accretion times for planets. It has not been developed to the same degree as the Safronov model and difficulties may well emerge. For example, the processes by which the terrestrial protoplanets achieved a state of initial gravitational stability permitting sedimentation of solids, after which their gaseous envelopes were dispersed by solar tidal action, appears somewhat contrived. It is possible, however, to conceive of a series of "hybrid models," representing various combinations of the Safronov and McCrea hypotheses, which might repay future investigation.

6.3 Temperature Regime and the Compositions of Planetesimals

The final chemical compositions of planets are likely to have been strongly influenced by the temperature–pressure–time paths followed within the nebula during the formation of planetesimals. Unfortunately, these parameters are poorly known. In these circumstances, it seems preferable to base our models as closely as possible on empirical evidence. Those infrared stars which are believed to be forming from collapsing interstellar gas clouds (Chapter 5) are particularly important in this respect. Protostars of this type consist of a hot, centrally condensed, contracting nucleus surrounded by a shell or cocoon of silicate dust particles radiating in the infrared at temperatures typically between 300 and 1000°K. Higher temperatures are doubtless reached in their interiors, but are not readily observable (Fig. 6.4).

Let us now reconsider the collapse of the circumstellar shell (assumed to be of solar bulk composition) to form a disk. Following previous discussion (Section 6.2), this requires firstly the coagulation of dust grains (initial size ~0.1 micron) to form small planetesimals with dimensions ranging perhaps from 10^{-2} to 10 cm. At this stage, the planetesimals become decoupled from the gas and attempt, initially, to follow Keplerian orbits about the sun. However, this results in strong dissipation between the planetesimals and the gas phase which is supported partly by its own internal pressure and may also be in a state of turbulent convection. Under the combined influence of self-gravitation perpendicular to the ecliptic plane and gaseous dissipation, the planetesimals spiral down to the ecliptic plane in a very short period (Safronov, 1972a; Huang, 1973). The paths taken by the planetesimals have a large vertical component, thereby cutting the gaseous isotherms at high angles. The overall result (Fig. 6.4) is that in a given zone where a planet may form, the solid material which ultimately settles into the ecliptic plane will have reached equilibrium with surrounding gas over a wide range of temperatures. We will see later that grain compositions are very sensitive

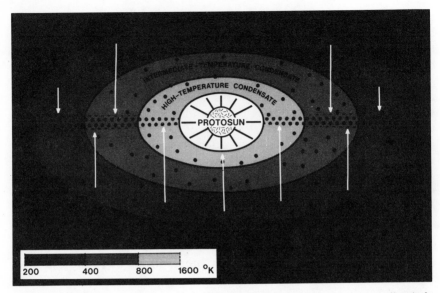

Figure 6.4 Diagram showing notional temperature distribution in "cocoon" nebula surrounding sun, and prior to collapse into discoidal configuration. Small first-generation planetesimals form within the shell and sink towards the ecliptic plane in the general direction indicated by the white arrows (note that the actual paths are more complex than this, as discussed in text). An important result is that at any given distance from the sun in the ecliptic plane, the solids that collect may have been subjected to a wide range of temperature and pressure conditions.

to temperatures over the range shown in Fig. 6.4, so that the chemical composition of the condensate which finally reaches the ecliptic plane may be highly heterogeneous. The degree of heterogeneity, however, will depend on several kinetic factors which cannot be readily quantified. These include (a) sizes of first generation planetesimals formed in the shell, before sedimentation into a disk, (b) the time (t_1) required for planetesimals to fall to the ecliptic plane, and (c) the time (t_2) required for planetesimals in the ecliptic plane to collect into asteroidal-sized bodies (second-generation planetesimals). Qualitatively, we can envisage three principal end results arising from the operation of these processes.

(a) The first generation of planetesimals formed in the shell are very small (e.g., $\ll 1$ mm), so that sinking toward the ecliptic plane is correspondingly slow. The nebula becomes increasingly opaque to solar radiation and cools efficiently as flattening towards a discoidal configuration occurs. Under these circumstances, the small planetesimals are likely to remain in chemical equilibrium with the surrounding gas. By the time the planetesimals reach the ecliptic plane, the nebula will be cold (e.g., $<300°K$ at 1 AU), as described by Safronov (1972a) and Schatzman (1967). The solid condensate will have remained in chemical equilibrium with surrounding gases at all stages, so that solids which finally collect into asteroidal-sized bodies (second-generation planetesimals) will reflect equilibration with solar gases *at relatively low temperatures*. A similar end result would be reached according to the model proposed by McCrea (1960).

(b) Efficient sticking and accretion of solid particles occurs within the circumstellar shell leading to the formation of relatively large (~ 10 cm) planetesimals *before the distended shell has collapsed to a disk*. These first-generation planetesimals sink rapidly to the ecliptic plane on a time scale (t_1) which is much shorter than the time scale (t_2) over which gravitational instability in the ecliptic plane causes the formation of asteroidal-sized bodies (second-generation planetesimals). In this case, the first-generation planetesimals will have been chemically isolated from surrounding gases at the positions where they accreted in the circumstellar shell. After sinking rapidly to the ecliptic plane, the resultant disk of solids will consist of an intimate mixture (at the 10 cm scale) of solids that have equilibrated with solar gases over a wide range of pressures and temperature (Fig. 6.4). Thus, the compositions of solids in the disk will be highly heterogeneous on a *small* scale (~ 10 cm) and this characteristic will be reflected in the asteroidal-sized intermediate bodies which form subsequently by gravitational instability.

(c) Consider an intermediate case, where the first-generation of planetesimals still form within the circumstellar shell but are smaller in size (perhaps from 1 mm to 1 cm) than for case (b) above. These planetesimals are nevertheless sufficiently large to have become chemically isolated from surrounding gases at the positions where they were formed in the circumstellar shell; however, they will sink much more slowly towards the ecliptic plane. In this case, we assume the time scale for gravitational instability in the ecliptic disk (t_2) to be short compared to the time scale (t_1) for sinking of planetesimals to the ecliptic plane. The planetesimals that arrive first at the ecliptic plane will therefore have separated from gases at the highest temperatures (Fig. 6.4) and the asteroidal bodies formed from these materials will consist of homogeneous high-temperature materials. As time passes, planetesimals that have equilibrated further from the ecliptic plane under lower pressures and temperatures will arrive at the ecliptic plane and rapidly form asteroid-sized bodies with compositions corresponding to lower-temperature precipitates. Thus, on this model, the compositions of asteroid-sized bodies forming in the ecliptic plane will vary widely from body to body, each individual body, however, being approximately chemically homogeneous and representing solids that have separated from the nebula within limited P–T fields.

The discussion to this point has emphasized the probable chemical heterogeneity among planetesimals reaching the invariant plane as being due to formation at different heights above the plane (Fig. 6.4). However, accretion theory (Wetherill, 1976a) also shows that a considerable degree of mixing of planetesimals originating at different radial distances in the plane will also occur as a result of increases in ellipticities and inclinations of orbits caused by gravitational perturbations from growing planetary nuclei (Fig. 6.3). Hartmann (1976) has shown that a considerable amount of the planetesimals swept up by any single terrestrial planet may have been derived by such perturbations from the "feeding zone" of another terrestrial planet, particularly during the later stages of accretion.

It should be noted, however, that despite these mixing processes, there will be an *overall trend* for the ratio of high-temperature condensates to low-temperature condensates to decrease with distance from the sun. This trend is of importance in explaining the compositional variations between terrestrial planets (Chapter 9).

The temperature distribution within a discoidal nebula surrounding the sun (at present luminosity) has been calculated by Safronov (1972a). He obtained temperatures of 115–41°K between Mercury and Mars, and 16–5°K between Jupiter and Neptune. Allowance for a transient stage of much higher solar luminosity would result in a greatly increased temperature in the innermost solar nebula near the orbit of Mercury, but because of high opacity caused by dust particles, temperature would fall off rapidly with distance from the sun. In the region where the earth was to accrete, temperatures within the flattened disk are unlikely to have exceeded 300°K (Kusaka et al., 1970).

Density Differences between Terrestrial and Outer Planets

The terrestrial planets, Mercury, Venus, Earth, and Mars, possess high densities (about 4.0–5.5 g/cm^3) as compared to the outer planets, Jupiter, Saturn, Uranus, and Neptune (0.7–1.7 g/cm^3). This basic difference is readily explained by the radial temperature gradient within the nebula outwards from the sun (see, for example, Safronov, 1972a). It appears that in the terrestrial planet region, temperatures were sufficiently high to prevent the condensation of more than small amounts of *ices* (Table 6.1). The dust particles which subsequently accreted into planetesimals and then into planets were thus comprised mainly of *rock* (silicates, metals, metallic oxides). In the vicinity of the outer planets, temperatures were sufficiently low to permit almost total condensation of both *rock* and *ice* components (Table 6.1), so that the planetary nuclei which collected in this region were much larger than the terrestrial planets. Uranus and Neptune represent this phase of development. The investigations of Safronov (1972a) and others show that in the region of the outer planets, the temperatures were not sufficiently low to permit the condensation of hydrogen and helium. In the cases of the giant planets Jupiter and Saturn, which now consist principally of hydrogen and helium, it appears that the planetary nuclei of rock plus ices may have been sufficiently large to cause gravitational instability in the solar nebula leading to capture of vast amounts of these elements as discussed in Section 5.4.

6.4 Gas–Solid Equilibria in the Nebula

Redox Conditions

The oxidation of iron in the dust particles of the primitive solar nebula is strongly influenced by the equilibrium (Latimer, 1950):

$$\tfrac{1}{4} Fe_3O_4 + H_2 = \tfrac{3}{4} Fe + H_2O; \qquad K = \frac{H_2O}{H_2}.$$

Within the nebula, K is fixed by the relative abundances of H_2 and H_2O, which yields a value for H_2O/H_2 close to 2×10^{-3}. The equilibrium constant K is also related to the free energy change ΔG for the above reaction by the well known expression

$$\Delta G_T = RTlnK.$$

With ΔG_T determined from thermodynamical data and K fixed by the H_2O/H_2 ratio of the nebula, the equilibrium temperature for the reaction is obtained. It is found to be 400°K. At temperatures below this, iron would occur in the oxidized state as magnetite, while above 400°K magnetite would be reduced to metallic iron. Under cool conditions, a substantial proportion of the oxidized iron would also occur as an FeO component of silicate minerals in the dust grains. The temperature required to reduce this oxidized iron to metal is higher than for magnetite because of the decrease in FeO activity caused by solid solution and compound formation. The relevant equilibria have been considered by Grossman (1972) and are shown in Figure 6.5. Essentially all oxidized iron is reduced to the metallic state by a temperature of 800°K.

It is thus seen that in the solar nebula, rather modest changes of temperature between 400 and 800°K are capable of producing extreme changes in the oxidation state of iron. We will find that this behavior is fundamental to all theories of planet formation.

In the Earth, Moon, and in most classes of meteorites, iron is present both in the metallic and oxidized states. This situation requires either that these bodies formed completely in a very narrow T, f_{0_2} field or that they consist of *mixtures* of components possessing variable degrees of oxidation and formed over much wider T, f_{0_2} ranges. Most theories of planet formation are more easily accomodated within the latter framework. In the context of the model shown in Figure 6.4, we see that the dust shell which collapses to form an equatorial disk contains material which has equilibrated with solar gases over a very wide temperature range. Thus, the planetesimals which ultimately collect in the ecliptic plane are likely to consist of material exhibiting a wide range of oxidation states [cases (b) and (c) of Section 6.3]. As noted earlier, the *mean* temperature of gas–solid equilibrium decreases with increasing distance from the sun, and is therefore paralleled by a corresponding

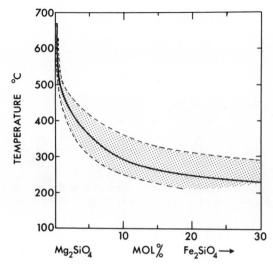

Figure 6.5 Equilibrium fayalite content of olivine as a function of temperature in a partially condensed system of solar composition. Shaded area represents uncertainty limits. (Based on data of Grossman, 1972.) (Diagram from Ringwood, 1975a, with permission.)

net increase in the proportion of oxidized to reduced iron in the ecliptic plane for these cases.

On the other hand, if solids equilibrated with nebula gases at relatively low temperatures as in the model depicted in Figure 6.2 [case (a) of Section 6.3], iron would be completely oxidized. Reduction to form metallic iron must then have occurred during the accretion process by which planets were formed (see, for example, Ringwood, 1966b). In this model, the energy source is likely to have been provided by the gravitational potential energy of accretion, so that we might expect a correlation between masses of planets and the degree of reduction of iron.

Condensation Sequence

The nature of the condensed phases separating from the nebula has been studied by many authors, recent treatments being provided by Grossman (1972), Larimer (1967), Grossman and Larimer (1974), Wasson and Wai (1976), and Wai and Wasson (1977). Formation of liquid phases in the nebula has been shown to require unacceptably high gas pressures (10^2–10^3 bar), so that, with the possible exception of the accretion of Mercury (Chapter 9), we are concerned with equilibrium between solids and gases. Most studies have taken the initial state to be a high-temperature gas of solar composition and have calculated, from the thermodynamic data, the sequence of solid phases that separate from the gas as temperature falls. Some of these results are summarized below. It should be noted that the same equilibria and sequence apply in the case of an initially low-temperature gas–solid system of solar composition, which is heated and subjected to increasing degrees of selective volatilization of solids as temperature rises.

We will now proceed to describe the condensation sequence of a gas of solar composition (at a total pressure of 10^{-4} bar), based on the results of the authors previously cited. These results are summarized in Table 6.2 and depicted in Figure 6.6. The first condensates consist of a group of refractory trace elements including Os, Re, and Zr. These condense well above 1679°K, the condensation point of Al_2O_3, which is the first condensate containing a major element. By 1500°K, all the Ti and most of the Ca have condensed as $CaTiO_3$ and $Ca_2Al_2SiO_7$, respectively. The rare earth elements, U, Pu, Th, Ta, and Nb, probably condense largely in solid solution in $CaTiO_3$. In contrast to Ca, Al, and Ti, less than 10 percent of the total Mg and Si are condensed until $CaMgSi_2O_6$ appears at 1387°K. Similarly, metallic Fe only begins to condense at 1375°K, carrying with it Ni and Co. At 1370°K, iron-free Mg_2SiO_4 appears, condensing most of the Mg. It later reacts with SiO in the vapor to form $MgSiO_3$, consuming all the remaining gaseous Si by 1200°K.

Below this temperature, Cu, Ge, and Ga condense in solid solution in the metal; Cr condenses as an oxide phase, while Mn condenses as either a silicate or a sulphide. At about 1200°K, Na, K, and Rb begin to condense in the form of solid solutions with previously condensed $CaAl_2Si_2O_8$. The alkali metals should be totally condensed by 1000°K, while Ag is totally condensed by 900°K. As discussed previously, metallic iron begins to oxidize significantly only after the temperature has fallen below 800°K, at which point Mg_2SiO_4 and $MgSiO_3$ contain about 0.5 mole-percent of the iron end-members of their respective solid solution series. At lower temperatures, their Fe^{2+} content rises rapidly (Fig. 6.5). Troilite (FeS) becomes stable at 650°K and forms by the reaction of gaseous H_2S with metallic iron.

The relatively volatile metals Pb, Bi, In, and Tl condense between 600° and 400°K, at which latter temperature magnetite also becomes stable. Below 350°K, magnesium silicates react with gaseous H_2O to form hydrated silicates, while below 200°K, argon, CH_4, $NH_4 \cdot H_2O$, and methane hydrate condense.

The very low temperature at which methane condenses renders it difficult to explain the occurrence of carbon compounds in meteorites and in the earth by mechanisms which involve the condensation and trapping of methane. At temperatures above about 700°K, carbon occurs predominantly as carbon monoxide. Below 500°K, the reaction

$$CO + 3H_2 \longrightarrow CH_4 + H_2O$$

proceeds strongly to the right. However, it is greatly impeded by kinetic difficulties and, in the presence of suitable catalysts (e.g., nickel–iron, magnetite and silicates), a wide range of condensed metastable hydrocarbon compounds possessing the general formula C_nH_{2n+2} condense, e.g.,

$$10CO + 21H_2 \longrightarrow C_{10}H_{22} + 10H_2O.$$

Table 6.2 Condensation temperatures and sequence of phases and elements separating from gas of solar composition at 10^{-4} atm. total pressure. Temperatures correspond to 50 percent condensation of a given element. (Based on Grossman and Larimer, 1974, Fig, 2; Grossman, 1972; Anders, 1968, Fig. 5; Wasson and Wai, 1976; and Wai and Wasson, 1977.)[a]

Element or compound	°K
Ca, Al, Ti oxides and silicates } Platinum metals, W, Mo, Ta, Zr, REE }	> 1400
Mg_2SiO_4	~ 1360
Fe–Ni metal	~ 1360
Remaining SiO_2 (as $MgSiO_3$)	1200–1350
Cr_2O_3	—
P	1290
Au	1230
Li	1225
Mn_2SiO_4	1190
As	1135
Cu	1118
Ga	1075
$(K,Na)AlSi_3O_8$	~ 1000
Ag	952
Sb	910
F	855
Ge	812
Sn	720
Zn	660–760
Se	684
Te	680
Cd	680
S (as FeS)	648
FeO (10% ss. as $[Mg_{0.9}Fe_{0.1}]_2SiO_4$)	~ 600
Pb	520
Bi	470
In	460
Tl	440
Fe_3O_4	400
NiO	—
H_2O (as hydrated silicates)	~ 300

[a] Significant differences appear in calculations by the authors cited. We have selected what appear to be the most reliable estimates based on recent thermodynamic data.

Equilibria of this type are well known and exploited industrially as Fischer–Tropsch reactions. Anders and coworkers have produced strong evidence that these reactions would produce a crop of complex organic compounds condensing upon grains present in the nebula at temperatures below 450°K. Marked resemblances between the compounds produced by these reactions and the organic compounds present in carbonaceous chondritic meteorites

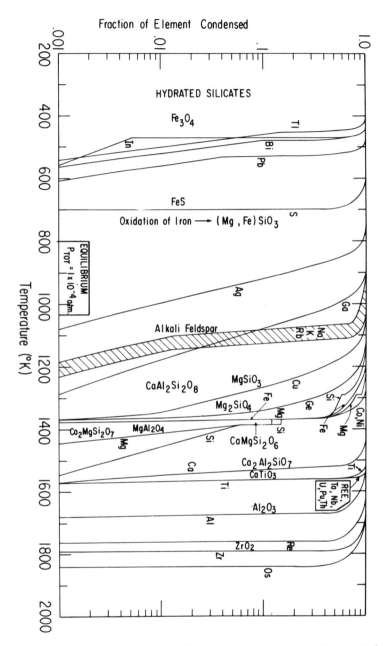

Figure 6.6 Condensation of the elements from a gas of solar composition at 10^{-4} atm. (From Grossman and Larimer, 1974, with permission.)

have been demonstrated by Studier et al. (1965, 1968, 1972) and Anders (1971). The degree of correspondence is increased if ammonia is present in the initial gas mixture, as would be the case in the solar nebula. Anders and his coworkers concluded that, most probably, the carbonaceous material present in the meteorites was initially condensed in this form, although the assemblage of organic compounds may well have been modified subsequently by mild thermal metamorphism within the meteoritic parent bodies. It seems quite likely that carbon was originally incorporated in the terrestrial planets via similar mechanisms. The discovery of a wide range of organic molecules forming within cool, dense, interstellar clouds of dust and gas (see, for example, R. D. Brown, 1974) is also suggestive of the operation of Fischer-Tropsch reactions. Miller and Urey have demonstrated that complex organic compounds can also be formed in gas mixtures of CH_4, NH_3, and H_2O by irradiation by ultraviolet and gamma rays (Miller, 1953). These mechanisms may also have contributed to the synthesis of carbonaceous compounds within the primordial solar nebula.

The effect of pressure upon the condensation behavior of some key components is shown in Figure 6.7. Except for forsterite and metallic iron, the relative condensation sequences do not vary significantly as the nebula pressure is changed from 1 to 10^{-6} atm. The exception is of some importance, however. At "high" pressures ($> 10^{-4}$ atm) iron condenses ahead of

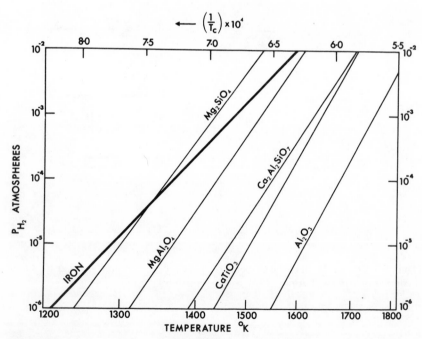

Figure 6.7 Pressure variation of condensation temperature for some major phases. (After Grossman and Larimer, 1974.)

forsterite, whereas these positions are reversed at lower pressures. Actually, the condensation temperatures are effectively identical (within experimental error) over the pressure range 10^{-3}–10^{-6} atm., which is the range of pressures to be expected throughout most of the terrestrial planet region. However, very close to the sun, near the orbit of Mercury, it is possible that higher nebula pressures may have prevailed, so that iron would have been less volatile than forsterite (Grossman and Larimer, 1974). This behavior may be of cosmochemical significance in connection with the high density of Mercury (Chapter 9).

The condensation behavior of magnesium and silicon are compared in Figure 6.8. Silicon is substantially more volatile than magnesium. Over 90 percent of the total magnesium condenses out as forsterite (Mg_2SiO_4) by 1290°K (10^{-4} atm.), while 40 percent of the total silicon remains in the vapor phase. As temperature falls, silicon in the vapor reacts with Mg_2SiO_4 to produce enstatite, the reaction approaching completion at about 1200°K. We saw in Chapter 2 that the volatilization of silicon away from enstatite to form forsterite may be of significance in explaining why the Mg/Si ratio in the Earth's mantle is apparently higher than the cosmic ratio.

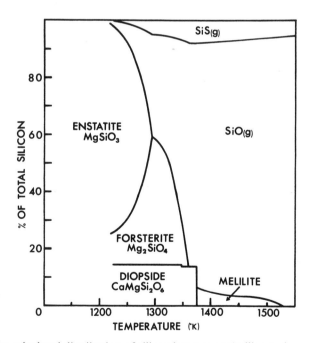

Figure 6.8 The calculated distribution of silicon between crystalline and vapor phases in a gas of solar composition at 10^{-4} atm. Nearly all magnesium has condensed at 1290°K (mainly as forsterite), whereas only 60 percent of the total silicon has condensed. Forsterite begins to react with silicon in the gas phase to form enstatite at 1290°K. This reaction continues with falling temperature until 95 percent total silicon is condensed at 1240°K. (Diagram kindly provided by L. Grossman.)

6.5 Chemical Fractionations among the Terrestrial Planets

The mean densities of terrestrial planets (including the Moon) are given in Table 6.3. They are seen to vary over a wide range. Part of this variation is caused by differential self-compression within the planets' own gravitational fields. However, when these effects are removed by calculating the densities at some common (low) pressure, the wide range in intrinsic densities remains (Table 6.3).

These density differences imply the existence of major differences in chemical composition among the terrestrial planets. Yet these bodies are believed to have been formed from the relatively involatile "rock" component (Table 6.1) of the solar nebula which was initially chemically homogeneous. How, then, were these fractionations of relatively involatile materials among the terrestrial planets achieved? This is one of the fundamental problems of cosmochemistry.

Metal–Silicate Physical Fractionation Hypothesis

Jeffreys (1937), Urey (1952), and many others have explained the varying densities of the terrestrial planets by assuming that they are composed of varying proportions of silicate ($\rho_o \sim 3.3$ g/cm^3) and nickel–iron ($\rho_o \sim 7.9$ g/cm^3) phases, each phase being essentially of constant composition. The low-pressure densities of planets would therefore increase with increasing metal/silicate ratios. Compositions of planets according to this model are given in Table 6.3. It is clear that the densities can be explained, at least in principle, on this basis. Urey (1952, 1957a,b) further hypothesized that fractionation of silicate particles from iron particles via *physical processes* had occurred in the solar nebula prior to accretion of planets, so that the ratio of silicate to iron varied widely throughout the nebula. Subsequent

Table 6.3 Mean densities of terrestrial planets at 10 kbar, assuming that they are composed of varying proportions of metal phase ($\rho_{10} = 7.9$ g/cm^3) and silicate phase ($\rho_{10} = 3.3$ g/cm^3)

Planet	Mean density	Mean density at 10 kb	Percent nickel–iron phase
Mercury	5.44	5.3	65
Venus	5.24	3.96	28.8
Earth	5.52	4.07	32.5
Moon	3.34	3.40	5
Mars	3.94	3.73	20

accretion of planets in specific regions reflected these preexisting metal–silicate inhomogeneities.

The high density of Mercury and the low density of the Moon, as compared to Mars, Earth, and Venus, clearly imply the occurrence of some kind of fractionation between iron and silicates. Moreover, during most of the 1960s it was widely believed that the abundance of iron in the sun relative to magnesium and silicon was about a factor of 5 lower than is found in Earth, Mars, Venus, and chondritic meteorites. This seemed to constitute decisive proof of the existence of gross iron–silicate fractionation between sun and nebula, and Urey's interpretation became widely accepted.

Nevertheless, certain difficulties emerged:

(a) Although various possibilities were explored, no really plausible mechanism was finally advanced for explaining the required physical fractionation of metal particles from silicate particles in the nebula.

(b) The composition of the silicate phase of chondritic meteorites was found to differ from that of the Earth's silicate mantle (Gast, 1960). Moreover, it is now known that the compositions of the silicate components of the Moon and Mars differ significantly from that of the Earth's mantle (Chapters 9 and 10). Thus, the silicate component of the nebula was not uniform in composition. *Chemical* fractionations, as distinct from *mechanical* fractionations, must be invoked to explain the inhomogeneity.

Urey's model has been extended in recent years by Anders (1968, 1971), Larimer and Anders (1967), Grossman (1972), and others. These authors assume that the inner solar nebula extending out to the asteroidal belt became hot enough ($\sim 2000°K$) to evaporate all the solids. On cooling, fractional condensation occurred as discussed in Section 6.4, with four or more principal groups of condensates: high-temperature condensates (Ca, Al, Ti, etc.), magnesium silicates, metallic iron, and low-temperature condensates (FeO, FeS, OH, volatile metals, etc.). These different batches of materials somehow became fractionated from each other by physical processes within the nebula, so that gross compositional differences were established in different regions of the nebula. These differences were reflected in the differing compositions of planets which accreted in localized regions of the nebula.

This hypothesis is directed at explaining not only the different abundances of iron relative to silicates, between the planets, but also the differing chemical compositions of the silicate phases in meteorites, Earth, and Moon. The four components can be mixed in arbitary proportions to explain the bulk composition of any planet (or meteorite). Moreover, the temperatures at which the low-temperature condensate separated from the nebula can also be varied, providing a wide range of possible compositions for this component. In more recent developments (Ganapathy and Anders, 1974), three more independent components were added to the mix. It is evident that the hypothesis provides an uncomfortably large number of degrees of freedom. This topic is discussed further in Chapters 9 and 11.

Redox Hypothesis

The difficulties encountered by the metal–silicate fractionation hypothesis in accounting for the varying densities of planets led Ringwood (1959)
to formulate an alternative hypothesis. He proposed that the relative
abundances of the common metals Fe, Mg, Si, Al, Ca, and Ni were the
same on Mars, Venus, and Earth and were identical to the relative abundances of these elements in chondritic meteorites and the sun. The differences in density between Mars, Venus, and Earth were attributed to differing
oxidation states, i.e., differing total amounts of oxygen. The effect of redox
state upon intrinsic density of a primordial "mix" of common metals is
seen in Figure 6.9. If all iron and nickel present in chondritic meteorites
were present as oxides, the density (at 10 kbar) would be 3.78 g/cm^3. As
FeO is reduced to elemental iron, the density increases until, with complete
reduction, the intrinsic density is 3.99 g/cm^3. Ringwood proposed that Mars
was completely oxidized, while in the Earth most of the iron was present
in the reduced state. Venus was thought to represent an intermediate state
of reduction (Fig. 6.9).

The objective of this hypothesis was to minimize the necessity for mechanical metal–silicate segregation processes occurring within the primordial
solar nebula, by invoking the operation of a simple *chemical process* involving redox equilibria. We have already seen (Section 6.4) that redox
states in the nebula are strongly dependent upon temperature and degree
of gas/solid segregation. Moreover, redox states may be markedly affected

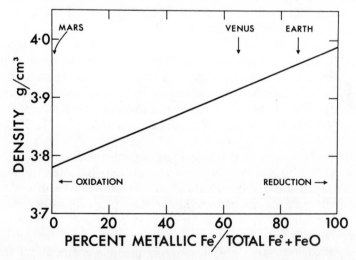

Figure 6.9 Relationship between density and oxidation state of primordial material
(represented by Type 1 carbonaceous chondrites on a volatile-free and sulphur-free
basis). Densities were calculated to correspond to mineral assemblages stable at 10 kbar.
Positions of planets in relation to redox states according to hypothesis of Ringwood
(1959) are indicated. (From Ringwood and Anderson 1977, with permission.)

by the conditions of planetary accretion (Ringwood 1966b). Thus, it is plausible to invoke redox state as a variable in explaining the densities of planets. Indeed, in view of the wide range of controlling factors, it would be remarkable if all the planets had finally achieved the same redox state.

The hypothesis was unable to explain the high density of Mercury, which was attributed (Bullen, 1952; Ringwood, 1960, 1966b) to its singular position in relation to the sun, so that it accreted at much higher temperatures than the other planets, causing loss of magnesium silicates. Although invoking additional processes (as discussed in Chapter 9), these extensions to the redox hypothesis did not seem unreasonable.

However, the low density of the Moon could not be reconciled with the hypothesis. If the latter were substantially valid, this implied that the Moon could not be regarded in any sense as a "terrestrial planet" formed by independent accretion within the solar nebula of the same material from which the other planets accreted. It must possess some entirely specialized and unique origin.

The redox hypothesis was not generally accepted during the 1960s, when it was widely believed that the abundance of iron in the sun (relative to silicon) was about a factor of 5 smaller than in the Earth and in chondrites. Although the mechanism of metal/silicate fractionation might be obscure, this large discrepancy persuaded most workers in the field that a very effective, large-scale metal/silicate fractionation process must have occurred in the solar nebula. Ringwood (1966b) preferred to believe that the solar iron abundance was in error, a prejudice which, fortunately, was confirmed in 1969 by Garz and coworkers, who discovered that a large revision was necessary in the oscillator strengths or iron lines (Garz et al., 1969). When the new oscillator strengths were employed, the revised abundance of iron in the sun was found to be similar to that in the Earth and chondritic meteorites (Fig. 2.2). We will return to this hypothesis in Chapter 9.

CHAPTER 7

Early Theories of Accretion of the Earth

7.1 Homogeneous Accretion

The most widely supported class of hypotheses of the Earth's origin during the period 1950–1970 maintained that the Earth accreted from an intimate mixture of silicate particles and metal particles, generally resembling chondritic meteorites (e.g., Urey, 1952, 1958, 1962a; Kuiper, 1952; Vino-gradov, 1961; Elsasser, 1963; Birch, 1965). The chondritic material was assumed to have been formed in the solar nebula by a complex series of chemical and physical processes that occurred prior to the accretion of planets.

According to this class of hypotheses, accretion of the Earth occurred over a sufficiently long period (10^7–10^8 years), so that its gravitational potential energy was efficiently radiated away and it formed in an initially "cool" and unmelted condition, with an average temperature of less than 1000°C. Subsequently, heating by long-lived radioactive elements occurred, leading to melting of the metal phase and its segregation into the core, much later (e.g., 10^9 years) than the primary accretion of the Earth. Thus, according to these models, not only did the Earth accrete initially from relatively homogeneous, well-mixed material, but also, for a considerable period thereafter, the Earth was approximately homogeneous, consisting of an intimate mixture of metal and silicate phases.

Homogeneous accretion hypotheses can explain, in principle, the major element composition of the Earth and its depletions in volatile elements, which are assumed to have resulted from the high-temperature processing that occurred in the solar nebula prior to accretion. These hypotheses also account for the inferred approximate chemical uniformity of the mantle subsequent to core formation and the small dispersions in abundances of many compatible siderophile, lithophile, and volatile minor elements in the mantle.

These early homogeneous accretion hypotheses did not, however, explain our inference (Section 3.4) that the core formed during, or very soon after, the accretion of the Earth: They also encountered some serious chemical difficulties:

(a) If the Earth had once consisted of an intimate mixture of iron and silicate phases which had been slowly heated until the metal melted and segregated, it is inevitable that local chemical equilibrium would have been established between metal and silicate phases. Yet the abundances of Ni, Co, Cu, Au, and many other siderophile elements in pyrolite are from 10 to > 100 times higher than can be explained by equilibrium partition between metal and silicate phases (Section 2.4).

(b) The Fe^{3+}/Fe^{2+} ratio of the upper mantle is ≥ 10 times higher than would be expected if this region had equilibrated with metallic iron, which subsequently separated into the core. In consequence, the volatiles degassed from the upper mantle are dominantly composed of CO_2 and H_2O, whereas if the mantle had once equilibrated with metallic iron, the gases evolved should be composed dominantly of CO and H_2 (Section 2.4).

(c) A related difficulty faced by homogeneous accretion hypotheses is the origin of the Earth's hydrosphere and atmosphere. According to these hypotheses, small amounts of volatile components such as H_2O and CO_2 were chemically bound and/or absorbed in the silicate component of the material from which the Earth accreted (see, for example, Brown, 1952). According to a more recent version (see, for example, Larimer and Anders, 1967), the volatiles were introduced via a small proportion of a well-mixed "carrier" similar to Type 1 carbonaceous chondrites. It is assumed that, with sufficiently slow accretion under cool conditions, the volatiles were trapped within the Earth. Subsequent heating of the Earth's interior by radioactivity led to magmatism accompanied by partial degassing—the principal species degassed being H_2O, CO_2, and N_2. These species accumulated to form the hydrosphere and atmosphere.

However, even if it were possible to trap some CO_2 and H_2O in this manner, these molecules would have been decomposed in the presence of excess metallic iron (Ringwood, 1975a, 1978b). As the intimate mixture of iron and silicate[1] was heated above 1000°C, leading to melting and core formation, any trapped water and carbon dioxide

---

[1] Even if the planetesimals from which the Earth accreted were initially chemically homogeneous on a scale of a few kilometers [as in case (c), Section 6.3], the planetesimals would have been extensively disintegrated and remixed by impact when they collided at high velocities with the accreting Earth.

would have been reduced according to the following simplified reactions (Ringwood, 1978b).

$$Fe + H_2O \quad \longrightarrow \quad FeO \text{ (in silicate)} + 2H_{\langle Fe\ soln\rangle}, \qquad \Delta v \text{ negative.}$$

$$2Fe + CO_2 \quad \longrightarrow \quad 2FeO \text{ (in silicate)} + C_{\langle Fe\ soln\rangle}, \qquad \Delta v \text{ negative.}$$

The hydrogen and carbon produced would both be soluble in excess iron, particularly at high pressure, and the free energies of these solutions would drive the equilibria to the right. The net effect would have been the decomposition of any trapped CO_2 and H_2O, the oxygen entering mantle silicates as FeO, while the C and H were removed from the mantle with the iron as it sank into the core. The homogeneous accretion model thus appears to offer an inadequate explanation of the Earth's hydrosphere and atmosphere.

These and other difficulties encountered by early homogeneous accretion models caused several authors to abandon them and explore other possibilities. Not the least of these difficulties was the complexity and contrived nature of the multistage chemical and physical fractionation processes that were assumed to have occurred in the solar nebula prior to accretion of the Earth and other terrestrial planets, (e.g. Anders, 1968, 1971, Ganapathy and Anders, 1974). Our conclusion (Chapter 4) that the proportions of "high-temperature condensates," iron, and magnesium silicates in the Earth are similar to those in the Sun and Type 1 carbonaceous casts considerable doubt on the need for invoking the operation of these complex processes. Moreover, they are not needed to explain existing data bearing upon the compositions of Mars and Venus (Chapter 9).

7.2 Single-Stage Hypothesis

In an attempt to avoid some of the difficulties in early versions of homogeneous accretion hypotheses, Ringwood (1959, 1960, 1966a,b, 1975a) proposed a simpler "single-stage" hypothesis of the Earth's origin, which was also extended to cover the origin of the terrestrial planets and the moon.

This hypothesis proposed that the Earth formed by accretion in an initially cool solar nebula [case (a), Section 6.3] from planetesimals appropriate in composition to the condensate to be expected on chemical grounds in such a nebula—i.e., highly oxidized primitive material containing carbonaceous compounds, water, and other volatiles, and resembling Type 1 carbonaceous chondrites (Column A, Table 8.1). It was maintained, furthermore, that reduction of oxidized iron to metal, loss of volatiles, and formation of the core occurred essentially simultaneously with the accretion process. The assumptions of intermediate stages of reduction and fractionation in the solar nebula prior to accretion were avoided, so that the hypothesis amounted to a single-stage process. A further key assumption was that accretion of most of the Earth occurred over a time interval smaller than 1 million years.

Formation of the Earth under the above boundary conditions is strongly influenced by the gravitational energy dissipated during accretion, which in turn controls the chemical equilibria in the accreting material. The accretional energy per gram is plotted against the radius of the growing Earth in Figure 7.1. It increases approximately as the square of the radius, reaching 15,000 cal/g during the final stages. The development of the Earth according to this model is shown in Figure 7.1. The stages referred to below are those depicted in this figure.

Stage I. During the early stages of accretion, the energy evolved is small and accretion is relatively slow. The temperature is accordingly low and is buffered by the latent heat of evaporation of volatile components (mainly H_2O) in the accreting material. Thus, during this stage, a cool, oxidized, volatile-rich nucleus of primordial material, perhaps about 10 percent of the mass of the Earth, is formed. Stage I is of key importance to the Earth's subsequent geochemical development, since it represents the stage at which volatiles, oxidized iron, and oxidized siderophile elements are trapped within the deep interior of the growing planet.

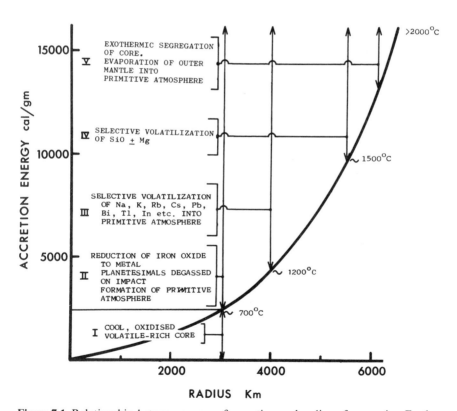

Figure 7.1 Relationship between energy of accretion and radius of a growing Earth-sized terrestrial planet. The principal stages of accretion are also shown in relation to the energy of accretion and approximate surface temperatures. (After Ringwood, 1970; with permission.)

Stage II. As the mass of the nucleus increases, the energy of infall of planetesimals becomes sufficient to cause strong transient heating on impact, leading to reduction of oxidized iron to metal by accompanying carbonaceous material. The heating also causes complete degassing, resulting in the generation of a primitive reducing atmosphere, mainly of CO and H_2.

Stage III. When the mass of the nucleus has increased to about one-fifth of the mass of the Earth, the near-surface accretion temperature has risen to about 1200°C. Reduction of iron oxide to metal now proceeds within the primitive reducing atmosphere (mainly CO and H_2), rather than at the solid surface of the nucleus. With further increase of temperature in the interval 1200–1500°C, reduction and volatilization of a number of relatively volatile elements (Cr, Mn, Na, K, Rb, Cs, F, Zn, etc.) into the primitive atmosphere occur. The material accreting thus consists of a mixture of metallic iron plus iron-free silicates, strongly depleted in volatile components.

Stage IV. As the mass of the nucleus and the rate of accretion increase still further, the temperature becomes sufficiently high (> 1500°C) to reduce and volatilize major components of the silicate phase of the infalling planetesimals. The most volatile major component is silicon (SiO). Consequently, the enstatite ($MgSiO_3$) component of the accreting material loses silicon and is replaced by forsterite (Mg_2SiO_4), which becomes the dominant accreting phase. At this stage, also, some silica becomes reduced to elemental silicon which enters the metal phase, forming a ferrosilicon alloy which continues to accrete.

Stage V. According to the model, the Earth develops "inside out," with a cool, oxidized nucleus and becoming successively more reduced and metal-rich towards the surface. This state is gravitationally unstable. As melting occurs near the surface, the metal segregates into bodies which are large enough to sink through the solid interior into the core. This process is highly exothermic, liberating over 600 cal/g for the entire Earth.

Core–Mantle Disequilibrium

The sinking of large bodies of metal into the core displaces the oxidized, volatile-rich material of the central cool nucleus upward into the mantle and also drives a forced convection within the latter. As a result of this strong stirring caused by core formation, the oxidized volatile-rich nucleus is mixed with the overlying degassed zone of iron-poor silicates. In this manner, oxidized iron (FeO + Fe_2O_3), siderophile elements (Ni, Co, Cu, Au, Re, etc.) and volatile elements and components (including H_2O and CO_2) become incorporated in the mantle. Equilibrium between sinking metal "drops" (cf. Elsasser, 1963) and surrounding silicates can be attained only by diffusion across the interfaces of sinking metal bodies. If the rate of sinking of metal is high compared to the rate of attainment of equilibrium by diffusion, the core which separates will not remain in equilibrium with the mantle. These, indeed, are just the conditions envisaged in the core formation mechanism of Elsasser (1963), in which the dimensions of the metallic "drops" are on the order of hundreds of kilometers. The occurrence of silicon in the core, coexisting with a mantle containing oxidized iron and siderophile elements, can then be understood on these grounds.

Further Development of Model

The single-stage model for the formation of the Earth involves the production of an enormous atmosphere (0.1–0.2 M_E) composed chiefly of H_2, H_2O, and CO, together with a few percent of volatilized metals. A critical requirement of the model is that the primitive atmosphere was completely lost at a very early stage of the Earth's history. In earlier versions of the model, Ringwood (1960, 1966a,b) envisaged that the massive primitive atmosphere accumulated until a late stage of accretion, when it was removed by some combination of high Earth-rotation rate, solar T-Tauri particle radiation, and turbulent interaction with the solar nebula. Urey (1960b) and many others, however, have correctly pointed out the difficulties of dissipating such a massive atmosphere, once it is allowed to accumulate.

Ringwood (1975a) revised his model to permit the primitive atmosphere to "blow off" continuously as the Earth accreted, so that at no stage was a *massive* atmosphere developed. The proposed mechanism required the turbulent mixing of hydrogen from the surrounding solar nebula into the primitive atmosphere, thereby lowering its mean molecular weight to about 3.5, which, together with a reasonable combination of high temperatures and high rotation rate, would permit the atmosphere to blow off continuously into the nebula.

Developing this model further, he suggested that the dynamic equilibrium between hydrogen derived from the solar nebula via turbulent mixing, and reduction products (H_2O, CO, H_2) produced during accretion, might have led to a steady-state situation corresponding to a mean H_2/H_2O ratio in the atmosphere near the Earth's surface of about 10 (as compared to the solar nebula value of about 1000). If it were possible to produce an atmosphere with this composition, then the relevant equilibria show that it would be possible to directly accrete FeS and FeO (as a component of silicates) at temperatures which were sufficiently high to evaporate Cr, Mn, Na, and other volatile elements. This model thus envisaged the possibility of a sulphur-rich core (instead of ferrosilicon), and with reduction of oxidized iron in the primordial material being controlled by hydrogen (derived from the nebula) rather than carbon (present originally in the low-temperature condensate). It must be admitted however, that the conditions assumed in order to achieve these objectives were highly specific and constrained, thereby lessening the plausibility of the model.

Difficulties Encountered by the Model

The author's "single-stage" hypothesis was developed partly to consider the consequences of planet formation from solid material which would be produced in equilibrium within a cold solar nebula (as advocated by Schmidt, 1958; Urey, 1959; Schatzman, 1967; and Safronov, 1972a) and partly in an attempt to avoid specific difficulties associated with existing homogeneous accretion hypotheses. The principal criticisms levelled against it are:

(a) Difficulty of disposing of gaseous reduction products (see, for example, Urey, 1960b; Lubimova, 1966; Levin, 1972a; Grossman, 1972);
(b) problems of forming a core out of chemical equilibrium with mantle (see, for example, Grossman, 1972; Brett, 1976);
(c) Short time scale required for accretion (Levin, 1972a).

In my opinion, difficulties (a) and (b) can be overcome, as in the revised model of Ringwood (1975a), although it must be admitted that the model is somewhat contrived. The problem of the short accretion time scale presupposes that accretion occurs according to the Schmidt–Safronov model (Figs. 6.2 and 6.3). This latter model, however, must be incomplete because it leads to impossible accretion time scales for Neptune, Uranus, and Mars (Section 8.2). The short accretional time scale required for the single-stage hypothesis might be achieved more readily by cosmogonic hypotheses of the type advocated by McCrea (1960, 1972) and McCrea and Williams (1965), as discussed in Section 6.2.

There are, however, additional geochemical difficulties encountered by the single-stage hypothesis that have become apparent to the author. These concern firstly the homogeneity of the mantle and the low dispersion in abundances of many siderophile, volatile, and lithophile elements (Chapter 2). It seems rather difficult to explain these small dispersions in terms of physical mixing of oxidized, volatile-rich material from the nucleus into the surrounding devolatilized, reduced mantle during core formation. It is scarcely credible that such a physical mixing process occurring in the solid state could be as efficient as is required.[2]

A second difficulty is that if all the oxidized nickel now present in the mantle had been introduced subsequent to core formation in the form of a low-temperature, metal-free condensate, then this condensate should also introduce the primordial abundances of all elements (Table 6.2) which are fully condensed above 400°K. (At temperatures higher than this, nickel would be present entirely in a metallic phase). Thus, for example, germanium,

---

[2] On the other hand, the abundances of some other trace siderophile elements (e.g., Au, Ir, Re, Os, and Pt) and also some primordial inert gases (Section 2.4) show extremely wide dispersions among mantle rocks, and may better be explained by inhomogeneous mixing of a primordial component comparatively rich in these elements, with mantle material previously equilibrated with metal, as envisaged by the single-stage model.

which condenses at 820°K, should be present in the low-temperature condensate in an amount equal to the primordial Ni/Ge ratio of 290 (by weight). However, the observed Ni/Ge ratio in the upper mantle is 1500. Moreover, on the basis of this model, the mantle should also contain the primordial Ni/Ir ratio of 20,000, whereas the observed ratio is about 850,000 (see also Table 7.1).

These difficulties, combined with the problems discussed earlier, necessarily challenge the viability of the single-stage model in its present stage of development. There are, however, certain elements of the model which are likely to be incorporated into any successful hypothesis of the origin of the Earth. These include (*i*) trapping of volatile components at an *early stage* of accretion when accretion energies are low; (*ii*) the emphasis upon chemical interactions between material accreting at the Earth's surface and the primitive atmosphere surrounding the Earth; (*iii*) the necessity for escape of a primitive atmosphere from the Earth's gravitational field; and (*iv*) the general concept that many of the overall chemical properties of the Earth are best explained in terms of the mixing of primitive, low-temperature oxidized, volatile-rich material (Component A, Table 8.1) with high-temperature, reduced, metal-rich, devolatilized material (Component B, Table 8.1).

These characteristics of the single-stage model will be encountered again in Chapter 8, where a compound model for the origin of the Earth is explored.

7.3 The Heterogeneous Accretion Hypothesis

An alternative approach which attempted to avoid the difficulties encountered by homogeneous accretion hypotheses proposed that the composition of the material from which the Earth accreted changed with time, so that the primary layered structure of the Earth was a direct result of the accretion process. This has become known as the heterogeneous accretion hypothesis, and has received a considerable degree of favorable attention in the literature over the last decade. Because of this, we will discuss it in some detail.

Actually, one of the earliest hypotheses in this category was proposed by Eucken (1944), who investigated the condensation behavior of hot solar gases which, for example, might have been torn out of the sun by tidal interaction with a passing star (Jeans, 1917). Eucken found that molten metallic iron would condense first, followed by silicates. He suggested that the gross structure of the Earth, with its metallic core surrounded by silicate mantle, was a consequence of direct sequential condensation in this manner. This concept has been further developed by Turekian and Clark (1969), Clark et al. (1972), Grossman (1972), Anders (1968, 1971), Ganapathy et al. (1970), Cameron (1973a), and D. L. Anderson et al. (1972) in the context of a specific theory of solar system origin advanced by Cameron (1963, 1969, 1973a) and Cameron and Pine (1973). According to this theory, the solar system developed from a rotating discoidal nebula (stellisk) amounting to two solar masses. Cameron showed that the contraction of a nebula of this mass would produce high temperatures (1000–1700°C) in the vicinity of the terrestrial planets (Fig. 5.2). Moreover, the dissipation time of the nebula would have been very short (10^3–10^4 years), and Cameron argued that most of the formation of the planets must have occurred over this brief time interval, before dissipation was complete.

Under the P,T, conditions reached in the inner regions of the solar nebula, the dust grains originally present in the parental cloud would have been completely evaporated. On cooling of the gas, the components would condense over a wide range of temperatures, following the condensation sequence given in Table 6.2. Turekian and Clark (1969) accordingly proposed that accretion of most of the Earth occurred simultaneously with condensation of the solar nebula over a period of about 10^4 years, resulting in gross stratification of the Earth, with an iron core surrounded by a mantle of magnesium silicates. The later stages of accretion occurred after the nebula had cooled to relatively low temperatures, so that the equilibrium condensate contained oxidized iron, iron sulphide, hydrated magnesium silicates, plus other volatiles, i.e., similar to Type 1 carbonaceous chondrites. Volatile-rich material of this nature, mixed with earlier high-temperature condensates which had failed to accrete previously, are believed to have accreted upon the Earth over a longer timescale (10^5–10^7 years) to produce the upper mantle–crust system.

An important feature of this heterogeneous accumulation model is that the upper mantle has never been in contact with the core. "It is a direct consequence of the short timescale of the main phase of accretion, which enables the core to settle to the centre and be surrounded by the lower mantle before the outer layers are added" (Clark et al., 1972). In this way, the heterogeneous accretion hypothesis provides an acceptable explanation of the high abundance of siderophile elements in the upper mantle, and its oxidation state and contents of volatiles. As noted earlier, the homogeneous accretion hypothesis appeared unable to account for these features.

In order to provide the required high temperatures in the nebula and the rapid accretion of planets, the model seems rather firmly committed to Cameron's specific hypothesis of a massive parental nebula amounting to two solar masses. A notable feature of the hypothesis is that nearly one solar mass must be removed from the system in the form of an intense solar wind (T-Tauri phase) combined with other dissipative mechanisms.

As noted in Section 5.2, Cameron's model has encountered serious difficulties and has been abandoned by its author. Cameron (1975a, 1977) and Cameron and Pollack (1976) are in the process of developing a considerably modified hypothesis involving a less-massive nebula and characterized by considerably lower temperatures. According to Cameron's (1977) new model, the maximum temperature attained in the nebula at the Earth's orbit was only 450°C. This would appear to remove the foundation of the heterogeneous accumulation hypothesis.

Application to the Earth: Some Specific Difficulties

The heterogeneous accretion model leads to an Earth zoned according to volatility (Fig. 7.2). Moreover, internal temperatures after accretion are well below the solidus at all depths. Since most of the metallic iron is concentrated towards the deep interior, the gravitational energy liberated by final segregation of the core is quite small and insufficient to cause strong heating of the Earth or to drive intense convective mixing processes within the mantle. It is a source of considerable mystery as to how this configuration of the Earth, after completion of accretion, could have evolved to its present observed structure. The two configurations could hardly be more different.

The strongly zoned initial structure in the heterogeneous accretion model cannot be reconciled with our conclusion concerning the homogeneity of the mantle (Section 2.2) unless some very efficient subsequent stirring process leading to homogeneity is postulated. Once a large-scale compositional stratification is established, it is extremely difficult to remove. Even if initially unstable—e.g., a dense layer overlying a less dense layer—the most likely to result would be a simple convective overturn, leading to reversal of the layers and a stable stratification.

The only process which might be invoked to homogenize the mantle is the segregation of metal phase into the core, as in the single-stage model (Section 7.2). However, since most of the iron is already concentrated towards the center in the heterogeneous accretion model, this mechanism appears inadequate. Moreover, as discussed earlier, a serious weakness of the previous single-stage model was its inability to explain the uniformity of distribution of compatible siderophile, lithophile, and volatile elements in the mantle by subsequent remixing processes connected with core formation. This weakness is also encountered by the heterogeneous accretion hypothesis, but to a much more extreme degree.

(b) Iron Content of the Lower Mantle. Equilibrium calculations (Fig. 6.5) show that significant quantities of FeO do not begin to enter the magnesium silicates condensing from solar gases until the temperature has fallen below 400°C. This is far below the temperature at which the lower mantle is believed to condense according to the heterogeneous accretion model (1000–2000°C). Accordingly, the model requires that the magnesium silicates in the lower mantle are essentially free of FeO (Fig. 7.2). The iron content of the lower mantle was discussed in Section 1.5. It was concluded that the FeO/(FeO + MgO) ratio of this region is similar to that of pyrolite and equal to 0.12. If the lower mantle contained no iron, its elastic ratio ($\theta = K/\rho$) distribution would be far too high in relation to its density distribution.

(c) Composition of Upper Mantle. According to the heterogeneous accretion model, the upper mantle, representing a low-temperature, lately-accreted component, might be expected to be strongly depleted in high-temperature

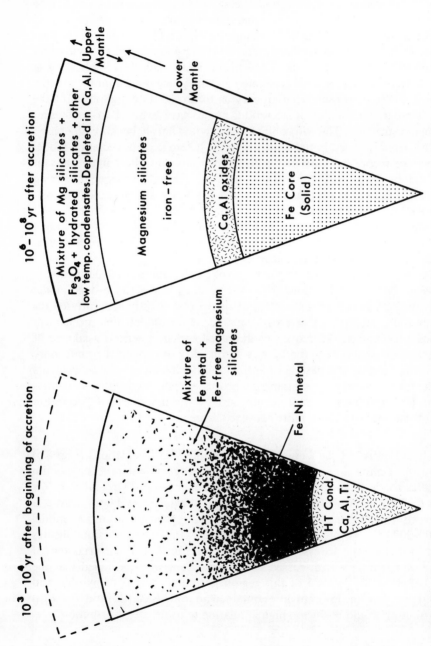

Figure 7.2 Zonal structure of Earth suggested by heterogeneous accumulation hypotheses of Turekian and Clark (1969) et al. (1972).

condensates such as Ca, Al, Ti, and REE, as compared to the major components Mg and Si. This does not appear to be the case. We concluded earlier that the abundances in pyrolite of Ca, Al, Ti, REE, and other members of this group, relative to Mg and Si, were similar to the primordial abundances. The model offers no explanation of this important relationship. The abundances of many volatile elements (e.g., sulphur) should also be *much* higher than is observed.

(d) **Composition of the Core.** Figure 3.2 shows that the Earth's core contains substantial quantities (10–15 percent) of light elements. Oxygen is believed to be the most likely candidate (Section 3.3), but significant amounts of sulphur and small amounts of H and C may also be present. Levin (1972a) has pointed out that the model does not provide any plausible means of introducing these light elements into the core. The required light-element components condense only at relatively low temperatures and are accreted to form the upper mantle *after* the core has been formed. The upper mantle is believed to have remained chemically isolated from the lower mantle and core.

(e) **Accretion Timescale.** The model requires that most of the Earth accreted on a time scale of 10^3–10^4 years. Such a short time scale poses severe dynamical problems that are unsolved (Levin, 1972a).

(f) **Siderophile and Volatile Element Abundances in the Upper Mantle.** During condensation of the solar nebula, nickel would have been present in the condensate as a metal until the temperature fell below 400°K, when NiO would have become stable. If the NiO now present in the upper mantle had been introduced subsequent to core segregation in the form of a low-temperature, metal-free condensate, then this condensate should also have introduced primordial ratios of all the elements that became fully condensed above approximately 400°K. Inspection of Table 7.1 demonstrates, however, that the ratios generally diverge widely from the primordial values.

Discussion

We conclude that the heterogeneous accumulation hypothesis provides a totally unsatisfactory explanation of the inferred properties of the Earth's mantle. It requires strong vertical zoning of (Ca + Al) relative to (Mg + Si), iron relative to magnesium, and other volatile components relative to involatile components. Evidence for these consequences is lacking. Indeed, the first-order conclusion arrived at in Chapter 2 related to the high degree of chemical homogeneity within the mantle, soon after the core had formed. Moreover, the model does not explain the presence of large amounts of light element(s) in the core or the proportions of siderophile and volatile elements present in

Table 7.1 Elements, which would have been fully condensed prior to oxidation of Ni in the solar nebula, are shown to occur in nonprimordial proportions in the Earth's upper mantle (ratios by weight)

	Earth's Upper Mantle	Primordial
Ni^a/Ir	85×10^4	2.0×10^4
Ni^a/Au	42×10^4	6.5×10^4
Ni^a/Ge	15×10^2	2.9×10^2
Cu/Au	85×10^2	7.2×10^2
F/Ge	62	2.2
Ni^a/S	7.2	0.2
Ni^a/Se	240×10^2	5.0×10^2
Ge/Ag	0.2×10^2	1.9×10^2
Ga/Au	15×10^2	0.6×10^2
Cu/Cd	9.7×10^2	1.7×10^2

[a] Ni has been selected for many of the ratios, since its abundances in the Earth's upper mantle and Type 1 carbonaceous chondrites are well established. Similar results would be obtained if either Co or Cu had been used as the standard of comparison. (From Delano and Ringwood, 1978a, with permission.)

the mantle. Additional geochemical difficulties encountered by this model have been discussed by Ringwood (1975a).

Origin of the D'' Layer

There is, however, one region of the Earth which might best be explained in terms of heterogeneous accretion. This is the lowermost 200 km of the mantle, which has long been known to exhibit anomalous seismic properties, including low or negative velocity gradients and a "lumpy" structure (Bullen calls this region the D'' layer).

The final stages of accretion of the Earth must have been much slower than the earlier stages, owing to depletions of planetesimals in the Earth's neighborhood (Safronov, 1972a). This low-density tail must have extended for a considerable period after the time of core formation and cooling of the surface. Indeed, the lunar highlands provide striking evidence of an intense bombardment between 4.6 and 3.9 billion years ago which must have been shared by the Earth.

These late, low-temperature condensates would presumably have contained the primordial abundance of iron, which may, however, have been highly oxidized. Such material, when it fell upon the Earth's surface, would have been degassed upon impact, but because of the relatively cool ambient conditions, iron would have remained oxidized, or become reoxidized. Thus, a cool, rigid crust of oxidized chondritic material would have collected. Because of its high iron content, it would be much denser than the mantle.

After a reasonably thick (10–20 km) cold plate of this material had collected, the consequent gravitational instability would cause the plate to subside into the mantle. Because of the large density contrast between mantle and plate, arising from the high iron content of the latter, the plate would sink directly to the base of the mantle and would accumulate above the core–mantle boundary. In such a way, a dense layer or iron-enriched material would be built up at the base of the mantle.

Because of its irregular distribution, controlled by the more or less random positions at which surface subduction had occurred, the layer might well have an uneven thickness which, by virtue of its high density, would lead to undulations of the core–mantle boundary. Moreover, the layer would possess chemical and dynamic instability, consequent upon the tendency of FeO from the layer to dissolve in the core at the interface. This would leave a thin transitional layer of lower density. This gravitationally unstable layer would ultimately segregate into a diapir (cf. salt domes) which would rise towards the top of the D'' layer, to be replaced by a fresh batch of iron-rich material. The net effect would be the development of small-scale heterogeneities over a long time scale, the latter being controlled ultimately by solid-state diffusion processes near the core–mantle boundary. The seismic characteristics of this region (Haddon and Cleary, 1974) might be explained by these processes.

CHAPTER 8

Homogeneous Accretion Revisited

8.1 Some General Considerations

The difficulties encountered by the single-stage and heterogeneous accretion models are sufficiently serious as to justify reconsideration of the homogeneous accretion hypothesis. It is conceivable that some of the earlier difficulties of this hypothesis might have been reduced or eliminated as a result of subsequent interpretations concerning the properties and composition of the Earth's core and its equilibrium with the mantle (Chapter 3).

In our discussion of the infrared cocoon protostar as a model for the early development of the solar system (Section 6.3), it was noted that solids may have become chemically isolated (as planetesimals) from the gas phase over a wide range of temperatures (Fig. 6.4). The planetesimals which spiral down to the ecliptic plane would thus show a great diversity of compositions, controlled by the condensation sequence of Table 6.2. This diversity is greatly increased by the gravitational perturbations of planetesimal orbits during accretion, which leads to a considerable degree of radial mixing. There would be a complete continuum from low-temperature condensates, similar to Cl chondrites, through planetesimals displaying increasing degrees of reduction of metallic iron and loss of volatiles, to relatively high-temperature condensates, totally reduced and strongly devolatilized, even to the extent of losing substantial amounts of silicon as SiO.

The subsequent homogeneous accretion of the Earth from this mixture of planetesimals readily explains some of our first-order conclusions concerning its composition, namely, the approximate chondritic relative abundances of major elements, Mg, Fe, Si, Ca, and Al, and of elements less volatile than Si. It explains, also, the depletions of volatile elements in the Earth and our conclusion that the degree of depletion increases approximately with volatility. Finally, it explains the inferrred high degree of homogeneity of the primordial mantle with respect to the distribution of crystallographically compatible lithophile, siderophile, and volatile elements.

Although it is important to recognize that the Earth probably accreted from a mixture of planetesimals possessing widely different thermal histories and, hence, compositions, some aspects of our discussion of the subsequent accretion of the Earth can be simplified if we consider the Earth to have accreted as a mixture of two principal components: A, an oxidized, volatile-rich primordial component similar to Cl chondrites; and B, a highly reduced, metal-rich, devolatilized material formed at high temperatures. Compositions of components A and B are given in Table 8.1.

In order to account for the abundances of non-siderophile volatile elements (e.g., Zn, Cl, Cs, and In) in the mantle, Ringwood (1977b) estimated that about 10 percent of component A would need to be mixed with 90 percent component B. On the other hand, in order to account for the large amount of oxygen believed to be present in the core, about 15 percent of

Table 8.1 Compositions of Orgueil Cl chondrite and of a reduced, devolatilized component condensing at elevated temperatures ($\sim 1000°C$) in solar nebula

	Orgueil[a]	High-temperature[b] component
	A	B
Metallic nickel–iron ($\sim 5\%$ Ni)	—	34.1
SiO_2	21.7	32.8
TiO_2	0.1	0.2
Al_2O_3	1.6	2.8
Cr_2O_3	0.35	0.2
MnO	0.2	0.1
FeO	22.9	—
NiO	1.2	—
MgO	15.2	27.7
CaO	1.2	2.3
Na_2O	0.7	—
K_2O	0.07	—
P_2O_5	0.3	—
Water	19.2	—
Organic compounds	9.7	—
Sulphur	5.7	—

[a] Wiik (1956)

[b] Based on devolatilized (including some Si, Cr, and Mn) Cl composition and with all oxidized iron and nickel reduced to the metallic state (after Ringwood, 1977b).

component A would be necessary (Ringwood, 1977b). The discrepancy, though not large, may be real. Anticipating the results of subsequent discussions, two explanations of the discrepancy can be advanced.

(a) A significant proportion of the volatiles in component A were evaporated into the atmosphere during accretion and lost from the Earth.

(b) A substantial degree of segregation of oxidized low-temperature dust particles from gas occurred prior to the cocoon star stage, so that the O/H ratio of the system greatly exceeded the solar ratio. Under these conditions, a significant amount of FeO may have accreted into silicates at relatively high temperatures (Ringwood, 1977b). In the author's opinion, the first explanation is more probable.

8.2 Accretion of the Earth

We now consider the accretion of the Earth from a mixture of planetesimals comprising about 15 percent of component A and 85 percent of component B. The accretion time scale is not a key factor in the following discussion, which could accomodate a "long" time scale of 10^7–10^8 years. The sizes of the initial planetesimal populations in the ecliptic plane of the nebula are believed to have ranged from a few kilometers downwards (Goldreich and Ward, 1973). This latter size also corresponded to that of the largest chemically homogeneous planetesimals, as in case (c) of Section 6.3. Accretion is believed to have proceeded as discussed in Section 6.2, whereby a population of intermediate-sized bodies with radii extending perhaps up to 1000 km was produced. The Earth thus formed from a hierarchy of intermediate bodies, planetesimals, and dust, as discussed by Safronov (1972a) and Wetherill (1976a).

During the early stages of accretion, the accretion energy was low (Fig. 7.1) and the original chemical heterogeneities among planetesimals over distances up to a few kilometers may have been preserved. This would permit the trapping of a significant proportion of highly volatile components introduced by component A at this stage. After the mass of the growing nucleus had increased to about $\frac{1}{10}$ of the present Earth mass, planetesimals and the intermediate-sized bodies would have arrived at the Earth's surface with sufficiently high velocities to cause strong transient heating and evaporation of volatiles (Fig. 7.1). Moreover, the impact velocities would have resulted in extensive disruption and fragmentation of incoming projectiles, accompanied by intense and repeated comminution. This would have caused intimate mixing of components A and B. This state of intimate mixing of A and B thus prevailed throughout most of the accretion of the Earth.

The temperature distribution during accretion depended on several variables, especially the time scale of accretion and the opacity of the atmosphere. For a wide range of plausible initial conditions, the *average* near-surface temperatures throughout most of the accretion of Earth would have exceeded the boiling point of water. Thus, water evaporated from component *A* during impact of planetesimals with the Earth would have accumulated in the primitive atmosphere, and, together with hydrogen (see below) would be its principal component. Smaller amounts of other gases including N_2, NH_3, inert gases, and CO (or CO_2, depending on temperature) would also accumulate in the primitive atmosphere. Moreover, transient high temperatures exceeding 2000°C would be reached during the entry of planetesimals and dust with high velocities into the primitive atmosphere. The interaction of these smaller bodies and particles with the primitive atmosphere would resemble that of meteors in the present atmosphere. Most would be totally evaporated and recondensed before they reached the surface, but would have produced extremely high transient temperatures. Intermediate-sized bodies such as the larger meteorites would have been extensively oblated at high temperatures, losing most of their mass before reaching the surface. The largest would penetrate the atmosphere without large proportionate loss of mass, becoming intensely shock heated and degassed upon impact with the solid Earth. A large proportion of these bodies would be vaporized and recondensed. Thus, only a small proportion of the incoming component *A* would retain its volatiles during atmospheric entry and impact with the growing Earth.

Our chosen mixture of 15 percent component *A* and 85 percent component *B* would contain the equivalent of 3 percent H_2O and 29 percent metallic nickel–iron (5 percent Ni). (The preponderance of metal phase is to be noted.) Thus, the principal chemical interaction of infalling planetesimals with the primitive atmosphere would be between metallic iron and water vapor. The equilibrium constants for the relevant reaction are given in Table 8.2.

It is immediately seen that most of the water vapor in the primitive atmosphere would be reduced by infalling excess metallic iron to form FeO (which enters ferromagnesian silicates in the accreting Earth) and hydrogen. In the presence of a young and active sun, the exosphere temperature of the Earth is likely to have been significantly higher than at present. Even now,

Table 8.2 Equilibrium constants for the simplified reaction $MgSiO_3 + Fe + H_2O = \frac{1}{2}Fe_2SiO_4 + \frac{1}{2}Mg_2SiO_4 + H_2$ corresponding to olivine and pyroxene compositions with FeO/(FeO + MgO) = 0.22. (Based on Mueller, 1964.)

°C	$K = P_{H_2}/P_{H_2O}$
400	132
600	35
800	17
1000	11
1200	7.6
1400	6

hydrogen is known to escape by thermal evaporation from the Earth's exosphere. It seems likely, then, that hydrogen produced by the above reaction would continue to escape from the atmosphere while the Earth accreted. This effect would be greatly enhanced by very high temperatures transiently reached when dust and planetesimals entered the primitive atmosphere at very high velocities.

A simple mass balance shows that most of the FeO finally incorporated in the Earth would have been produced by reaction of metallic iron in component B with water vapor from component A. The proportion of FeO introduced directly from component A was much smaller (Ringwood, 1977b).

Although most volatiles from component A were transiently evaporated into the atmosphere during entry into the atmosphere, only the most highly volatile components—e.g., H_2, H_2O, CH_4, CO_2, CO, N_2, NH_3, some H_2S, and inert gases remained in the atmosphere. Components of lesser volatility, e.g., most of the depleted metals (alkali metals, Pb, Tl, Bi, etc.) from component A simply recondensed in the primitive atmosphere and were incorporated into the solid Earth. It seems likely that much of the sulphur in component A was likewise ultimately condensed as FeS, given the redox state of the primitive atmosphere (see further discussion by Ringwood, 1977b).

8.3 Formation of the Core

According to our model, the Earth accreted as an intimate mixture of silicates (containing FeO) and metallic iron, together with small amounts of FeS and volatile components. The internal temperature distribution during accretion lay between curves S and R of Figure 3.5.

Evidence discussed in Section 3.4 shows that the core formed very soon after accretion. Thus, the temperature distribution soon after accretion must have been sufficiently high so that the melting point of iron alloys was exceeded. Following earlier discussions, this required that a substantial proportion of the Earth was accreted in the form of intermediate-sized bodies (Safronov, 1972a; Wetherill, 1976a). On the other hand, we have concluded that most of the Earth's mantle remained unmelted during accretion and core formation. In order to permit core formation within a homogeneously accreted Earth under these conditions, an additional component(s) must have been present which was capable of lowering the melting point of the metal phase, but not the silicate. The properties of the Earth's core indeed imply the presence of large amounts of a component possessing a low atomic weight.

We have argued earlier that this component is likely to have consisted dominantly of oxygen (as FeO), although a small amount of FeS was also likely to have been present. The effects of FeS in lowering the solidus temperature of the metal phase have been widely discussed (e.g., Murthy and Hall, 1970) and continue to play a significant role in the present model. However, the dominant role in lowering the melting point of the metal phase is believed to have been played by FeO, an effect which is displayed only at very high pressures.

It is likely that at very high pressures, the system Fe-FeO may behave analogously to the system Fe-FeS at low pressures (Section 3.4). Thus, an extensive degree of solubility of FeO in molten iron is expected to cause a large decrease in the Fe–FeO solidus (eutectic) temperatures as compared to the melting point of pure iron. For this reason, melting of the metal phase leading to segregation of large bodies of liquid (Fe–FeO) solution is likely to have commenced first at considerable depth (1500–2000 km) within the Earth (in contrast to the Elsasser, (1963) model wherein the core-segregation process commenced near the surface). The effect of dissolved FeO on the melting point of the metallic alloy also ensured that the metal phase melted at a much lower temperature than the silicate phase at the high pressures attained deep within the Earth's interior.

The commencement of core segregation at great depths caused acute convective instability in the overlying regions. This was accentuated because of the decrease in FeO/(FeO + MgO) ratio of the silicate–oxide phases with depth as FeO entered the metal phase. After segregation of metal from a localized region, the residual silicate-oxide phase rose convectively or diapirically upward, to be replaced by a new batch of dense FeO-rich silicate-and-metal mixture from the upper mantle. This new batch quickly re-equilibrated. FeO (or Fe_3O_4 h.p.p.) migrated from the silicate–oxide phase into the molten metal phase, which segregated and sank into the core. The FeO content of the residual silicate–oxide phase was thus established by high-pressure equilibrium with (Fe–FeO) metal phase deep within the Earth. Ultimately, the entire volume of the Earth was convectively recycled in this manner and all of the metal entered the core.

8.4 Siderophile Elements and Oxidation State within the Mantle

The relatively high abundances of siderophile elements and oxidized species (Fe^{3+}, CO_2, H_2O) in the upper mantle show clearly that this region could not have equilibrated with pure metallic iron (Section 2.4). This was a principal reason for rejecting the early version of homogeneous accretion models. We concluded in Section 3.3 that the core probably contains about 10 weight-percent oxygen and also a small amount of sulphur in solution. It seems quite likely that the partition equilibria for siderophile elements between the iron–oxygen solution and silicates might differ considerably from the case for pure iron silicates, and that this factor might ultimately be responsible for the high siderophile-element abundances in the mantle and its highly oxidized state.

Consider the case where mantle silicates containing FeO are in equilibrium with pure metallic iron:

$$2Fe + O_2 = 2FeO.$$

The equilibrium constant $K = a_{FeO}^2/a_{Fe}^2 \cdot f_1(O_2)$, where a_{FeO} is the activity of FeO in the mantle, a_{Fe} is the activity of iron (equal to unity), and f_1 is the fugacity of oxygen. Thus, $K = a_{FeO}^2/f_1(O_2)$. Now consider the case where the core contains a large quantity of dissolved FeO, so that the activity of Fe is lowered, while, a_{FeO} in the mantle remains fixed. The fugacity of oxygen changes to $f_2(O_2)$ and the equilibrium constant is $a_{FeO}^2/a_{Fe}^2 \cdot f_2(O_2)$. Equating the equilibrium constants, cancelling a_{FeO}^2 terms and rearranging, we have

$$\frac{f_2}{f_1}(O_2) = \frac{1}{a_{Fe}^2}$$

The activity of iron in the Fe–FeO solution may well be considerably decreased compared to its pure state. Applying the above equation, it seems quite possible that the oxygen fugacity (f_2) in regions of the Earth's mantle which had equilibrated with (Fe–FeO) solutions prior to and during core formation could have been 1 or 1.5 log units higher than in the case (f_1) when solution of FeO did not occur.

This increase in oxidation state may have been enhanced by another effect. We have already (Section 3.3) considered Mao's suggestion that in the deep mantle, FeO may partially disproportionate to form a high pressure phase of Fe_3O_4 plus metallic iron. This assemblage would be stable only under subsolidus conditions. Once the melting temperature required for the formation of the iron–oxygen solution in the deep mantle had been exceeded substantially, the Fe_3O_4 (h.p.p.) phase would have been totally dissolved in excess liquid metallic iron and would ultimately have been carried into the core. However, under subsolidus conditions *prior* to core segregation, the lower mantle phase—e.g., $(Mg_{0.88}Fe_{0.12})\,SiO_3$ perovskite—would have equilibrated with Fe_3O_4 (h.p.p.) and would therefore contain a small proportion of Fe_2O_3, which would possess limited solid solubility in the perovskite phase. After the metallic (Fe–O) solution had segregated into the core, the Fe^{3+}-bearing perovskite would be returned to the upper mantle via convection, thereby accounting for the presence of a small amount of Fe^{3+} in this region and a greatly increased oxygen fugacity.

It seems possible that under the much higher oxygen fugacities in the mantle caused by a combination of the effects discussed above, oxidized species such as OH^{-1} and CO_3^{-2} may have been stable throughout most of the Earth during the formation of the core. Likewise, the higher $f(O_2)$ conditions would be expected to cause considerable increases in the equilibrium amounts of siderophile elements which entered the silicate phases of the mantle (Mysen and Kushiro, 1976; see also Fig. 11.5). The high abundances of these elements might then be attributed to equilibrium partition between (Fe–O) metallic liquid and silicates.

The apparent uniformity in distribution of several compatible siderophile elements in the upper mantle may thus represent a combination of equilibrium partitions between metal and silicate phases in the deep mantle, combined with convective mixing during and subsequent to core formation. The equilibrium model might also provide an explanation for other puzzling characteristics of the mantle not readily explained by the heterogeneous accretion model, e.g., the observations that the relative abundances of many siderophile elements in the upper mantle—e.g., Ni/Ge and Ni/Ir— deviate strongly from the primordial relative abundances (Table 7.1).

The abundances of gold and platinum-group elements appear to display very large dispersions in mantle rocks, and are unlike the compatible siderophiles in this respect. It is not clear whether these dispersions can be explained by crystal chemical factors arising from mantle differentiation processes, combined with near-surface metasomatism. It is possible that these dispersions might arise from incomplete homogenization of chemical heterogeneities on the scale of a few kilometers introduced into the Earth during the earliest stages of accretion when accretion energies were low. It will be recalled (Section 6.3) that in this earliest stage of accretion, planetesimals possessing contrasting chemical compositions might have possessed sizes ranging up to a few kilometers. These may not have been completely disintegrated and remixed when they impacted the growing Earth, during the early stages of accretion, while it was still small. It is only during the more mature phase of accretion that all material falling upon the Earth became thoroughly remixed.

A similar factor might have been responsible for the very large dispersions in the abundances of primordial inert gases occurring in the glassy selvedges of oceanic tholeiites (Craig and Lupton, 1976).

Iron-Oxide Content of the Mantle

We have already remarked upon the striking degree of uniformity of $FeO/(FeO + MgO)$ ratios among mantle-derived alpine ultramafics and lherzolite xenoliths in alkali basalts all over the world. These ratios lie mostly within the range 0.07–0.12 (molecular), and appear to be applicable throughout most of the volume of the mantle (Section 1.5). It is tempting to seek an explanation for these small variances in $FeO/(FeO + MgO)$ ratios in terms of similar processes to those which were responsible for inferred small variances of compatible siderophile elements in the upper mantle.

We concluded earlier that if the Earth accreted homogeneously from a mixture of metal and silicate, the solubility of FeO in the liquid metal phase would increase with depth, once the solidus of the Fe–FeO system was reached. Accordingly, the $FeO/(FeO + MgO)$ ratio of the coexisting silicates would tend to decrease with depth. A similar situation would prevail if FeO (from silicate phases) were disproportionated under pressure to form Fe_3O_4 (h.p.p.) + metallic iron. These reactions would not proceed to completion, however, since the effect of increasing temperature with depth would tend to preserve some FeO in solid solution in silicates, stabilized by the free energy of mixing. It is possible that the FeO/(FeO + MgO) ratios of 0.12, corresponding to the least fractionated mantle silicates (pyrolite) reflected an equilibrium between the effect of pressure tending to decrease the ratio and temperature tending to increase it.

Dynamical factors connected with the process of core formation may also have played a role. We have already noted that the core formation process involved strong convection throughout the Earth. It is possible that there was a limited depth interval in which metallic iron–oxygen alloy segregated into bodies large enough to sink into the core. Such a limited depth interval may have been established by a balance between the effect of pressure tending to increase the solubility of FeO in molten Fe, thereby decreasing the minimum melting point of the eutectic and the normal effect of pressure in increasing the melting points of end-member components.

If metal-segregation initially occurred within a limited depth interval (e.g., 1500–2000 km), then the FeO/(FeO + MgO) ratio of silicates in this region would have been fixed within fairly narrow bounds. This ratio would be imposed upon the remainder of the mantle by convection associated with core formation as discussed in Section 3.4.

8.5 Atmosphere and Hydrosphere

Escape of Primitive Atmosphere

Most of the low-temperature Cl component A accreting upon the Earth was completely degassed upon impact. Assuming that the mixture from which the Earth accreted contained 15 percent of component A, the Earth should contain about 400 ppm of nitrogen, most of which would have entered the primitive atmosphere. However, the actual amount of nitrogen in the present atmosphere corresponds to less than 1 ppm for the bulk Earth. It seems, therefore, that a sizeable primitive atmosphere must have escaped from the Earth during or soon after accretion, if a substantial amount of component A had been accreted. Consideration of the mass balances of other volatiles in the accreting material and in the present Earth would lead to the same conclusion.

It has frequently been argued (e.g., Urey, 1960b, 1962a) that escape of high-molecular-weight gases from the Earth's gravitational field is not possible. This argument has been used both as a boundary condition for the formation of the Earth and as an objection to Ringwood's single-stage hypothesis, which required the large-scale escape of high-molecular-weight gases during accretion of the Earth.

Geochemical considerations indicate rather strongly that the escape of a primitive atmosphere from the Earth indeed occurred at an early stage. It is believed that most of the present atmosphere and hydrosphere are of secondary origin, having developed gradually over geological time by degassing of the mantle, connected with processes of magmatism (Rubey, 1951). The abundances of H_2O and N_2 in primitive basalt magmas and ultramafics imply that volatiles presently locked in the mantle exceed the amounts released to the atmosphere and hydrosphere (Ringwood, 1966a, 1975a). Immediately after its formation, before the present secondary atmosphere and hydrosphere had developed substantially, the proportion of volatiles—chiefly H_2O, CO_2, and N_2—locked inside the Earth would have been far greater.

It is most difficult to understand how the Earth could have been formed in this condition. Referring to Figure 7.1, we see that when the mass of the Earth had grown beyond one-tenth of its present value, accreting solids arrived with such high velocities that they were subjected to strong transient heating, accompanied by melting and degassing. By the time the Earth had reached one-quarter of its present mass, accreting matter was largely volatilized during impact. Beyond this stage, nearly all the accreting material was strongly degassed. At the conclusion of accretion, amounts of degassed volatiles present in the atmosphere and hydrosphere exceeded, on any reasonable assumptions, the amount of volatiles trapped in the interior. These considerations strongly indicate that a primitive atmosphere much larger than the present secondary atmosphere and hydrosphere was present after the Earth had formed and that some mechanism for the escape of the primitive atmosphere must have existed.

It can readily be shown that atmosphere loss on the scale required cannot occur via the Jeans–Spitzer mechanism of evaporative escape of single atoms from the exosphere. The key physical process required is that of continuous atmospheric blow-off during accretion. Öpik (1963a) pointed out that blowing off of a planetary atmosphere is governed by an escape parameter B, which represents the ratio of gravitational potential energy to thermal energy of the molecules:

$$\bar{B} = \frac{GM\bar{m}}{RkT},$$

where G is the gravitational constant, M is the mass of the planet inside the spherical surface with radius R, $\bar{m}$ is the *mean* mass of the gas molecules, k is Boltzmann's constant, and T is absolute temperature. For a diatomic gas, the thermal energy equals $2.5\,kT$ per molecule; if this exceeds the gravitational energy, the top of the atmosphere blows off into interplanetary space. Thus, for escape of this kind, which is indiscriminate as to molecular species and depends only on the mean properties of the gas,

$$\frac{GM\bar{m}}{RkT} = \bar{B} < 2.5.$$

In the case of the primitive Earth rotating with angular velocity ω, the effective gravitational energy is reduced and the condition for blow-off becomes

$$\bar{B} = \frac{\bar{m}}{kT}\left(\frac{GM}{R} - \omega^2 R^2\right) < 2.5.$$

Clearly, blow-off is favored by increasing R, T, and ω, and decreasing $\bar{m}$.

Ringwood (1975a) estimated that blow-off of a primitive terrestrial atmosphere would occur during accretion providing the Earth were rotating with a period of 4–5 hours (as implied if the Moon were derived from the Earth), the atmospheric temperature were 1800–2000°K (comparable to the present exosphere temperature when the sun is in an active phase), and the atmosphere possessed a mean molecular weight $\bar{m}$ of 3.5. According to his model, this low value of $\bar{m}$ was produced by turbulent mixing of the primitive terrestrial atmosphere with hydrogen derived from the primordial solar nebula. The model therefore required that accretion of the Earth occurred *before* the primitive, hydrogen-rich solar nebula had been dissipated by the sun as it passed through the T-Tauri phase. Since Ringwood's model envisaged that the Earth accreted in 10^6 years of less (in order to obtain high surface temperatures from accretional energy) this requirement seemed reasonable.

However, the situation is different if one considers a long time scale for accretion of the Earth, e.g., 10^8 years, since in all probability, the T-Tauri phase of the sun would have occurred long before the accretion of the Earth had been completed. In this case, it would not be possible to lower the mean molecular weight of the terrestrial atmosphere adequately by mixing with hydrogen from the solar nebula, and hence blow-off would not occur.

This problem is circumvented in the present model, whereby the required supply of hydrogen is produced by the reaction of metallic iron in component B with water from component A:

$$\mathrm{Fe} + \mathrm{H_2O} = \mathrm{FeO} + \mathrm{H_2}.$$

This reaction would have occurred mostly in the primitive terrestrial atmosphere as metallic iron from component B fell with high velocity into an atmosphere of water vapor plus other volatiles evaporated from component A. The relevant equilibrium constants for the reaction (Table 8.2) show that in the lower regions of the atmosphere, the H_2/H_2O ratio was probably in the vicinity of 100. Because of the great preponderance of hydrogen in the system, the mean molecular weight $\bar{m}$ was only slightly greater than 2. Under these circumstances, providing high temperatures (1500–2000°K) were maintained in the upper atmosphere, the primitive atmosphere would have blown off continually. Thus, hydrogen produced by this reaction would continually have "washed" other volatile components, including N_2, CO, CO_2, NH_3, inert gases, and volatile metals, back into space during accretion. For this reason, the total amount of component A which was accreted by the Earth may have been significantly higher than estimates based on the content of volatile elements Zn, In, Cs, and Cl now present in the mantle (Section 2.4). A substantial amount of these elements may have been lost from the Earth by the above mechanism. This may account for the discrepancy between the amount of component A implied by the present inventory of volatile elements in the mantle (10 percent) and the amount required (15 percent) to explain the total FeO present in the mantle and core.

The required high temperature in the upper atmosphere could readily have been produced by the early sun providing it was in a somewhat more active state than today. This appears quite likely. Transient high temperatures would also be produced by the falling of small planetesimals into the atmosphere. It is possible that atmosphere escape may have occurred via a large number of small "puffs" as limited volumes were heated to extremely high temperatures by planetesimal infalls. The phenomena would be similar to those produced by the infall of meteors and meteorites, but on a vastly greater scale.

We conclude from this discussion that the loss of a sizeable primitive atmosphere from the Earth does not appear to present any severe problems, whether accretion occurs on a time scale of 10^6 or 10^8 years.

Incorporation of H_2O and CO_2 into the Mantle

The mechanism by which small but significant amounts of water and CO_2 became incorporated into the mantle is not as simple as it might seem. We have already noted that nearly all of the water in low-temperature component A would have been evaporated into the atmosphere during accretion and reduced to FeO and H_2 by the excess metallic iron present in component B.

A small proportion of volatile-rich, low-temperature condensates would have survived passage through the atmosphere to become incorporated within the solid body of the accreting planet. We see this process today with the sporadic arrival at the Earth's surface of carbonaceous chondrites. These represent only small proportions of the original parent bodies, most of which were burnt up during their passage at high velocities through the atmosphere. But even this small amount of water, initially buried in the Earth during accretion, was destined to become decomposed because of the great excess of metallic iron with which it was in intimate contact. The water would have been reduced to form iron oxide, which combined with the silicate phases present, while the hydrogen dissolved in the excess metallic iron forming an interstitial Fe–H solution. With the ultimate melting of iron and its segregation into the core, all of the water present in the upper mantle would have been decomposed, and the hydrogen carried into the core.

A similar situation would have occurred with CO_2, most of which would have been reduced either in the primitive atmosphere or by excess iron in the upper mantle, while the resultant carbon also formed interstitial solutions with iron. Thus, there is a problem in explaining why, indeed, the Earth possesses any water and carbon dioxide.

It is possible that the formation of these components is due to a reversal in the deep mantle of some of the equilibria just discussed. When metallic iron reduced CO_2 and H_2O in the primitive terrestrial atmosphere, a small amount of hydrogen and carbon entered the iron to form interstitial solutions. Within the upper mantle, still more hydrogen and carbon entered the metal phase as a result of the reduction of OH^{-1} and CO_3^{-2} ions just discussed.

In Section 3.3, we concluded that at very high pressures in the deep mantle, iron and oxide would be extensively soluble above the solidus. The entry of oxygen into the iron alloy already containing significant amounts of elemental carbon and hydrogen would tend to reverse the earlier equilibria by reoxidizing the dissolved hydrogen and carbon. The effect of the solubility of oxygen would be to cause a large increase in the thermodynamic activity of H_2O and CO_2 in the system, so that these molecules would be reformed and then migrate to the silicate phase, where they would become fixed as minor components of the mantle.

8.6 Conclusion

The present extension of the homogeneous accretion hypothesis appears capable of explaining several difficulties encountered by earlier versions of this model. These include the high abundances of siderophile elements in the mantle, the redox state of the mantle, the nature of the light elements in the core and the escape of a primitive atmosphere during accretion. It also has the major advantage of explaining the homogeneity of distribution of lithophile, siderophile, and volatile compatable elements in the mantle and the bulk chemical composition of the Earth.

The viability of the present model depends to a considerable extent upon our conclusions that the compatible elements are indeed relatively uniformly distributed throughout the mantle and that oxygen is the principal light element in the core. The evidence in favor of these conclusions is strong but not yet finally decisive. Fortunately, this key evidence is in principle procurable. If the above conclusions are finally confirmed, the homogeneous model will become most attractive and consistent with a wide range of geochemical and geophysical data.

PART III

THE MOON AND PLANETS

CHAPTER 9

The Terrestrial Planets

9.1 Introduction

The terrestrial planets possess markedly different intrinsic densities (Table 6.3), implying the existence of corresponding differences in chemical composition. Possible reasons for the compositional differences which were considered in Section 6.5, involved (a) fractionation of metallic iron from silicates in the nebula prior to accretion, and/or (b) differences in oxidation state of iron between planets. Having previously considered evidence bearing on the bulk composition of the Earth (Chapters 1–4), general processes involving accretion of planets in the solar nebula (Chapters 5 and 6), and theories of the origin of the Earth (Chapters 7 and 8), we will now consider compositions of the terrestrial planets in the light of the above discussions. It is not our objective to provide a comprehensive review of the chemistry, constitution, and evolution of each of these planets, but rather to discuss those aspects that have a general bearing on the compositions and origins of the Earth and Moon.

9.2 Mars

The density of Mars is 3.94 g/cm^3; its mass is only about one-tenth that of the Earth. When corrected for the effect of self-compression in its own gravitational field, this density is substantially smaller than the decompressed densities of the Earth and Venus (Table 6.3). The traditional explanation of this density deficit (Jeffreys, 1937; Urey, 1952; MacDonald, 1962) was that Mars consisted of a metallic iron phase and a silicate phase, similar in composition, respectively, to the Earth's core and mantle. However, the lower density of Mars implied that it possessed less iron than the Earth (Table 6.3). This was taken to support the hypothesis (Urey, 1952) that major physical fractionation of metallic iron from silicates had occurred in the solar nebula prior to accretion of the planets.

An alternative hypothesis was advanced by Ringwood (1959, 1966b), who suggested that the total amounts of major metals (Fe, Si, Mg, Al, and Ca) were the same in Mars, the Earth, Venus, the Sun, and Type 1 carbonaceous chondrites. The lower density of Mars was attributed to its more oxidized state (Fig. 6.9). These alternative models have an essential bearing on general problems such as the chemical evolution of the solar system and it is important to differentiate between them. Because of the markedly different bulk physical properties of Mars compared to other planets, it offers the possibility of distinguishing between these two interpretations. A property which turns out to have a key bearing on the topic is the moment of inertia of Mars.

Moment of Inertia

If Mars were in perfect hydrostatic equilibrium, it would possess a moment of inertia coefficient, I/MR^2, of 0.376. However, the observed deviation of the optical flattening (5.86×10^{-3}) from the dynamical flattening (5.24×10^{-3}), as well as large differences in surface elevation, imply that the hydrostatic assumption is not correct and that an adjustment of the moment of inertia coefficient may be required (Binder and Davis, 1973).

The nature of this correction depends partly upon the possible existence of a differentiated crust and the degree to which it is isostatically compensated. From the observed relationships between surface gravity and elevation, Phillips and Saunders (1975) concluded that the older regions of the planet (as indicated by crater counts) were underlain by a relatively low-density crust in isostatic equilibrium. The depth of isostatic compensation was estimated to be up to 100 km. However, their analysis also showed that the younger regions of the crust, including the Tharsis Plateau, were only partially compensated at this depth. Isostatic compensation might have been achieved at greater depths or, alternatively, part of the uncompensated load might have been supported by the strength of the Martian interior.

Lamar (1962) proposed that the difference between dynamical flattening and optical flattening could be accounted for by the existence of an increase in crustal thickness from pole to equator. Binder and Davis (1973) examined the effect of such an isostatically compensated crust on I/MR^2 and concluded that a small reduction, from 0.376 to about 0.372, would be caused. However, these authors used an earlier estimate of the optical flattening, which has since been superceded by more accurate measurements based on spacecraft occultations. These later measurements (quoted above) are much closer to the dynamical flattening, so that the correction to I/MR^2 would be smaller than estimated by Binder and Davis.

On the other hand, if the surface relief on Mars were not isostatically compensated, a larger correction to I/MR^2 could be necessary. Reasenberg (1977) assumed that the enormous mass comprising the entire Tharsis Plateau represented a largely uncompensated load on the crust, supported by the strength of the interior. On the basis of this model, he obtained a moment of inertia coefficient of 0.365. However, Reasenberg's model appears questionable in view of the evidence presented by Phillips and Saunders (1975) that the Tharsis Plateau is partially compensated at a depth of 100 km and could be fully compensated at greater depths, a conclusion also reached by Lambeck (1979). Moreover, Lambeck has shown that the stresses implied by Reasenberg's model are probably in excess of the long-term strength of the Martian mantle. He finds that if Mars has a similar rheology to the Earth, the non-hydrostatic component of the flattening would imply a lower limit to I/MR^2 of 0.374. In a recent discussion of the moment of inertia of Mars, Cook (1977) also concluded that corrections to the dynamical value arising from nonhydrostatic components of the gravitational potential were probably small. In the light of evidence currently available, it seems most likely that the moment of inertia coefficient of Mars is close to 0.374.*

Internal Constitution

Rapid advances during the late 1960s (Section 1.4) in knowledge of the crystallographic and elastic properties of high-pressure mineral phases and their stability relationships made it possible to calculate the mean density and moment of inertia of different compositional models of Mars with considerable precision. Ringwood and Clark (1971) accordingly investigated the properties of a series of models possessing bulk compositions similar to Type 1 carbonaceous chondrites (minus volatiles). The models contained all of the iron in the oxidized state and had varying Fe^{2+}/Fe^{3+} ratios. The most plausible of the (low-pressure) mineral assemblages investigated consisted of olivine (60 percent) + pyroxene (19 percent) + magnetite (21 percent). The $FeO/(FeO + MgO)$ molecular ratio of olivine and pyroxenes was 0.32, the mineral assemblages being generally similar to that displayed by the Karoonda chondrite. When this mineral assemblage was subjected to the $P-T$ field corresponding to the Martian interior, two major series of phase transitions to denser structures occurred at depths of about 1200 and 2000 km. The calculated density of a chemically homogeneous Martian model of this

---

* *Note added in proof*: However, a subsequent study of the topic by Kaula (1979) using a different model from that adopted by all of the above workers leads to an estimate of 0.365 for I/MR^2, similar to that of Reasenberg. Kaula's model assumed that non-hydrostatic contributions to the moments-of-inertia are axially symmetric about the principal axis in the direction of Tharsis. To the extent that the various alternative solutions are model dependent, the question cannot be regarded as closed. See footnote on page 141.

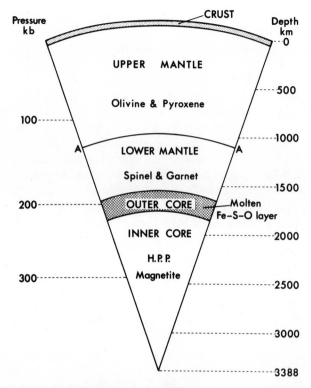

Figure 9.1 Section through Mars based on model which assumes that the composition corresponds to that of a differentiated, devolatilized Type 1 carbonaceous chondrite (from Ringwood and Clark, 1971).

composition was 3.97 g/cm^3, in good agreement with observation; but the moment of inertia coefficient was 0.391, much larger than observed. This model was then varied by assuming that Mars had extensively melted and differentiated, thereby permitting the high-pressure form of magnetite to segregate into the core (radius 1640 km), as indicated in Figure 9.1. The calculated density of this model was 3.94 g/cm and I/MR^2 was 0.373, in excellent agreement with the observed properties of Mars.

The above model is undoubtedly oversimplified. It was recognized that if Mars had differentiated sufficiently to form a large core, then a low-density crust would probably be present. This would have the effect of reducing I/MR^2, but not below acceptable limits. There is also the possibility of changing the core-to-mantle ratio by varying the Fe^{3+}/Fe^{2+} ratio of the model.

It is likely that some sulphur would also have been accreted. For example, the Karoonda carbonaceous chondrite contains about 4 percent of (Ni, Fe)S. Thus, a liquid Fe–S–O layer would probably also be present, as indicated in Figure 9.1. Ringwood and Clark (1971) concluded that it would not be possible to replace the magnetite core by troilite (FeS) because this would lead to an unacceptably low density. However, subsequently discovered transitions of FeS to denser states under high pressures (Ahrens, 1976) weaken this conclusion and permit the Fe–S–O layer to be much larger than indicated in Figure 9.1.

The available information is obviously insufficient to specify a unique model of Mars. Nevertheless, it can be firmly concluded that a differentiated, oxidized chondritic model of the type considered by Clark and Ringwood (1971) is capable of explaining the density and moment of inertia of the planet. The essential characteristic of this model is that the Martian mantle [FeO/(FeO + MgO) ~ 0.35] contains much more FeO than the Earth's mantle,* for which FeO/FeO + MgO ~ 0.12 (Section 1.3). However, the relative abundances of the principal metals Fe, Mg, Si, Ca, and Al in Mars, on the basis of this model, are similar to their abundances in the bulk Earth.

Urey's (1952) model is also capable of explaining the density and moment of inertia of Mars. This is facilitated by the additional degree of freedom, whereby the ratio of iron to silicates can be varied arbitrarily in order to explain the density. The model, however, has some further implications which can be tested. According to this model, the composition of the silicate phase in Mars is similar to that of the Earth's mantle. However, the density of the Martian mantle, as implied by the moment of inertia, is found to be substantially higher than that of the Earth's mantle (at corresponding pressures). The Martian mantle thus contains much more total iron than the Earth's, and according to the model, much of this iron must be present in the *metallic state*, dispersed throughout the mantle (Urey, 1952, 1957a; MacDonald, 1962). Evidence relating to this property will be discussed in the next subsection.

The internal structure of Mars has been discussed by several other authors (e.g., Kovach and Anderson, 1965; Reynolds and Summers, 1969; Binder, 1969; D. L. Anderson, 1972a; Binder and Davis, 1973; Johnston et al., 1974; Johnston and Toksöz, 1977). In several models proposed by these authors, the proportion of sulphur in the Martian core is treated as an independent variable. In general, the favored models tend to be intermediate between those of Urey (1952) and Ringwood and Clark (1971). Most of these studies concluded that the FeO/(FeO + MgO) ratio of the Martian mantle was between 0.2 and 0.4, and thus substantially higher than that of the Earth's

---

* *Note added in proof*: If however, the moment of inertia coefficient of Mars is indeed closer to 0.365, as recently suggested by Kaula (1979), the density of the mantle will be somewhat decreased while the size and/or density of the core will be increased. Johnston and Toksöz (1977) have calculated Martian models with $I/MR^2 = 0.365$ and find that the FeO/(MgO + FeO) ratio of the silicates in the mantle would be about 0.25, which is still twice as high as that of the Earth's mantle. The bulk composition of the planet can then be well matched by a model intermediate between those Models I and II of Clark and Ringwood (1971). In this case, the core would still consist of an Fe–O–S alloy but would contain a larger amount of iron oxide than the case considered in the text. The basic conclusion that the bulk composition of Mars can be matched by an oxidized chondritic composition is not affected by Kaula's downward revision in I/MR^2.

mantle (Section 6.5), while the Martian core (mostly believed to be comprised of an Fe–S alloy) was proportionately much smaller in size than the Earth's core. Thus, Ringwood's (1959, 1966b) interpretation that Mars is more oxidized than Earth has been widely adopted; however, the cosmochemical implications have been largely ignored. For example, several of the above authors concluded that Mars possesses a substantially smaller overall Fe/(Mg + Si) ratio than the Earth, thereby implying the operation of a process within the nebula by which metallic iron was fractionated from silicates (Urey, 1952). It is now evident that the above conclusion relating to Fe/(Mg + Si) ratios is unjustified, being based on an overestimate by some of these authors (e.g., Binder, 1969; D. L. Anderson, 1972a) of the FeO abundance in the Earth's lower mantle (Section 1.5).

Discussion

Additional lines of evidence provide further support for the interpretation that the Martian mantle is richer in FeO than the Earth's mantle and does not contain large amounts of dispersed metallic iron:

(a) Viking analyses of Martian soils (Toulmin et al., 1977) show that they are highly oxidized and rich in iron oxides (~ 19 percent Fe_2O_3). Toulmin et al. (1977) concluded that these soils had been formed by weathering of an FeO-rich basaltic crust, probably of global extent, which had in turn been derived by differentiation from an FeO-rich mantle.

(b) The Martian soil at the Viking lander sites contains from 1 to 7 percent of a highly magnetic mineral (Hargraves et al., 1977). Ferric oxide is also a component of the soil, demonstrating the existence of a redox state which would not permit metallic iron to be identified with this magnetic mineral. The preferred explanation is that the magnetic component consists of maghemite (γ-Fe_2O_3) — (Hargraves et al., 1977). Moreover, the aerosol component of the Martian atmosphere was interpreted by Pollack et al. (1977a) to contain about 10 percent of magnetite.

(c) The principal volatile species degassed from the Martian interior consist of CO_2 and H_2O. If the Martian mantle contained large amounts of dispersed metallic iron, the degassed species would consist predominantly of CO and H_2 (Section 2.4). The formation of a highly oxidized surficial crust of global extent from a mantle containing abundant dispersed metallic iron would be extremely difficult to explain.

Chemical Evolution

The major differentiation within Mars to form a crust, mantle, and core implies the occurrence of extensive internal melting processes. Toksöz and Hsui (1978) point out that the heavily cratered regions of the crust must have formed during the first half billion years, prior to cessation of the "late-heavy" bombardment of the Moon about 3.9×10^9 years ago. It seems likely that the core also formed during this major differentiation. The extensive occurrences of inferred volcanic rocks with a wide range of ages imply that differentiation continued throughout most of Martian history (Mutch and Head, 1975).

An analysis of soil from one of the landing sites is given in Table 9.1. In view of the similarity of the soil compositions from two widely separated sites and its probable aeolean origin, it seems that this composition is likely to be of global significance. Toulmin et al. (1977) interpret the composition as representing a mixture of clay minerals, sulphates, carbonates, and iron oxides. The overall mafic nature of the soils, combined with the low abundances of K, Rb, and Zr, strongly imply that the source rocks were also mafic in nature. Highly differentiated rocks which predominate in the terrestrial crust, such as granites and their weathering products, could constitute only a very small proportion, at most, of this material. The Martian crust is thus inferred to be dominantly mafic in nature. The large absolute amount of iron and the high $FeO/(FeO + MgO)$ ratio reinforce our previous inference that the Martian mantle is richer in FeO than the terrestrial mantle. This is also supported by the inferred low viscosity of Martian volcanic magmas (Schonfeld, 1977b).

The low abundance of potassium in the Martian soil is particularly significant. A basalt formed by partial melting of ultramafic material possessing

Table 9.1 Chemical composition (as oxides) of Martian soil from S1 site (Toulmin et al., 1977). A similar composition was found at S2 site.

	Wt. %
SiO_2	44.7
Al_2O_3	5.7
Fe_2O_3	18.2
MgO	8.3
CaO	5.6
K_2O	<0.3
TiO_2	0.9
SO_3	7.7
Cl	0.7
	91.8[a]
Rb ppm	≤30
Sr ppm	60
Y ppm	70
Zr ppm	30

[a] The missing components probably consist mainly of water and carbonates.

the chondritic abundance of potassium would contain at least 0.5 percent K_2O. The upper limit for K_2O (Table 9.1) thus suggests that Mars (like the Earth and Venus) is depleted in potassium relative to chondrites. This is supported by the orbital gamma-ray measurements carried out by the Mars 5 spacecraft, which yielded a K/U ratio for the Martian surface that was comparable with the corresponding ratios for the Earth and Venus (Surkov and Fedoseyev, 1977).

The discussion in the previous subsection would imply a generally similar absolute abundance of uranium in Mars to that in the Earth and carbonaceous chondrites. The inferred deficiency in potassium, a key heat-producing element, accordingly has important implications for the thermal history of the planet. In order to account for the early differentiation of crust, mantle, and core, the internal temperatures in Mars must have been quite high soon after the planet accreted. Partial conservation of gravitational energy of accretion, as discussed by Safronov (1972a) and Wetherill (1976) seems to be the most likely source of this early heat input. According to the model depicted in Figure 9.1, early segregation of the core would have been facilitated by the low minimum melting point in the system Fe–S–O (ternary minimum is 920°C compared to 988°C for Fe–FeS).

The physiographic features of Mars do not suggest the occurrence of plate tectonic processes analogous to those which predominate on the Earth (Mutch and Saunders, 1976). This may be due to the presence of a strong lithosphere, proportionately thicker than that of the Earth. The absence or paucity of granitic rocks can then, therefore, readily be explained. On the Earth, such rocks are formed by a multistage petrogenetic process involving subduction of lithosphere plates and remelting (see, for example, Ringwood, 1975a). In the absence of these processes on Mars, the widespread occurrence of granitic rocks is not to be expected.

Volatiles

The Viking results on the compositions of Martian volatiles present several enigmas. The amount of degassing of Mars, as indicated by the inventories of inert gases, N_2, H_2O, and CO_2 (even allowing for large crustal reservoirs of H_2O and CO_2), is surprisingly low, particularly in view of the large degree of global differentiation which is inferred to have occurred. For example, the amount of argon-36 in the Martian atmosphere (per gram of total planetary mass) is proportionately only about 1 percent of that in the Earth's atmosphere (Biemann et al., 1976). This could suggest (a) that Mars is intrinsically depleted in volatiles compared to the Earth, (b) that a large proportion of the Martian atmosphere escaped from the planet at an early stage, or (c) that Mars had been degassed much less efficiently than the Earth.

Currently, there is insufficient evidence to decide finally between these alternatives. However, in view of the decisive evidence that Mars is more oxidized than the Earth, and our interpretation (Section 6.4) that the oxidized components of planets accreted at low temperatures, the author finds it difficult to believe that Mars was intrinsically depleted in volatiles relative to the Earth when it accreted. This skepticism is reinforced by the large amounts of the relatively volatile species, Cl and S, which occur in the Martian soil. The existence of a large difference in isotopic composition of nitrogen between the atmosphere of Mars and the Earth strongly suggests that a considerable proportion of the Martian atmosphere escaped early in its history (McElroy et al., 1976). It is doubtful, however, whether this factor alone can explain the ^{36}Ar deficiency. It is easier to believe that Mars may not have been degassed as efficiently as the Earth. The loss of primary volatiles from the Earth's mantle is closely related to the process of plate tectonics—in particular, partial melting of mantle rocks at shallow depths beneath mid-oceanic ridges to form a basaltic oceanic crust, accompanied by degassing. However, we have already noted that terrestrial-type, large-scale plate tectonics do not appear to have occurred on Mars. Consequently the rate of degassing might well have been much smaller.

The ^{40}Ar/^{36}Ar ratio on Mars is about 10 times higher than the terrestrial ratio. Superficially, this might suggest a higher content of ^{40}K on Mars, but this does not appear likely for reasons given previously. Tentatively, it might be attributed to extensive early loss of degassed ^{36}Ar, as implied by the nitrogen isotope data (McElroy et al., 1976).

The isotopic ratios of oxygen and carbon are similar in the atmospheres of Mars and Earth. This similarity, in contrast to the differences in isotopic composition of nitrogen, is attributed to exchange with large reservoirs of H_2O and CO_2 present in the outer crust (McElroy et al., 1976). The relative abundances of Ne, Ar, Kr, and Xe are very similar in the Martian and terrestrial atmospheres and, except for Xe, are very similar to the "planetary" component of chondritic meteorites. Xenon is depleted both in Mars and the Earth, presumably owing to adsorbtion by clay minerals and sediments (Owen et al., 1977). However, ^{129}Xe derived by radioactive decay of ^{129}I is relatively more abundant on Mars.

If Mars is indeed depleted in potassium relative to chondrites, as suggested by the foregoing evidence, then it is reasonable to expect a corresponding depletion in sulphur, since the latter element is much more volatile than potassium under most conditions which might prevail in the solar nebula. For this reason, the author suspects that the model depicted in Figure 9.1 is probably more applicable than a model in which the Martian core is composed predominantly of iron sulphide (FeS).

Accretion of Mars

According to the views expressed by Ringwood (1959, 1966b), Ringwood and Clark (1971) and Lewis (1973), Mars accreted at relatively low temperatures in the solar nebula, thereby accounting for its highly oxidized state, governed by the equilibria discussed in Section 6.4. According to this model, however,

Mars should possess the chondritic abundance of potassium, which does not seem to be the case. The inferred depletion of potassium suggests, rather, that Mars accreted as a mixture of high-temperature condensate rich in metallic iron and depleted in volatiles (including K) with low-temperature condensate similar to Cl chondrites (Table 8.1), in a generally similar manner to that advocated in the case of the Earth (Chapter 8). However, to account for the more highly oxidized state of Mars, about 30 percent of low-temperature condensate would be necessary, as compared to 10–15 percent in the case of the Earth (Chapter 8). The water contained in this larger proportion of low-temperature condensate would be sufficient to oxidize all the free metallic iron in the high-temperature condensate, as discussed earlier. The occurrence of a larger amount of low-temperature condensate in Mars is readily explicable on the basis of the nebula model depicted in Figure 6.4.

9.3 Venus

In many respects (Table 9.2), Venus can be regarded as a twin planet of the Earth. A large part of the density difference between the two planets is caused by differing degrees of self-compression in the respective planetary gravitational fields, consequent upon the smaller mass of Venus. If Venus were identical to the Earth in composition and structure, possessing the same mass ratios of core-to-mantle-to-crust, and the interior temperatures were the same at corresponding pressures, then, for the observed mass (Table 9.2), Venus would possess a mean density of 5.34 g/cm^3 [compared to the observed value of 5.24 g/cm^3 (Ringwood and Anderson, 1977)]. Thus, the actual density of Venus is 1.9 percent smaller than that of an Earthlike model. This difference, though small, is well determined, since it is based on application of the same empirically derived equations of state to both planets. Small errors in the particular equations of state employed would not affect the density differential, which is obtained simply by determining what the mean density of the Earth would be if gravity $[g = GM(r)/r^2]$ were reduced to the Venus value. An allowance for the higher near-surface temperatures of Venus would reduce this density differential to 1.7 percent.

Table 9.2 Observed and inferred properties of Venus and the Earth (from Ringwood and Anderson, 1977).

	Venus	Earth
Mass ($\times 10^{27}$g)	4.871	5.976
Radius (km)	6054	6371
Mean Density (g/cm^3)	5.24	5.517
Proportionate mass of metallic core	23	32.5
FeO/(FeO + MgO) ratio in mantle (mole fraction)	0.24	0.12

Lewis (1972, 1973) suggested that the smaller density of Venus was caused by the formation of Venus at higher temperatures than the Earth because of its closer proximity to the sun. In consequence, sulphur failed to condense on Venus, while the silicates on that planet were devoid of FeO. In the case of the Earth, assumed lower temperatures of formation permitted large amounts of sulphur to enter the core and substantial amounts of FeO to enter the silicates of the mantle. However, Ringwood and Anderson (1977) showed that this hypothesis was quantitatively inadequate to account for the density differential. Moreover, it disagrees with several other observations and inferences:

(a) our conclusion that sulphur is *not* the principal light element in the Earth's core (Section 3.4);
(b) the observations implying that the clouds of Venus are composed of sulphuric acid (Young, 1973);
(c) similar K/U ratios in the surface rocks of both planets (Surkov, 1977);
(d) the similar amounts of total CO_2 present in Venus and Earth (Sagan, 1962), which suggest that both planets may possess similar mantle redox states (Ringwood and Anderson, 1977).

In the light of this unfavorable evidence, Lewis' hypothesis must be regarded with considerable reserve.

The slightly smaller intrinsic density of Venus compared to Earth could be explained in principle by Urey's (1952) hypothesis of physical metal/silicate fractionation in the solar nebula prior to accretion. It could be assumed that Venus contained a smaller amount of metallic iron than the Earth (Table 6.3). We note, however, that this hypothesis was unsatisfactory when applied to explain the relatively low intrinsic density of Mars (Section 9.2).

An alternative hypothesis favoured by Ringwood (1959, 1966b) proposes that Venus is more oxidized than the Earth (Figure 6.9). The density deficit could be explained if the Venusian mantle possessed an FeO/(FeO + MgO) ratio of about 0.24 and a metallic core amounting to only 23 percent of the mass of the planet. This would permit Venus to contain relative abundances of major metallic elements similar to those of the Earth, Mars, and Type 1 carbonaceous chondrites. An hypothesis of this type has the further advantage of economy of assumptions, since it invokes a known process. Large variations in redox state can be demonstrated to exist between the Earth, the Moon, Mars, and the various classes of meteorites.

Accretion Conditions

The lower intrinsic density of Venus and the inference that this is caused by a higher oxidation state than the Earth are anomalous in relation to the overall planetary trend for density to decrease with increasing distance from the Sun and the expectation, based on Figure 6.4, that the proportions of low-temperature condensates in planets would increase with distance from the

Sun. Ringwood and Anderson (1977) suggested that this "anomaly" might be caused by the smaller mass and consequent smaller gravitational accretional energy, in conjunction with the single-stage autoreduction hypothesis discussed in Section 7.2. However, in view of the weaknesses of the single-stage model as outlined in that section, this explanation is no longer convincing.

The anomalous density of Venus implies that accretion conditions were more complicated than those suggested by Figure 6.4. A possible explanation may derive from considerations advanced by Kaula and Bigeleisen (1975), who showed that if the growth of Jupiter had been far advanced before the corresponding accretion of the terrestrial planets, large amounts of planetesimals (dominantly low-temperature condensates) from the Jupiter region would have been gravitationally scattered into the region of the solar nebula where the terrestrial planets were accreting. There is no reason to suppose that accretion of all the planets was precisely isochronous. It is possible that the Earth accreted more rapidly than Venus and that a disproportionate share of low-temperature planetesimals from the Jupiter region became incorporated, via collisions, in the planetesimal cloud from which Venus was accreting.

The available geochemical data indicates considerable similarities between the Earth and Venus. The potassium/uranium ratios measured by the Venera 8, 9, and 10 missions were generally similar to the terrestrial ratio of 10^4, implying a corresponding degree of depletion of potassium (and probably other volatiles) in Venus as compared to the primordial abundances. Absolute abundances of K (4 percent), U (2.2 ppm), and Th (6.5 ppm) discovered by Venera 8 implied the occurrence of a granitic rock (Surkov, 1977). On the Earth, such rocks are generally formed by multistage petrological processing involving plate tectonics. On the other hand, the K, U, and Th abundances of samples measured by Venera 9 and Venera 10 were consistent with basaltic rocks, as suggested also by their bulk densities of about 2.8 g/cm^3 (Surkov, 1977).

Volatiles

The abundances of carbon dioxide, nitrogen, and argon in the Venusian atmosphere are of considerable interest. These are similar to the corresponding terrestrial abundances on a weight-for-weight basis, providing that the CO_2 locked up in terrestrial sedimentary carbonate rocks is included (Sagan, 1962; Surkov, 1977). Thus, the volatile inventory (including K) and the degree of outgassing in both planets appear to have been generally similar.

The outstanding difference is the present paucity of water (~ 0.1 percent) in the Venusian atmosphere (Florensky et al., 1977). The inventories of other volatiles strongly suggest that Venus accreted under generally similar conditions to the Earth and therefore it would be expected that a comparable amount of water would have been trapped in the interior, later to be released by degassing. There are two pieces of evidence which suggest that Venus was not, or is not now, as "dry" as it appears. Firstly, it seems likely that carbon was originally accreted in planets in the form of condensed hydrocarbons (Section 6.4) present in the low-temperature component A (Table 8.1). J. Walker et al. (1970) point out that the conditions under which the hydrocarbons are oxidized to carbon dioxide in the planetary interior would necessarily generate a comparable amount of water, e.g.,

$$\text{"}CH_2\text{"} + 3 Fe_3O_4 \longrightarrow CO_2 + H_2O + 9 FeO.$$

Second, the potassium and uranium abundances observed at the surface of Venus by Venera 8 suggest the occurrences of a granitic rock. On the Earth, the formation of granites is believed generally to involve the presence of high water-vapor pressures.

The deficiency of water in the Venusian atmosphere may well have been caused by a "runaway greenhouse" effect accompanied by efficient photodissociation of water vapor and escape of hydrogen (see, for example, Sagan, 1967; Dayhoff et al., 1967; Ingersoll, 1969). The oxygen may have been consumed by reaction with FeO in surface rocks to form Fe_2O_3 and with FeS to form sulphates. The plausibility of the latter reaction is supported by the identification of sulphuric acid as a major component of the Venusian clouds (Sill, 1972; Young, 1973). A process of plate tectonics on Venus, resulting in continued exposure to the atmosphere of fresh rocks derived from the mantle, and disposal of oxidized products into the mantle via crustal subduction would have facilitated these reactions.*

9.4 Mercury

The high density of Mercury, 5.45 g/cm³, clearly implies that this planet contains much more iron, relative to silicates, than the other terrestrial planets. The models of Siegfried and Solomon (1974) imply an iron content of about 60 percent. Thus, Urey's hypothesis that fractionation of iron

---

* *Note added in proof*: The above preliminary data on the composition of the Venusian atmosphere have been confirmed in general by the results from the Pioneer Venus mission. The composition of the lower atmosphere was given to be CO_2–96.4%, N_2–3.4%, H_2O–0.14% and O_2–69 ppm, Ar–19 ppm, Ne–4.3 ppm and SO_2–186 ppm (Oyama et al., Science **203**, 802–804, 1979). Further strong evidence for the occurrence of H_2SO_4 in the clouds was obtained and the presence of additional sulphur-bearing species (including pure sulphur) was inferred. A surprising aspect of the results was the high concentrations of primordial [36]Ar and [20]Ne in the atmosphere, implying that these species are 200–300 times as abundant as on the Earth. About half of the total argon on Venus consisted of [36]Ar.

It has been speculated that these "excess" inert gases may have been captured from the solar wind, this process being facilitated by the weak Venusian magnetic field (Zahn et al., Science **203**, 768–769, 1979). Alternatively, they may be a consequence of specialized conditions under which Venus accreted. It has been suggested above that the accretions of Venus and Earth were not isochronous and that the feeding zone of Venus received a larger share of "low-temperature condensate" as a result of scattering of late planetesimals by Jupiter after accretion of Earth had been completed.

relative to silicates was an important process in the solar nebula is confirmed in the case of Mercury, if not for the other planets.

The anomalously high iron content of Mercury has generally been attributed to some process controlled by the planet's proximity to the sun. However, the nature of this process is speculative and poorly understood. Bullen (1952) proposed that all of the terrestrial planets formed initially with the same Fe/Si ratio. Owing to strong heating of the surface caused by its proximity to the newly formed sun, Mercury subsequently lost part of its silicate mantle by evaporation, after an iron core had segregated. However, this model makes rather extreme demands upon the properties of the assumed superluminous early phase of evolution of the sun, prior to its arrival on the Main Sequence. Current models of star formation do not predict luminosities nearly as large as are required to evaporate and remove silicates from the surface of Mercury (Larson, 1972a,b).

Ringwood (1966b) suggested that the high density of Mercury was caused by accretion within the solar nebula in a high-temperature environment caused by proximity to the Sun. As a result, most of the magnesian silicates failed to condense with the accreting material, so that Mercury became enriched in metallic iron, with a mantle enriched in Ca and Al silicates and devoid of oxidized iron and alkalis. D. L. Anderson (1972b), Grossman (1972), and Lewis (1973) proposed essentially similar models. These models do not require temperatures as high as those of Bullen's model discussed above.

The main problem with this model arises from the generally similar condensation temperatures of magnesium silicates and iron in the solar nebula (Grossman, 1972) (Fig. 6.7). The nebula gas pressure at the orbit of Mercury was probably about 10^{-3} atm. (Weidenschilling, 1978). At this pressure, the "condensation temperature" of Fe is only 30°C higher than that of Mg_2SiO_4 (Fig. 6.7). Actually, both components condense over a range of temperatures and Weidenschilling (1978) has shown that it is quite difficult to formulate a P, T environment within the zone of accretion of Mercury, in which iron can become fractionated from Mg_2SiO_4 as efficiently as is required to explain the high iron content of the planet.

Weidenschilling (1977) has pointed out that the low mass of Mercury may be an even more anomalous feature than its high density. From the empirical relationships between surface density and distance from the Sun in the primitive solar nebula (Fig. 6.1), the mass of Mercury would be expected to be similar to those of Venus and the Earth, whereas its actual mass amounts to only 5 percent or so of the mass of the Earth. Weidenschilling suggests that the high density and low mass are consequences of a single process which occurred in this region of the nebula, namely, aerodynamic drag. This arises because the gas is supported against the Sun's gravity by a combination of its own internal pressure and its orbital velocity. Solid bodies, on the other hand, are not supported by the pressure gradient, but rotate at the Kepler velocity which necessarily differs from the orbital velocity of the gas. The solid bodies are therefore subjected to gas drag which causes them to spiral towards the

Sun. Weidenschilling supposes that most of the solid matter in Mercury's accretion zone was thereby driven into the Sun, thus explaining its small mass.

Other factors being equal, planetesimals of low density are driven into the Sun more efficiently than higher density planetesimals by this aerodynamic fractionation mechanism. In view of the large proportion of total mass lost, Weidenschilling points out that the fractionation mechanism need only have been slightly more effective for silicate-rich than for iron-rich planetesimals, in order to account for the high density of the residual material from which Mercury accreted.

This model has some attractive features, although its effectiveness depends critically upon several assumed initial conditions, such as the size distribution of planetesimals. Also, it is far from clear how the mechanism worked so efficiently in the region of Mercury, whereas it appears not to have been significant in the zone of accretion of Venus, which has a *lower* intrinsic density than the Earth.

At the present stage, it appears that neither of the two principal models invoked for explaining the high density of Mercury—the one relying on temperature-controlled chemical fractionation and the other on a physical metal/silicate fractionation mechanism—is free from difficulties.

Differentiation and Thermal History

The surface of Mercury is similar in many respects to that of the Moon. A large proportion of the planet is heavily cratered and probably possesses an age generally similar to that of the lunar highlands ($3.9-4.4 \times 10^9$ years). The reflectance spectra of the surface materials of Mercury are also similar to those of mature soils from anorthositic regions of the lunar highlands containing less than 6 percent FeO (Adams and McCord, 1978). These authors have inferred that Mercury possesses an anorthositic crust similar to that of the Moon. This conclusion is also supported by the combined ultraviolet-reflection spectral studies of Carver et al. (1975) and Wu and Broadfoot (1977). Parallel observations of the albedo of Mercury at optical wavelengths show, however, that its surface contains less TiO_2 than the lunar surface, and that the FeO content of surface soils on Mercury lies between 3 and 6 percent FeO, significantly smaller than that of the average lunar soil (Hapke, 1977).

The role of volcanism on Mercury is currently under debate. Strom (1977) and Strom et al. (1975a) believe that the extensive Mercurian intercrater plains are probably of volcanic origin and were formed during the later stages of the heavy bombardment which was responsible for the major crustal features. However, Wilhelms (1976) has argued against this interpretation.

The occurrence of a global magnetic field in Mercury (Ness et al., 1976) has been interpreted by most investigators to imply the existence of a differentiated liquid metallic core, presently convecting and generating a magnetic field by dynamo action (see, for example, Solomon, 1976; Cassen et al., 1976). If all the iron had differentiated to form a core, its radius would be about 75 percent of the present radius of Mercury, implying that Mercury had undergone a global differentiation process. This would be consistent with the interpretation that the crust possesses an anorthositic composition (Adams and McCord, 1978) and that extensive volcanism had occurred on the planet (Strom, 1977). Because of the exothermic nature of the core separation process (Siegfried and Solomon, 1974), and the intense internal convection involved, it is probable that core formation was completed early in the history of Mercury, prior to termination of the phase of heavy meteoritic bombardment which was mainly responsible for the present surface features. Moreover, core formation would be accompanied by an expansion of the planet, causing widespread tensional features (Solomon, 1976), whereas observations indicate that the lithosphere has been in a state of compression throughout much of the history of the planet (Strom, 1977; Strom et al, 1975a).

Early planetary differentiation and formation of a core requires that a substantial proportion of the volume of the planet was at elevated temperatures ($> 1000°C$) soon after its formation (Solomon, 1976), similar in this respect to the Moon. Accretional heating, possibly combined with high ambient temperatures in the solar nebula, appear to be the most plausible sources of this heat (Cassen, 1977). There are some problems in understanding how a core which formed so early would still be molten today, and capable of generating a magnetic field (Solomon, 1976). However, they are probably not insuperable, since they depend rather critically upon assumptions as to the distributions of heat sources in the mantle and the role of solid-state convection (Cassen, 1977; Cassen et al, 1976).

The existence of 3–6 percent of FeO in the Mercurian crust (Hapke, 1977), if finally confirmed, would have some important implications. It is not readily reconciled with simple condensation models such as those of Ringwood (1966b) and Lewis (1972, 1973), which predict that the materials from which Mercury accreted, separated from the nebula at high temperatures, and were therefore free of FeO (Fig. 6.5). It suggests that a more appropriate model might be based on mixing a high-temperature condensate with a *small* amount of low-temperature condensate, as inferred for the case of the Earth (Chapter 8) and implicit in the cocoon nebula model of Figure 6.4. In the case of Mercury, the proportion of low-temperature condensate was much smaller than that which accreted on the Earth, while the high-temperature condensate separated from the nebula at much higher temperatures and was therefore depleted in magnesium silicates compared to metallic iron. The final accretion of Mercury must then have occurred under relatively cool ambient conditions and/or in an environment in which the H_2/H_2O ratio was much smaller than in the solar nebula (Ringwood, 1977b), so that FeO remained stable.

The inference that a finite but relatively small amount of FeO is present in Mercurian surface materials also has a bearing on Weidenschilling's aerodynamic model. In its simplest form, this model would nôt predict that the silicates of Mercury would be poorer in FeO than the silicates of the other terrestrial planets. The difficulty may, however, be circumvented by the introduction of additional assumptions.*

9.5 Asteroids and Meteorites

Introduction

There is a vast literature dealing with the subjects of asteroids and meteorites. The present very brief discussion is directed towards those aspects of meteorite and asteroid evolution which have been considered by some authors to have a bearing on the formation of the terrestrial planets.

The asteroids total about 5×10^5 small bodies (≤ 1.6 km in diameter) (Gehrels et al., 1971) orbiting mainly between Mars and Jupiter (Fig. 5.1). The largest asteroid is Ceres with a diameter of about 770 km; Ceres comprises about half of the total mass of material in the asteroid belt, which is estimated to be about 0.4×10^{-3} of the Earth's mass (Gehrels et al., 1971). Formerly it was thought that these bodies were derived from the remains of a disintegrated planet. However, it is now generally accepted that they represent, rather, an aborted planet. The gravitational field of neighboring massive Jupiter would have caused strong perturbations of the orbits of planetesimals in the region and, in addition, would have delivered an intense flux of high-velocity planetesimals from its own zone of growth into the asteroid–Mars region. These processes are believed to have prevented the accretion of a planet (in the case of the asteroids) and also removed most of the mass of the material in this region (see, for example, Kaula and Bigeleisen, 1975; Weidenschilling, 1975; Fig. 6.1).

Until the recent advent of space exploration, meteorites falling upon the Earth provided our only accessible samples of extraterrestrial material. Hence it was only natural that traditional theories of formation of the planets have been heavily based on attempts to understand the chemical evolution of these bodies. We shall be interested to enquire whether this view is still appropriate.

In the past, it had been widely assumed that most meteorites were derived from the asteroid belt and were deflected into Earth-crossing orbits by

---

* *Note added in proof:* McCord and Clark (*Lunar and Planetary Science* **10**, 789–791, 1979) have recently observed an absorption band in the Mercury spectrum near 0.9 μm and attribute this to d-shell transitions in Fe^{2+} ions. The position of the absorption band is best explained by the occurrence of Fe^{2+} ions in orthopyroxene. McCord and Clark conclude that the surface of Mercury contains about the same amount of FeO and of orthopyroxene as the lunar Apollo 16 site, which averages about 5.5% of FeO. The important implications of this observation have been discussed above. It provides further support for the mixing model of Fig. 6-4 and its applicability to the conditions of accretion of Mercury.

collisions. However, it was subsequently realized that collisional deflections of this kind required excessive impulses which would have caused all meteorites arriving at the Earth to be intensely shocked, contrary to observation. Problems also arose in explaining their cosmic-ray exposure ages. A considerable amount of subsequent research into these problems is beginning to produce some clarification. Wetherill and Williams (1978) have demonstrated the possibility of transferring asteroids into Earth-crossing orbits on acceptable time scales via rather subtle gravitational and resonance effects. They have made a plausible case that the differentiated meteorites (irons, stony-irons, and achondrites) may be derived from parent bodies in the inner regions of the asteroid belt. Moreover, it now seems likely that the ordinary chondrites are derived from the relatively small families of Mars- and Earth-crossing (Amor and Apollo) asteroids as argued by Anders (1964) and Wetherill (1976b). These interpretations are supported by spectrophotometric studies of asteroid surfaces, to be discussed subsequently.

The Earth- and Mars-crossing Apollo and Amor asteroids have half-lives against collisions with these planets only on the order of 10^7 years, so that there must be a source capable of placing new bodies into these orbits. Öpik (1963b) has argued for a cometary source, whereas Anders (1964) maintains that the source was situated in the inner regions of the asteroid belt. Wetherill (1976b) originally inclined towards a cometary source, but recently (1978) has appeared willing to consider asteroids as possible sources.

Although the meteorites are discussed later in this section, it will be convenient at this stage to describe their classification and abundances. Many classifications of meteorites have been devised, the most satisfactory being that of Prior (1920). The principal groups, together with their major constituents, are listed in Table 9.3. This classification is incomplete; however, it includes 99 percent of known falls. Observed falls of meteorites on the Earth's surface are given in Table 9.4. The chondrites are seen to be by far the largest group, and are further subdivided as in Table 9.5.

Asteroids

During recent years, major progress has been made towards interpreting the surface compositions of asteroids from spectrophotometric observations. This topic has been reviewed by Chapman (1976) and Gaffey and McCord (1977). Two major groups were recognized. The first, characterized by low albedos (≤ 0.09), strong negative polarizations at small angles (> 1.1 percent), and relatively flat, featureless spectral reflectance curves, was called by Chapman "C type" owing to similarity with carbonaceous chondrites. The second group, characterized by higher albedos (≥ 0.09), weaker negative polarizations (0.4–1.0 percent), and reddish, sometimes featured spectral curves, were called "S type" indicative of similarity to meteorites containing abundant silicate minerals (Chapman, 1976).

Table 9.3 Classification of meteorites (Prior, 1920).

Irons (*siderites*)	Iron containing from 4 to 35 percent of nickel, together with small amounts of Co, S, P, and C.
Stony-irons (*siderolites*)	
Pallasites	Nickel–iron, olivine.
Mesosiderites	Nickel–iron, hypersthene, calcic plagioclase, olivine.
Stones (*aerolites*)	
Chondrites	Olivine, orthopyroxene, nickel–iron, sodic plagioclase. (Posses a characteristic spheroidal structure.)
Achondrites	Do not possess the spheroidal structure. Poor in metal. Subdivided into:
	Calcium-rich achondrites (eucrites and howardites). Composed dominantly of pigeonite, hypersthene, and calcic plagioclase.
	Calcium-poor achondrites (diogenites and aubrites). Composed dominantly of hypersthene (d.) and enstatite (a.).

A detailed study of these general classes was carried out by Gaffey and McCord on a population of 65 asteroids. Forty percent of these were found to possess optical properties similar to C1 and C2 chondrites, while another 15 percent possessed properties indicating that they were composed of a mafic silicate assemblage (olivine $\pm$ pyroxene), together with a relatively abundant opaque phase (carbon and/or magnetite). The properties of these asteroids are matched best by C3 carbonaceous chondrites. Ten percent of Gaffey and McCord's sample corresponded best to a mixture of silicate minerals with a smaller (but still substantial) proportion of opaque minerals. This class included the largest asteroids, Ceres and Pallas. The only known meteoritic class with the appropriate mineral assemblage to explain the

Table 9.4 Observed falls of principal classes of meteorites (Prior, 1953). Total falls = 632.

Class of meteorite	Percent
Irons	7
Stony-irons	2
Chondrites	82
Calcium-rich achondrites	5
Calcium-poor achondrites	4

Table 9.5 Classification of chondrites (simplified)[a].

Class	Group	Major minerals	Percent of observed chondrite falls
Enstatite chondrites	E	Enstatite (Fs_0) nickel–iron (containing Si), sodic plagioclase, troilite, minor graphite	1.8
Ordinary chondrites	H	Olivine, bronzite, (Fs_{15-22}) nickel–iron, sodic plag., troilite	38.1
	L (metal depleted)	Olivine, hypersthene (Fs_{22-28}) nickel–iron, sodic plag., troilite	46.2
	LL (stronger metal depletion)	Olivine, hypersthene (Fs_{28-31}) nickel–iron, sodic plag., troilite	8.5
Carbonaceous chondrites	C1	Low-temperature mineral assemblage (LTA): hydrated silicates, carbonaceous compounds, magnetite, sulphur	0.8
	C2	LTA about 60 percent. Remainder: olivines and pyroxenes of heterogeneous composition.	2.3
	C3	LTA about 30 percent. Remainder: heterogeneous olivines and pyroxenes, (Fe,Ni)S	2.3

[a] See Wasson (1974) for a more comprehensive classification, particularly including metamorphic grades.

optical properties is represented by the highly metamorphosed C3 chondrites,[1] Karoonda (olivine + magnetite).

An important feature of the observations is that the surfaces of 65 percent of the asteroids, including the largest, appear to be composed of materials resembling various classes of highly oxidized, essentially metal-free carbonaceous chondrites.

Of the remaining asteroids, Gaffey and McCord found that another 30 percent probably contained substantial proportions of metallic nickel–iron, together with a range of different silicate components. They do not, however,

[1] Referring to the simplified classification given in Table 9.5.

resemble ordinary chondrites. Chapman (1976) suggests that many of these may be pallasites (olivine + metal) but Gaffey and McCord do not consider this identification to be firm, since many of these asteroids apparently contain abundant pyroxenes, unlike pallasites.

The above studies represent a nearly complete observational sampling of main belt asteroids with diameters above 80 km and include significant sampling of asteroids with diameters down to 40 km. A remarkable fact is that ordinary chondritic mineral assemblages appear to be absent or extremely rare as surface materials on main-belt asteroids, despite their great preponderance among observed meteoritic falls on Earth (Table 9.4). However, among the relatively small families of Apollo and Amor asteroids, which have highly eccentric Mars- and Earth-crossing orbits, surface properties indicate that ordinary chondritic compositions are probably present (e.g., Eros and Toro), while carbonaceous chondrite types are rare. It seems that the ordinary chondrites, which have so greatly influenced cosmochemical speculations because of their abundance amongst observed falls, may be quantitatively very rare and derived from a small group of parent bodies with highly specialized orbits.

Chapman and Davis (1975) have suggested, on the basis of a study of the size distributions and collision histories of asteroids, that C-type asteroids were originally about 100 times as abundant as the differentiated S asteroids. They estimate that the total mass of material in this region may have approached the mass of Mars. Because of the very low mechanical strength of C-type asteroids, the vast majority were rapidly destroyed by collisions. Most large S-type asteroids are interpreted as the iron or stony-iron cores of differentiated asteroids, of which Vesta is the only intact survivor. These S-asteroid cores have survived because of their much greater mechanical strengths.

A further observation of cosmogonic significance is that, whereas the most abundant class of differentiated stony meteorites is comprised of eucrites and howardites, only one asteroid (Vesta) has been identified which has optical properties commensurate with this class.

In general terms, it seems that most of the S-group asteroids may correspond to classes of metal-silicate meteorites known to fall upon the Earth (Chapman, 1976; Gaffey and McCord, 1977). The mineral assemblages of these meteorites are indicative of melting and differentiation processes occurring in parent bodies under moderately reducing conditions. The interesting feature is that these (differentiated?) meteorites tend to be concentrated in the inner regions of the asteroid belt, although even here the more primitive C asteroids are comparable in abundance. Serious problems arise, therefore, in understanding how one set of asteroids became melted and differentiated, whereas most of the others escaped this fate, or alternatively were heated sufficiently to metamorphose them but not to cause melting.

One speculative possibility is that asteroids accreted over a time scale that was long compared to the half life of ^{26}Al (7.4×10^5 years) which was present during the early stages of formation of the solar nebula (Gray and

Compston, 1974; Lee and Papanastassiou, 1974; Lee et al., 1977). Asteroids which accreted relatively early may then have had sufficient radiogenic heating from ^{26}Al to cause melting and differentiation, whereas asteroids accreting after the decay of most of the ^{26}Al would have escaped this fate. This explanation does not account directly, however, for the concentration of differentiated asteroids in the inner regions of the asteroid belt.

An alternative suggestion (Wetherill, 1977a) is that differentiated asteroids might represent planetesimals left behind after formation of the Earth. These residual planetesimals were then deflected by complex gravitational perturbations into the inner region of the asteroid belt. However, in view of the difficulties encountered in bringing meteorites to the Earth from the asteroid belt, Wetherill relegates this suggestion to the highly speculative category. It should also be recalled that differentiated asteroids are believed to exist in the outer regions of the asteroid belt (Gaffey and McCord, 1977).

Meteorites

Meteorites have preserved an important chemical and isotopic record of processes that occurred during the earliest stages of the solar system's evolution. It is widely assumed that the *chondrites* are the most primitive class of meteorites, owing to a general similarity between their compositions and the corresponding solar composition, combined with the belief that chondrites have escaped the complexities caused by igneous fractionation processes of the kind which affected the achondrites. Unfortunately, however, there is a great deal of controversy concerning most aspects of the origin of chondrites, the causes of their chemical fractionations, and the nature of their remarkable petrological and mineralogical characteristics.

Despite this controversy, there is general agreement that the Type 1 carbonaceous chondrites represent the most primitive chondritic class. The chemical and mineralogical characteristics of these objects suggest that they could have been formed by the accretion at low temperatures of primordial oxidized dust to form small parent bodies which were subjected to mild recrystallization and low-grade metamorphism. The Type II and III carbonaceous chondrites are readily interpreted as disequilibrium physical mixtures of primitive C1 material with heterogeneous high-temperature minerals, mainly olivines and pyroxenes (Ringwood, 1963; Anders, 1964). It is clear, however, that the history of the ordinary and enstatite chondrites has been much more complicated (see, for example, Wasson, 1972). There is a growing consensus (shared by the author), that many of the remarkable textural and petrological characteristics of ordinary chondrites are caused by impact phenomena and mixing processes in the regoliths of one or more parent bodies (see, for example, Urey and Craig, 1953; Fredriksson, 1969; Fredriksson et al., 1973; Kurat et al., 1972). However, this view is disputed by some (e.g., Wood and McSween, 1978).

The chemical fractionations displayed by chondrites are also extremely complex (see, for example, Anders, 1968, 1971; Wasson, 1972, 1974). Four major processes seem to have operated:

(1) fractionation of refractory elements (Ca, Al, Ti, etc.; see Table 6.2) causing these to be increasingly depleted in ordinary and enstatite chondrites relative to Cl chondrites (Si normalized).
(2) fractionation of siderophile elements leading to marked depletions of this group in the cases of L and LL ordinary chondrites;
(3) substantial depletions ($\times 2$–$\times 10$) of moderately volatile elements (e.g., Cu, Ga, and Zn; see Table 6.2) in ordinary chondrites;
(4) large depletions ($\times 10$–$\times 1000$) of highly volatile elements (e.g., Pb, Tl, and Bi) in ordinary chondrites.

Abundances of many elements in ordinary chondrites are shown in relation to their primordial abundances (as displayed by Cl chondrites) in Figure 9.2. This may be compared with condensation temperatures of elements and compounds from a gas phase of solar composition (Table 6.2). The latter provide a measure of comparative volatilities of the relevant elements and compounds. There is seen to be a strong trend for the degree of depletion of a given element in ordinary chondrites to increase with increasing volatility. However, this trend is not shared by the "high-temperature

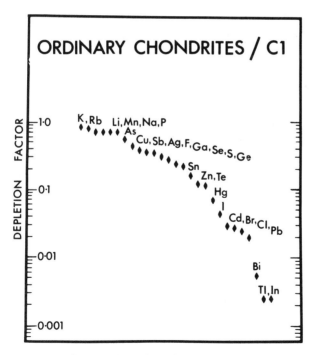

Figure 9.2 Comparison of abundances of volatile elements in ordinary chondrites with corresponding abundances in Cl chondrites. (From Ringwood and Kesson, 1977; with permission.)

condensates," which, as a group, are depleted by about 40 percent in ordinary chondrites, compared to C1 chondrites (Larimer and Anders, 1970).

Detailed interpretations of these relationships have been proposed by Larimer and Anders (1967, 1970), Anders (1968, 1971, 1977), Ganapathy and Anders (1974), Grossman and Larimer (1974), Wasson (1972, 1974), and Wasson and Chou (1974). It is argued by Ganapathy and Anders (1974) and Anders (1977) that chondrites are mixtures of seven distinct components, each of them formed in the solar nebula under differing thermal conditions. These seven components were then mixed in appropriate proportions to reproduce the observed chondritic abundance patterns (Fig. 9.2). These components comprise (1) high-temperature condensates (Ca, Al, Ti, etc,), (2) Mg silicates, which may contain some FeO, (3) metallic nickel–iron, (4) sulphide phase, FeS, (5) low-temperature condensate similar to C1 meteorites, and (6), (7) remelted portions of components (2) and (3), respectively, partially depleted in volatiles.

There is little doubt that Anders and his colleagues have demonstrated the important roles played by selective condensation/volatilization and mixing processes in producing the chemical compositions of chondrites. However, because of the great flexibility of the model, including its seven degrees of freedom and several other ad hoc elements, the significance of the apparent agreement between the model and the observed compositions of chondrites is somewhat ambiguous. Indeed, Wasson and Chou (1974), Wasson (1972, 1974), and Dodd (1969, 1974) have drawn attention to an array of mineralo- logical, chemical, and thermal difficulties encountered by the model, and have offered alternative explanations and interpretations.

One of the assumptions shared by many modern cosmochemists concerns the relations that are taken to exist between the processes responsible for the formation of chondrites and the large-scale evolution of the parental solar nebula. While it is accepted that chondrites formed *within* the solar nebula, the further assumption that the detailed chemical and physical fractionations recorded by chondrites provide information about the *large-scale pressure, temperature and compositional evolution* of the inner regions of the solar nebula have never been adequately justified, as far as the author is aware. This viewpoint is usually taken to be axiomatic; the solar nebula is then characterized in terms of these processes, and the consequent consistency between properties of chondrites and properties of the nebula is held to confirm the reality of the models. A degree of circularity in these arguments is evident. This extends also to many current interpretations of the significance of the high-temperature inclusions from the Allende meteorite. While there is no doubt that these record important gas–liquid–solid fractionations which occurred within the nebula at a very early stage of its history, there are many other features of these inclusions that show that the fractionations must have occurred in highly specialized environments which, probably, were not characteristic of large regions of the nebula. These include the compara- tively rapid cooling rates experienced by the inclusions, as shown by their textures, presence of ferrous iron (indicating a much higher f_{O_2} than would be present in the nebula), and the extremely complex condensation history, as manifested by the distribution and mineralogy of high- and low-temperature assemblages within individual inclusions (Lovering and Wark, 1977).

In the author's opinion, a more likely environment for producing the Allende high-temperature inclusions might have been provided by high-velocity impacts of primordial "dust–ice" planetesimals, similar to comet nuclei, upon the surfaces of large asteroids or among themselves (see, for example, Kaula and Bigeleisen, 1975). Impacts of this nature must have been common during the early evolution of the solar system and would have provided rapidly cooling gas spheres of nonsolar (H_2-depleted) composition in which rapid disequilibrium condensation of solids would occur. It has already been noted that a number of contemporary workers now favor the view that many of the characteristic properties of chondrites arise from repeated mixing and impact events occurring in the regoliths of asteroid-size parent bodies.

These, however, represent individual viewpoints. There is, as yet, no consensus despite the enormous amount of research which has been carried out in this field.

9.6 Cosmogonic Models Based on Chondrites

It is understandable that, in the past, our concepts of planetary formation have been heavily influenced by evidence derived from meteorites, particularly chondrites. For an extended period, nearly all our knowledge of extraterrestrial chemistry and mineralogy was based on these objects which, moreover, recorded processes that occurred during the earliest history of the solar system, long before the oldest rocks on the Earth were formed.

Perhaps the most explicit rationale of those who base their cosmogony upon meteorites is given by Ganapathy and Anders (1974):

> The fundamental assumption is that the inner planets were made by exactly the same processes as the chondrites. For the chondrites, it has been obvious for some time that they are mixtures of about half a dozen components that formed in a cooling solar nebula. We assume that the Earth and Moon (and other terrestrial planets) were made from the same components but in different proportions. The problem then reduces to estimating these proportions from appropriate geochemical constraints.

The author believes that the above philosophy is no longer acceptable for the following reasons.

(a) The processes by which chondrites were formed are not yet established. The specific model advocated by Ganapathy and Anders (above) is opposed by many workers of equal competence.

(b) In the past, emphasis on the significance of ordinary chondrites has rested heavily on their great preponderance amongst observed falls, which suggested a corresponding preponderance of this material in their source region, presumably the asteroid belt. However, it is now believed that parent bodies with the properties of ordinary chondrites are exceedingly rare in the main asteroid belt, where carbonaceous chondritic material seems to predominate. It is likely that ordinary chondrites are derived from the quantitatively insignificant Apollo asteroids, whose origin is as yet uncertain. In the light of these considerations, it seems premature to base theories of cosmogony on these cosmogonically rare objects.

(c) Terrestrial planets such as the Earth and Venus are several orders of magnitude larger than the parent bodies of ordinary chondrites. Even if the highly specific fractionation processes responsible for the compositions of ordinary chondrites had occurred on a wider scale in the nebula, it seems quite likely that they would have been averaged out and obscured as a result of the very much larger scale of accumulation and mixing processes which formed the terrestrial planets. Wetherill (1975) and Hartmann (1976) have demonstrated the occurrence of extensive overlapping between the feeding zones from which terrestrial planets formed, and that large-scale mixing of planetesimals probably occurred throughout the inner solar system.

(d) As was discussed extensively in Part I, the observed bulk composition of the Earth is readily explained without recourse to the seven-component chondritic model. The situation is similar in the cases of Mars and Venus. In view of the similarity in major element abundances between Cl chondrites, the Earth, and the Sun, the chondritic fractionation model is quite superfluous.

(e) We will find in Chapter 11 that the seven-component chondritic model provides a decidedly inadequate explanation of the composition of the Moon.

CHAPTER 10

Constitution and Composition of the Moon

10.1 Internal Structure

A set of four seismometers was placed on the moon by the Apollo astronauts. These were used to study seismic waves generated by impacts of Saturn S4B booster rockets and ascent stages of the lunar modules at known sites, as well as meteoroid impacts on the lunar surface and moonquakes generated deep in the lunar interior. Interpretations of data obtained by the two principal groups of investigators are in good overall agreement, although differences in detail exist.

Beneath the Apollo array, seismic P velocity increases from 4 km/sec a few kilometers below the surface, to 6 km/sec at 20 km, where it then jumps to 6.7 km/sec (Toksöz et al., 1974). Below 20 km, velocity is nearly constant, increasing slightly to about 6.8 km/sec at a depth of 55 km (Dainty et al., 1974). A major increase of velocity from 6.8 to 8.0 km/sec occurs between 55 and 60 km, marking the base of the crust and the beginning of the lunar mantle (Fig. 10.1). Gravity observations combined with measurements of topography show that the crust is generally in approximate isostatic equilibrium, the principal exception to this statement being provided by the mascons (excess mass concentrations) associated with circular mare basins (see, for example, Kaula, 1975). The gravity and topography observations also imply that large variations in crustal thickness occur, ranging from 30–35 km beneath mascons to 90–110 km in the farside highlands (Bills and Ferrari, 1977). The average crustal thickness is greater on the farside than on the nearside, causing a 2.5 km displacement of the center of mass from the center of figure (Kaula, 1975).

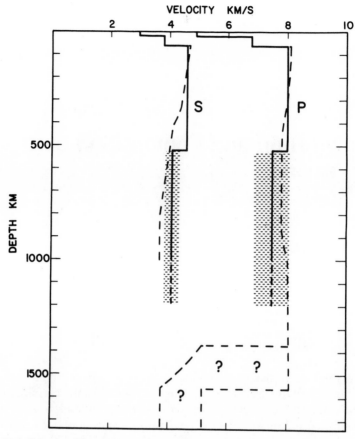

Figure 10.1 Seismic velocity distributions within the Moon according to Nakamura et al. (1976a,b) (broken line) and Goins et al. (1977) solid line. Formal uncertainties in the latter distribution are indicated by shading.

Seismic velocity distributions within the mantle strongly imply the presence of chemical zoning. Goins et al. (1977) recognize an upper mantle extending to a depth of about 500 km, characterized by uniform P and S wave velocities of 8.0 and 4.6 km/sec, respectively, and a Poisson's ratio of 0.25. Below this layer, P and S seismic velocities decrease abruptly to about 7.7 and 3.8 km/sec, respectively, yielding a Poisson's ratio of 0.34. This decrease marks the beginning of the lower mantle, which is also characterized by constant velocities extending to at least 1000 km (Fig. 10.1). Goins et al. (1977) point out that the velocities in the lower mantle possess substantial uncertainties (as shown in Fig. 10.1), and that the boundary between the upper and lower mantle could be continuous rather than discontinuous. However, the high value of 0.34 for Poisson's ratio in the lower mantle is believed to be well established. Note that the proportional uncertainty in S wave velocity is substantially smaller than that for P waves.

Nakamura et al. (1967a,b) also infer a layered structure for the lunar mantle, but the boundary between their upper and lower mantles occurs between 300 and 400 km and is characterized by a continuous velocity distribution. In the upper mantle, between 60 and 300 km, P wave velocity decreases from about 8.1 to 7.9 km/sec, while S wave velocity decreases from about 4.7 to 4.4 km/sec. Below 300 and extending to 400 km, the velocity gradient for S waves becomes still more negative, dropping to 4.15 km/sec at 400 km. It remains negative until 800 km (3.85 km/sec). In contrast, P wave velocity is almost constant at 7.9 km/sec in the 300–800 km interval. Nakamura and coworkers obtain a value of 0.250 $\pm$ 0.025 for Poisson's ratio in the upper mantle and 0.36 $\pm$ 0.02 in the lower mantle.

Attenuation of seismic waves has been studied by both groups. For P waves, Goins et al. (1977) obtained Q (quality factor) values of about 5000 for the upper mantle and 1500 for the lower mantle. Nakamura et al. (1976a,b) obtained a Q value for S waves of about 4000 in the upper mantle. These upper mantle values are very high by terrestrial standards and imply that this region is well below the solidus and contains no partial melt or volatiles (Tittman et al., 1976). Moreover, the Q value of 1500 for P waves in the lower mantle is still high by terrestrial standards (Dainty et al., 1976). This property, combined with occurrences of moonquakes to depths of 950 km, strongly suggests that the lower mantle to this depth does not contain an interstitial melt phase and is also water free.

Normal transmission of P waves is observed to depths of at least 1400 km, whereas S wave transmission in this region cannot be detected. Thus, the interval between 1000 and 1400 km must be characterized by strong attenuation of S waves. Nakamura et al. (1976b) suggested that this might be caused by a state of partial melting. Their explanation, however, may not be unique. D. L. Anderson (personal communication) has pointed out that the attenuation could be caused by processes not involving partial melting.

Nakamura et al. (1974) proposed the existence of a low-velocity core (V_p 3.7–5.1 km/sec) with a minimum radius of 170 km. They emphasize that this suggestion is very tentative, being based on an analysis of signals from only one farside meteoroid event (Nakamura et al., 1974, 1976b). An important conclusion, however, is that the maximum radius of a possible lunar core is about 350 km, this limit being set by observations of normal P wave transmissions to this depth. The mass of a possible metallic core of this size would be less than 2 percent of the mass of the moon. A similar limit is set by magnetometer results (Wiskerchen and Sonett, 1977).

The internal structure of the moon as revealed by seismic investigations is depicted in Figure 10.2.

Some Constraints on the Lunar Density Distribution

The mean density of the moon is 3.344 $\pm$ 0.002 g/cm^3 (Bills and Ferrari, 1977). This value is very similar to that of the Earth's upper mantle. This similarity may, of course, be coincidental. On the other hand, it may be of

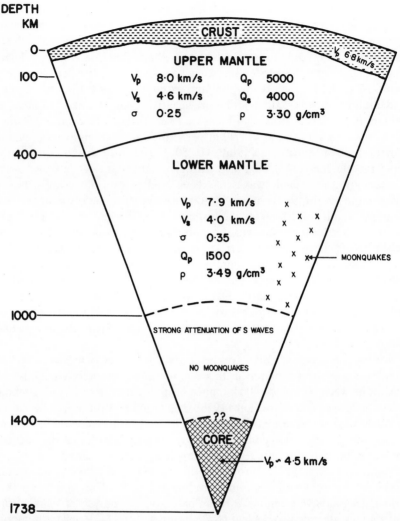

Figure 10.2 Internal structure of the Moon. Based on Nakamura et al. (1974, 1976a,b) and Goins et al. (1977). σ—Poisson's ratio; Q—quality factor (reciprocal of seismic attenuation); ρ—density.

profound genetic significance. Reasons for favoring this latter interpretation will be advanced in Chapters 11 and 12.

Estimates of the coefficient of moment of inertia (I/MR^2) have shown encouraging signs of convergence in recent years. The following values have been obtained for this coefficient: 0.392 ± 0.003 (Sinclair et al., 1976), 0.391 ± 0.002 (Blackshear and Gapcynski, 1977), and 0.3931 ± 0.0021 (Bills and Ferrari, 1977). In the light of these studies, a reasonable current estimate would be 0.392 ± 0.002. Since the coefficient of moment of inertia of a sphere of uniform density is 0.4, the lunar value implies the existence of a substantial increase of density with depth within the lunar interior.

The mean thickness of the lunar crust has been estimated as 60 km (Kaula et al., 1974), 70 km (Bills and Ferrari, 1977), and 75 km (Kaula, 1977), and these authors have estimated mean densities ranging from 2.86 to 3.07 g/cm^3. The densities and crustal thicknesses are not independent of each other— isostatic considerations show that high mean crustal densities imply greater mean crustal thicknesses. However, the problem of determining each of these quantities independently is complicated by the occurrence of porosity in lunar crustal rocks. This is shown by the observation that seismic velocities in the crust are lower than can be explained on the basis of inferred mineralogy. This behavior is clearly displayed in the upper crust (Fig. 10.1) but also exists to a lesser degree in the lower crust (Liebermann and Ringwood, 1976). The degree of porosity is also likely to be a function of thermal history and heat flow. It is possible that the mean porosity of the thick farside crust is appreciably higher than on the nearside, where the abundances of radioactive elements tend to be higher. This would result in higher subsurface temperatures, contributing to densification and loss of porosity at shallower depths. A further complication is the evidence of regional variations in chemical composition (see, for example, Schonfeld, 1977a), which would be accompanied by corresponding density variations. The author believes that a mean crustal density of 3.07 is likely to be too high because of lack of allowances for porosity, while the lower limit of 2.86 seems too low on petrological grounds. Most probably, the mean density of the crust is between 2.9 and 3.0 g/cm^3.

If we take the crust to have a mean thickness of 70 km and a density of 2.95 g/cm^3, while the deeper regions of the moon are assumed to be homogenous, then the moment of inertia coefficient of this lunar model would be 0.396 while the density of the homogeneous interior would be 3.40 g/cm^3. If, in addition, we allow the presence of an iron core with the *maximum* radius permitted by seismic observations (350 km), then the moment of inertia coefficient would be 0.393 while the density of the (homogeneous) lunar mantle would be 3.35 g/cm^3. The moment of inertia coefficient of this model is thus in essential agreement with observation. On the other hand, if the moon does not possess a core, then the density of the mantle must increase substantially with depth. Since the effects of thermal expansion and compressibility approximately cancel themselves out within the moon (Kaula et al., 1974), a lunar model lacking a metallic core implies an inhomogeneous mantle. If we take a two-layer mantle model plus crust (Fig. 10.2), the mean density and moment of inertia would be satisfied if the mean density of the upper mantle (70–400 km) was 3.30 g/cm^3, while the mean density of the lower mantle (below 400 km) was 3.49 g/cm^3.

The seismic velocity distributions (Fig. 10.1) strongly support the second inhomogeneous model. The observed substantial decreases of velocities with depth, whether continuous or discontinuous, are too large to be explained by the effects of acceptable temperature gradients (see, for example, Goins et al., 1977). Moreover, the high Q values and occurrence of moonquakes as deep as 1000 km preclude the existence of partial melting in this region of the lower mantle. Partial melting cannot therefore be invoked to explain the decrease of seismic velocities between 400 and 1000 km (Dainty et al., 1976). The large difference in Poisson's ratio between the upper and lower mantle may also be indicative of chemical and petrological zoning. Alternatively, it might be caused by high temperatures in this region (Langseth, personal communication).

The occurrence of decreases in seismic velocities with depth combined with an increase of density with depth can be explained only if the iron content of the lower lunar mantle is appreciably higher than that of the upper mantle. A contributing factor to the velocity decrease could also be an increase in the pyroxene/olivine ratio with depth, as inferred in Section 10.3. Note also that the occurrence of phase transformations to denser mineral assemblages involving the formation of garnet (Ringwood and Essene, 1970) would necessarily cause velocities to *increase* with depth in a homogeneous mantle providing the lower mantle contained at least a small amount (e.g., 4 percent) of alumina, as appears likely.

We conclude that the lunar mantle is probably chemically inhomogeneous and that the content of iron (presumably as FeO) in the lower mantle is higher than in the upper mantle. This conclusion has also been reached by Kaula et al. (1974), Nakamura et al. (1974), and Goins et al. (1977). Nevertheless, the presence of a very small metallic core, amounting to less than 2 percent of the lunar mass and smaller than 200 km in radius, cannot finally be eliminated solely on physical grounds. A small core of this kind may ultimately be required to explain the observed magnetizations of lunar rocks (Runcorn, 1977), although its existence is not favored by geochemical arguments (Section 11.6).

10.2 The Lunar Crust—Upper Mantle System

The lunar crust consists of two major petrologic provinces—the anorthositic highlands and the basaltic mare regions. Although the latter occupy a large area on the nearside of the moon, their average thickness is quite small and their total volume is estimated to be 1 percent or less that of the lunar crust (Head, 1976).

The predominant rock types occurring in the upper regions of the lunar highland crust consist of a suite of breccias with compositions ranging between anorthositic gabbro, gabbroic anorthosite, and anorthosite. Wänke et al. (1974, 1975) have demonstrated that, after minor corrections for the presence of KREEP component and meteoritic nickel–iron, the highland breccia compositions lie upon well-defined mixing lines, with pure anorthosite as an end component (Fig. 10.3). Mixing relationships between different components of highland breccias have also been studied by Taylor and Jakeš (1974) and Taylor and Bence (1975), who have used the observed breccia compositions together with orbital XRF and gamma-ray data on abundances of Ca, Al, and Th to estimate the mean composition of the upper layer of the

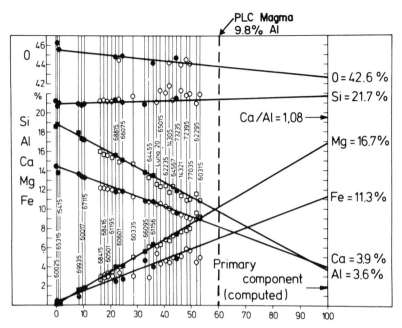

Figure 10.3 Diagram from Wänke et al. (1976, with permission), showing that highlands rock compositions can be interpreted in terms of the mixing of two components, one consisting of pure anorthosite and the other representing a "primary" component. Wänke and coworkers propose that the latter consists of material from which the Moon accreted. An alternative view, favored herein, is that the "primary" component consists of the parental magma from which the anorthositic cumulates crystallized.

lunar highlands (Table 10.1, column 1). This composition corresponds to that of an anorthositic gabbro. The high MgO/(MgO + FeO) ratio and the substantial Cr content should be noted. These features have been interpreted to imply that the lunar crust has a substantial "primitive" component and has not been subjected to *extensive* fractional crystallization which would result in a lower mean MgO/(MgO + FeO) ratio and a lower mean Cr content (Wänke et al., 1974, 1976; Taylor and Jakes, 1974; D. Walker et al., 1975a). It should be noted, however, that this latter generalization applies only to the *average* composition of the lunar crust. The orbital x-ray fluorescence and gamma-ray experiments (see, for example, Arnold et al., 1977; Metzger et al., 1977; Bielefield et al., 1977) reveal the existence of a considerable degree of regional chemical heterogeneity in the lunar highlands. These heterogeneities may reflect the operation of widespread igneous fractionation processes. For some elements, a striking correlation of chemical heterogeneities with crustal thickness exists (Metzger et al., 1977; Schonfeld, 1977a).

On the Earth, anorthositic rocks are often formed during the crystallization of large stratiform intrusions of basaltic magma, e.g., the Bushveld and Stillwater complexes (E. D. Jackson, 1967). In these complexes, dense olivine and pyroxene crystals sank during crystallization of the magmas to form

Table 10.1 Derivation of composition of basaltic magma multiply saturated with plagioclase and olivine, which may have been parental to the lunar crust (PLC magma)

	I Mean composition of upper crust (Taylor and Jakeš, 1974)	II Column I minus 57% of plagioclase An_{95} PLC magma
SiO_2	44.8	47.3
TiO_2	0.55	0.88
Al_2O_3	24.6	18.4
Cr_2O_3	0.1	0.13
FeO	6.6	9.4
MgO	8.6	12.3
CaO	14.2	11.3
Na_2O	0.45	0.16
$\dfrac{MgO}{(MgO + FeO)}$	0.70	0.70

thick basal layers, while plagioclase, which possesses a density similar to that of the parental magma, became concentrated in thick layers overlying these basal mafic cumulates. It is widely believed that the anorthositic suite of the lunar crust formed in an analogous manner from a huge ocean of mafic or ultramafic magma (see, for example, Wood, 1970, 1975; D. Walker et al., 1975a). However, because plagioclase has been shown to *float* in lunar magmas of appropriate composition (D. Walker and Hays, 1977), whereas it sinks in terrestrial mafic magmas (E. D. Jackson, 1967), segregation of plagioclase to form a thick layer may have been relatively more efficient on the Moon. Presumably, a regular stratiform structure was formed originally. However, in near-surface regions, this stratiform structure has been destroyed by subsequent intensive meteoritic bombardment, which has caused extensive vertical mixing of major rock types.

Lunar anorthositic rocks possessing compositions similar to the mean near-surface lunar crust composition (Table 10.1) display P velocities of 6.7–6.9 km/sec at confining pressures of 5–10 kbar (see, for example, H. Wang et al., 1973). These values agree closely with the seismically observed velocity in the lower crust of 6.8 km/sec. The agreement has been widely interpreted to imply that the entire lunar crust possesses a composition similar to that of the observed mean near-surface composition (see, for example, S. R. Taylor and Bence, 1975).

This conclusion, however, is unwarranted. Liebermann and Ringwood (1976) measured the P velocity of pure anorthite and, using existing velocity data for other relevant minerals, demonstrated that a pore-free gabbroic anorthosite containing 69 percent plagioclase (An_{95}), with orthopyroxene > olivine $\gg$ clinopyroxene $\gg$ ilmenite (mean $MgO/(MgO + FeO)$ ratio = 0.72), as advocated by Taylor and Bence (1975), would have a V_P velocity of 7.4 km/sec. Similar calculations showed that a lunar gabbro (40 percent plagioclase An_{95}, 50 percent orthopyroxene, 10 percent olivine) would have $V_P = 7.5$ km/sec. These results show firstly that P velocity is insensitive to large changes in the relative proportions of plagioclase to pyroxenes in lunar gabbroic anorthosite–anorthositic gabbro–gabbro compositions. Secondly, the intrinsic velocities of these rocks (7.4–7.5 km/sec) are substantially higher than the observed lower crust velocities of 6.7–6.8 km/sec. The discrepancy is most likely due to shock damage and microfracturing in the rocks of the lower crust, caused by large meteoritic impacts and cratering. The observed lunar velocities, therefore, do not justify the conclusion that the lower crust is of gabbroic anorthosite composition. The basic ambiguity in matching seismic velocities to rock compositions would equally permit the lower crust to consist of a mafic gabbro containing 40 percent or less of anorthite. In this case, the bulk composition of the whole crust could be equivalent to that of a normal gabbro (basalt) containing about 50 percent of plagioclase and 18 percent of Al_2O_3. The latter interpretation would be consistent with the hypothesis that the crust represents a former gabbroic magma derived by extensive partial melting of the moon's upper mantle. During crystallization of this parental magma, a limited amount of plagioclase elutriation occurred, resulting in a relative enrichment of plagioclase in the upper crust, and a corresponding depletion in the lower crust, as compared with the initial composition of the parent magma (Liebermann and Ringwood, 1976).

Analogous behavior is commonly observed in terrestrial mafic stratiform intrusions. It is plausible that it should also have occurred on the moon. In view of the efficient differentiation which has occurred to form the bulk lunar crust, it would be surprising if this differentiation had not continued *within* the lunar crust, resulting in a significant degree of enrichment of plagioclase in the upper 20–30 km.

A petrologically zoned model of the crust, as discussed above, is supported by studies of depths of the origin of major rock types in the crust. Ryder and Wood (1977) have pointed out that the rocks excavated by the largest and most deeply penetrating meteoroid impacts generally contain much less normative plagioclase than the anorthositic rocks widely exposed at the surface. They proposed a model in which the proportion of plagioclase relative to ferromagnesian minerals decreased with depth throughout the crust. A similar model was suggested by Charette et al. (1977). The major element bulk composition for these crustal models are similar to that derived by an entirely different method in the following subsection.[1]

---

[1] S. R. Taylor (1973a) has suggested that subsequent remixing by cratering processes would have homogenized any initially nonuniform lunar crust. However, this appears unlikely in view of the strong lateral compositional heterogeneities in the highland crust as revealed by orbital x-ray and gamma-ray spectroscopy (Adler et al., 1973; Metzger et al., 1974). While it seems likely that the top 10 km or so of the crust have been mixed and overturned by saturation bombardment of meteoroids making 50–100 km diameter craters, the depths of the original craters represented by the much rarer large ring basins (>200 km radius) and the degree of mixing caused by these events are poorly known. Head et al. (1975) believe that the maximum depth of excavation was only 20 km. A detailed Monte Carlo simulation of the cratering history of the moon by Hörz et al. (1976) showed that less than 5 percent of the lunar surface would have been cratered to depths exceeding 15 km by a bombardment history sufficient to produce all the craters observed in the highlands.

Nature of the Lunar Crust's Parental Magma (PLC Magma)

The formation of the lunar crust is generally believed to have involved a large-scale melting and differentiation process which affected an outer zone of the moon some hundreds of kilometers thick (see, for example, D. Wood, 1970; D. Walker et al., 1975a). This major episode of melting and differentiation occurred within a time interval 4.4–4.6×10^9 years ago, during or very soon after the formation of the moon (Tera and Wasserburg, 1974). It is possible that an outer layer, perhaps 400 km thick, was *totally* melted, thereby forming an ultramafic parent magma (see, for example, D. Walker et al., 1975a). However, consideration of available energy sources and heat balances, and the large temperature interval between the liquidus and solidus (Ringwood, 1976), indicate a greater likelihood of producing extensive (e.g., 20–30 percent) *partial* melting of the outer regions of the moon. This would produce a magma of basaltic bulk composition (parental to the lunar crust) overlying a thick zone of residual, refractory ferromagnesian minerals. Even in the event that total melting of the outer moon occurred, thereby forming an ultramafic magma, extensive fractional crystallization of olivine + pyroxene would be necessary before the Al_2O_3 content of the residual magma was sufficiently high (17–19 percent Al_2O_3) to precipitate plagioclase (Ringwood, 1976). Thus, in this case also, by the time that the plagioclase-rich lunar upper crust began to form, the parental magma must have been of a general mafic or basaltic composition.

Schonfeld (1975) and others have pointed out that the composition of the upper lunar crust can be interpreted in terms of a mush of cumulus plagioclase crystals plus trapped intercumulus parental gabbroic magma from which the plagioclase had crystallized. This is a simple and attractive concept. The high $MgO/(MgO + FeO)$ ratio and Cr contents of the lunar crust suggest that the parental mafic liquid had not evolved extensively via fractional crystallization when it became trapped in the plagioclase cumulate.

Melting relationships of relevant lunar gabbroic anorthosite compositions at low pressures (< 5 kbar) have shown that plagioclase crystallizes over a wide temperature interval before being joined at a cotectic by olivine and/or pyroxene (Kesson and Ringwood, 1976a). The cotectic liquid is of an overall basaltic composition. In the case of the lunar crust, we obtain the composition of the parental basaltic composition by removing increasing amounts of liquidus plagioclase (An_{95}) from the mean upper crust composition (Table 10.1), and determining the stage at which the residual liquid becomes multiply saturated at its liquidus by plagioclase and a ferromagnesian mineral (olivine and/or pyroxene). This composition has been experimentally determined (Table 10.1, column 2). The liquidus phases at the cotectic at atmospheric pressure are plagioclase (An ~ 95) and olivine (Fo_{88}). This composition is believed to approximate that of the magma parental to the lunar crust and, in terms of the previous discussion, to represent the bulk composition of the lunar crust. Thus, it may be of major volumetric and petrogenetic significance.

Table 10.2 Comparison of composition of basaltic magma, which could have been parental to the lunar crust, with composition of a typical primitive terrestrial oceanic tholeiite modified by partial loss of volatile components

	I Primitive terrestrial oceanic tholeiite[a]	II Parental lunar crust magma	III Column I minus $(7\% \, SiO_2 + 1.8\% \, Na_2O)$[b]
SiO_2	50.3	47.3	47.5
TiO_2	0.7	0.9	0.8
Al_2O_3	16.6	18.4	18.2
FeO	8.0	9.4	8.8
MgO	10.2	12.3	11.2
CaO	13.2	11.3	14.5
Na_2O	2.0	0.2	0.2
$\dfrac{MgO}{MgO + FeO}$	0.69	0.70	0.69
$\dfrac{\sum REE}{Chondrites}$	~ 9	~ 10	~ 10

[a] Sample DSDP 3-18 from Frey et al. (1974).
[b] This is equivalent to 14 percent of the total SiO_2 and 90 percent of the Na_2O in the composition given in column I.

It is of interest to compare the composition of this magma with that of primitive terrestrial oceanic tholeiites, which represent the most abundant rock type erupted at the Earth's surface. This comparison is made in Table 10.2. It is well known that lunar basalts are depleted in volatile elements relative to terrestrial basalts. Of the major elements shown in Table 10.2, column I, the most volatile are Na and Si (Grossman, 1972). In Table 10.2, column III, we have arbitrarily removed 14 percent of the total SiO_2 and 90 percent of the total Na_2O from column I. We see that the composition of the residual modified terrestrial tholeiite is very similar to that of the lunar parental highland magma.[2]

The resemblance between the magma parental to the lunar crust and devolatilized terrestrial oceanic tholeiites extends also to key trace elements such as the rare earths. Hubbard et al. (1971) calculated the REE abundances of the parental basaltic magma from which the anorthosite 15415 had crystallized, using experimental plagioclase-liquid partition coefficients. The parental liquid was found to possess chondritic relative abundances of the REE elements at about 10 times the absolute chondritic abundances (Fig. 10.4). It is similar in this respect to many primitive terrestrial tholeiites including the example chosen in Table 10.2. Hubbard et al. (1971) also showed that other low-K lunar anorthosites possessing positive Eu anomalies had

[2] The higher content of Cr_2O_3 in the PLC magma (0.13 percent), compared to the terrestrial tholeiite (0.05 percent Cr_2O_3), can be attributed to the effect of differing oxygen fugacity conditions on the partition of chromium between magma and residual olivines and pyroxenes.

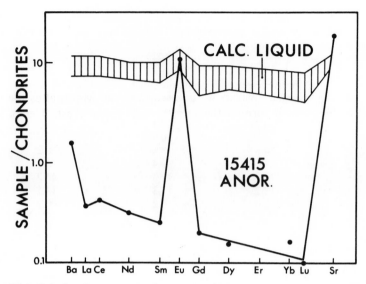

Figure 10.4 Calculated rare earth element abundances in parental (basaltic) liquid from which lunar anorthosite 15415 is believed to have crystallized (after Hubbard et al., 1971).

likewise crystallized from parental liquids with generally similar REE abundances. Similar results were obtained by Laul et al. (1974) and McCallum et al. (1975).

The close resemblance in major element and REE abundances between the most abundant kind of primitive lunar basaltic magma and the most abundant class of primitive terrestrial basaltic magma, modified only by the partial loss of two of the most volatile components (Na_2O and SiO_2), is believed to be of considerable genetic significance. We will return to this point subsequently.

Wänke's "Primary Matter" Hypothesis

Wänke et al. (1974, 1975, 1976, 1977, 1978b) have proposed an alternative explanation of lunar highland geochemistry which, nevertheless, is quite closely related to the one which we have just described. On the basis of Figure 10.3, Wänke et al. showed that lunar highland rock compositions can be interpreted in terms of the mixing of two components, one consisting of pure anorthosite and the other representing "primary matter," which is comparatively rich in Mg and Cr.

The composition of this latter component was obtained from the mixing lines of Figure 10.3, extrapolated to the composition at which the chondritic Ca/Al ratio of 1.08 was reached. In a subsequent, more detailed study using the same approach, Wänke et al. (1977) preferred to use the Al/Sc ratio for this purpose. The latter composition so obtained was also found to possess chondritic Ca/Al and Mg/Cr ratios, accompanied by a subchondritic Mg/Si ratio. Moreoever, Wänke et al. (1978b) demonstrated a good correlation between Ni and Mg in selected highland rocks.

In the light of these and other geochemical properties, Wänke et al. interpreted this second component of Figure 10.3 as representing, in large part, "primary matter" which was accreted by the Moon after the early differentiation which formed the anorthositic upper crust. The primary matter was then mixed with anorthositic crustal material as a consequence of meteoritic impact processes, thereby explaining the trends shown in Figure 10.3. According to Wänke et al's. interpretation, the composition of the "primary component" should be representative of that of the bulk Moon.

In later discussions, Wänke et al. (1977, 1978b) recognized that this simple model required revision in order to account for the subchondritic Mg/Si ratio. They proposed that during the early major differentiation which formed the crust, a large amount of olivine separated from the parental magma and became incorporated in the lunar mantle. In order to obtain the true composition of this "primary matter," a correction for this missing olivine should be applied. When this was carried out, a chondritic Mg/Si ratio was obtained for "primary matter," the composition of which is given in Table 10.5.

Wänke et al.'s (1977, 1978b) revised interpretation of lunar highland geochemistry is quite closely related to the interpretation given by the author in the previous subsection. In particular, it leads to a similar estimate of the lunar bulk composition (Table 10.5); its implications with respect to the origin of the Moon are identical with those drawn by the author (Chapters 11 and 12).

The two models are not mutually exclusive. I prefer the interpretation of highland geochemistry given in the previous subsection on the grounds of simplicity and economy of assumptions. Subsequent discussions of the petrological and geochemical development of the Moon is accordingly based on Ringwood's (1977a) model. However, a great deal of this discussion could readily be adapted to the Wänke et al. (1977, 1978b) model.

The Lunar Upper Mantle

The composition of the (PLC) magma which is believed to be parental to the lunar crust is given in Table 10.1. Olivine (Fo_{88}) plus plagioclase crystallize simultaneously on its liquidus at 1250°C (in vacuum). The absence of a negative europium anomaly (Fig. 10.4) implies that plagioclase was not a residual phase in the source region, and that the magma had not crystallized substantial amounts of plagioclase after segregating from its source region. Olivine, however, remains on the liquidus to about 7 kbar (Kesson and Ringwood, 1976a) and was probably present as a residual phase in the source region. It is possible, therefore, that the composition of the PLC magma has been modified by crystallization of olivine at relatively shallow depths and that the primary magma was richer in normative olivine than the composition given in Table 10.1.

The primary PLC magma may have formed by extensive partial melting of the outer few hundred kilometers of the moon. After segregating from the residual phases in its former source region, the vast ocean of primary PLC magma would have precipitated olivine until plagioclase saturation was reached. Most probably, therefore, a layer of olivine cumulates underlies the crust (Fig. 10.5).

If the PLC magma segregated from residual refractory ferromagnesian phases at average pressures less than 7 kbar (150 km), the experimental phase equilibria show that the residual phase consisted of olivine (Fo_{88}). At this pressure, reconnaissance runs showed that olivine is joined by ortho-

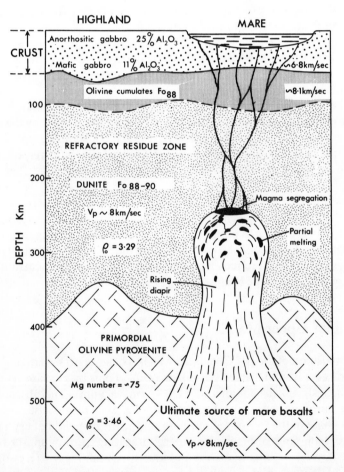

Figure 10.5 Structure of the Moon based on petrological studies of the origins of highland rocks and low-Ti mare basalts. Small corrections have been applied to the ambient densities and P velocities to allow for the effect of high P and T in the lunar interior.

pyroxene (Kesson and Ringwood, 1976a). For a model primary PLC magma containing ≥ 10 percent more normative olivine than the composition given in Table 10.1, olivine remains on the liquidus to pressures exceeding 15 kbar and is joined by aluminous orthopyroxene, which crystallizes over a wide pressure interval. These relations suggest that the residual, refractory phases remaining in the source region after segregation of the primary PLC magma consisted of olivine alone, or of olivine + orthopyroxene.

The PLC magma possesses a CaO/Al_2O_3 ratio of 0.61, substantially smaller than the chondritic ratio of 0.8. If the source region possessed a near-chondritic CaO/Al_2O_3 ratio, as suggested by its near-chondritic relative REE abundances (Fig. 10.4), then the residual phases remaining behind after magma generation must have possessed a high net CaO/Al_2O_3 ratio. However, the liquidus orthopyroxenes observed in PLC and (PLC + olivine) compositions have CaO/Al_2O_3 ratios smaller than 0.4. The presence of orthopyroxene as a residual phase would exacerbate the discrepancy with the chondritic CaO/Al_2O_3 ratio. Orthopyroxene does not, therefore, appear likely to be a major residual phase in the source region. On the other hand, olivine containing 0.4–0.5 percent CaO but no Al_2O_3, is a required residual phase and would modify the CaO/Al_2O_3 in the desired direction, although not sufficiently far as to produce an overall chondritic ratio.

If the latter were characteristic of the bulk system from which the lunar crust was derived, then large quantities of clinopyroxene with $CaO \gg Al_2O_3$ must be assumed to be residual in the source region. Although this possibility cannot yet be finally excluded, it seems improbable in the light of existing phase equilibria data.

It seems likely, therefore, that the residual, refractory component remaining in the lunar mantle after extraction of the primary PLC magma consisted of pure dunite similar to, or somewhat more magnesian than, the observed liquidus PLC magma olivine (Fo_{88}, 0.5 percent CaO). If we assume that the PLC magma represented a 25 percent melt of its source region (comparable to the most primitive oceanic tholeiites in the Earth's mantle), then the CaO/Al_2O_3 ratio of the bulk system would be 0.70—significantly, but not excessively, below the chondritic ratio of 0.8. The bulk system would contain 4.6 percent Al_2O_3.

The mean composition of the lunar crust and upper mantle on the above assumptions is given in Table 10.5, column II. It is seen to be quite similar to the pyrolite composition of the Earth's upper mantle. The principal differences are the lower SiO_2 content of the Moon, implying a higher olivine/pyroxene ratio, and its lower CaO/Al_2O_3 ratio.

The density of the dunitic (Fo_{88}) lunar upper mantle at a depth of 100 km and a corresponding temperature probably in the vicinity of 600°C (Keihm and Langseth, 1977) would be 3.29 g/cm^3, while its P-wave velocity at this depth would be 8.1 km/sec (Birch, 1969). These values, based entirely on petrologic reasoning, agree almost exactly with the seismic velocity and density of the upper mantle obtained by completely independent physical methods (Fig. 10.2). The agreement is most satisfactory.

10.3 Mare Basalts and the Lower Mantle

Petrogenesis

One of the most significant results arising from the Apollo project was the demonstration that the lunar maria are composed of rocks resembling terrestrial basalts. Moreover, we have seen in Section 10.2 that the lunar crust was ultimately derived by differentiation from a parental magma ocean of basaltic affinities. In recent years, important progress has been made towards understanding the origin of terrestrial basaltic rocks. It is now widely accepted that they formed by partial melting of an ultramafic source rock (pyrolite) in the Earth's mantle. The observed spectrum of basaltic compositions is interpreted primarily in terms of the degree of partial melting of the source material, the depth of partial melting, and the subsequent crystallization history during ascent to the surface. High-pressure, high-temperature investigations of the crystallization behavior of terrestrial basalts using the methods of experimental petrology have been successfully employed to constrain the nature of their source regions (Green and Ringwood, 1967; Green, 1970a; Ringwood, 1975a).

It is natural to attempt to understand the origin of lunar basalts in terms of processes analogous to those that have operated in the petrogenesis of terrestrial basalts and, likewise, to use experimental petrology to provide information on the nature of their source regions.

Early hypotheses of mare basalt petrogenesis based on experimental petrology proposed that all classes of mare basalts had formed by varying degrees of partial melting of a common olivine–pyroxenite source at depths of 150–500 km (see, for example, Ringwood and Essene, 1970). Although this straightforward single-stage hypothesis provided an adequate explanation of many aspects of the major element chemistry of mare basalts and their source regions, it was unable to provide satisfactory explanations of other geochemical characteristics of high-Ti and low-Ti mare basalts, e.g., their REE patterns, Eu anomalies, TiO_2 contents, and isotope systematics.

To meet these difficulties, a second class of hypotheses was developed (see, for example, Schnetzler and Philpotts, 1971) which maintained that mare basalts could have formed by the remelting of chemically and mineralogically inhomogeneous olivine + pyroxene $\pm$ ilmenite cumulates formed during the early differentiation of the moon around 4.6–4.4 $\times$ 10^9 years ago, as the complement of the plagioclase-rich crust. Specifically, it was suggested (for example, by D. Walker et al., 1975b) that low-Ti basalts had formed by the partial melting of early olivine + pyroxene cumulates at considerable depths (200–400 km), whereas the high-Ti basalts formed by partial melting of a zone of late-stage olivine + pyroxene + ilmenite cumulates at relatively shallow depths (around 100 km). The cumulate remelting hypothesis offers, in principle, an explanation for several important geochemical characteristics of mare basalts, e.g., their diverse TiO_2 contents, the two-stage history recorded by Sm–Nd and Pb–U isotopes and of the Eu anomalies. However, this hypothesis still encounters a number of fatal difficulties. For example, it is unable to explain the observation that the least fractionated high-Ti and low-Ti basalts have similar Cr contents and MgO/(MgO + FeO) ratios. This and other shortcomings have been discussed in detail by Ringwood (1976) and Ringwood and Kesson (1976). Further discussion of this topic is given in Section 10.4.

Although the cumulate-remelting hypothesis is not acceptable in its present form, its success in explaining several key features of the trace-element and isotopic geochemistry of mare basalts strongly suggests that some form of cumulate reprocessing was involved in mare basalt petrogenesis. This raises the possibility of combining the more attractive aspects of the single-stage, uniform, primordial source-region hypothesis with those of the cumulate-remelting hypothesis.

Hubbard and Minear (1975) pointed out that most objections to both previous classes of hypotheses could be eliminated or minimized if mare basalts were interpreted as hybrid magmas produced by the interaction between early (4.4–4.6 × 10^9 years) cumulates and later (3.0–4.0 × 10^9 years) liquids formed by partial melting of the primordial, deep lunar interior. This hypothesis was reformulated by Kesson and Ringwood (1976a) to avoid difficulties encountered by the Hubbard–Minear model.

Ringwood and Kesson (1976) adopted as their starting point the widely held belief that during or soon after its formation, the outer regions of the Moon (to a depth of a few hundred kilometers) were extensively melted and differentiated, whereas the deeper primordial interior remained unmelted at this stage. Accordingly, the temperature of the latter region continued to rise owing to radioactive heating. Experimental studies (Ringwood, 1976) showed that the temperature gradient in the cumulates was much higher than the adiabatic gradient and hence this region was convectively unstable. Moreover, the convective instability within the cumulate sequence was further enhanced by the normal effects of fractionation which caused the later cumulates to be richer in iron and titanium and therefore more dense. The initial presence of a few percent of interstitial liquid in this otherwise solid material would also have reduced the viscosity. The cumulate layer in the Moon's upper mantle was thus in a state of acute gravitational instability, and convective overturn was inevitable. This had the effect of transporting highly fractionated cumulates into the primordial deep interior, where thorough assimilation and hybridization occurred. Mare basalts are believed to have formed by the subsequent partial melting of these contaminated regions.

A key implication of this hypothesis is that high-Ti basalts were derived from more highly contaminated (or hybridized) local source regions than low-Ti basalts. Although the latter do contain a small but important component derived from the early (4.4–4.6 × 10^9 years) differentiation process, they are nevertheless mainly composed of material derived from the primordial deep lunar interior. It follows that low-Ti basalts are better able to

provide information on the composition of the primordial lunar interior than
are high-Ti basalts.

This general interpretation is supported by several specific lines of evidence.
Rare-earth element (REE) distributions for some representative high- and
low-Ti basalts are shown in Figure 10.6. Two features which are attributable
to participation of 4.6–4.4 billion-year-old differentiated cumulates are the
magnitude of the europium anomalies and the depletion of light REE.
Depletions of light REE are generally much smaller in low-Ti basalts than in
high-Ti basalts. Likewise, the magnitude of europium anomalies are generally
much greater in high-Ti basalts than in low-Ti basalts, although there is a
degree of overlap between some differentiated Apollo 12 low-Ti basalts and
some of the more primitive of the Apollo 17 high-Ti basalts. Another feature
implying a more complex history for high-Ti basalts is their low nickel
contents (less than 10 ppm) as compared to low-Ti basalts, which contain up
to 120 ppm Ni (180 ppm Ni in Apollo 15 Green Glass).

Figure 10.6 shows that there is a general trend for the size of the Eu-
anomaly to decrease as the absolute abundances of the remaining trivalent
REE decrease. The trend is shown more clearly in Figure 10.7, which is based

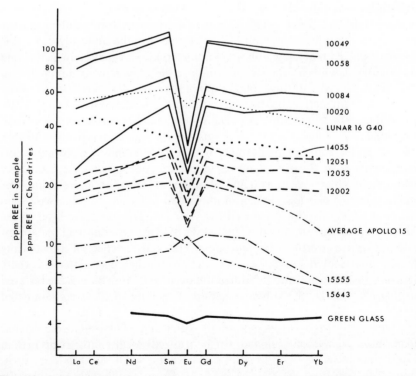

Figure 10.6 Abundances of rare earths in a representative suite of mare basalt samples.
Samples 10049, 10058, 10084, and 10020 are high-titanium samples. Remainder are in
the low-titanium category.

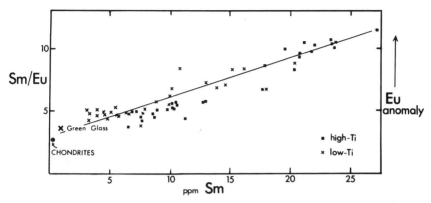

Figure 10.7 Graph showing samarium/europium ratios versus samarium concentrations for Apollo 11, 12, 15, and 17 basalts.

on similar diagrams presented by Haskin et al. (1970) and Helmke et al. (1973). The overall linear relationships between Sm/Eu ratios (a measure of the Eu anomaly) and Sm contents is quite impressive. This may be caused by the combination of two factors: (a) varying degrees of mixing between a primordial component from the deep interior (no Eu anomaly) and a fractionated component from the subcrustal cumulates possessing a deep Eu anomaly, and (b) varying degrees of partial melting of the source region, negatively correlated with the degrees of concentration of incompatible elements (trivalent REE in this case) in the resultant magmas.

Figures 10.6 and 10.7 clearly demonstrate the more primitive nature of most low-Ti basalts (particularly from Apollo 15) as compared to most high-Ti basalts. The fact that the Sm/Eu ratio versus Sm regression passes very close to Green Glass and to the chondritic ratio is of considerable significance, implying that the "primitive" component of mare basalts from the deep primordial interior did not possess an Eu anomaly. Likewise, the "flat" pattern of trivalent REE in Green Glass implies the existence of chondritic relative abundances in its source region.

Vaniman and Papike (1977) identified a new type of basalt characterized by very low titanium contents (VLT-type, <1 percent TiO_2) and a wide range of MgO/(MgO + FeO) ratios in thin sections from the Apollo 17 drill core. They concluded that the VLT basaltic suite were genetically related to Green Glass. The basalts recovered from Mare Crisium by Luna 24 also belonged to this suite, containing about 1.0 percent TiO_2 (Barsukov, et al. 1977). They were relatively fractionated, possessing high Al_2O_3 and low MgO/(MgO + FeO) ratios. However, studies of the Luna 24 soils revealed the presence of high-magnesium, VLT olivine vitrophyres (Coish and Taylor, 1977). Rare earths of most Luna 24 soils were present at about 10 × chondritic levels (Blanchard et al., 1977). The patterns were flatter than for most Apollo 15 rocks and the Eu anomalies were generally smaller. It seems that the VLT suite has been derived from a comparatively primitive source region.

Spectral studies (Andre et al., 1977; Schonfeld and Bielefield, 1977) suggest that high-magnesium, VLT basalts related to Green Glass might occur widely among the lunar maria.

Source Region of Low-Titanium Basalts

In view of the preceding discussion, which led to the conclusion that high-Ti basalts have had a much more complex history than low-Ti basalts, we have chosen to employ compositional data and phase relationships displayed by low-Ti basalts alone in an attempt to constrain the composition of the primordial deep lunar interior.

Low-Ti basalts, which best qualify as primitive magmas, are those possessing relatively high Cr and Mg contents and high contents of normative olivine, and which can furthermore be demonstrated to have reached the lunar surface essentially as liquids containing negligible or small amounts of previously crystallized phenocrysts. Rocks in this category include 12009 (Green et al., 1971a), 12002 (D. Walker et al., 1976), and 15555 (Kesson, 1975). These rocks contain 10–20 times the chondritic abundances of REE and other incompatible, involatile elements and are similar in this respect to oceanic tholeiites. The relative abundances of trivalent REE and other incompatible, involatile elements are, moreover, quite close to chondritic relative abundances, although significant deviations, including the Eu anomaly (small in the case of 15555), testify to a degree of complexity in the chemical history of the source region prior to magma generation.

The Apollo 15 Green Glass is believed to possess considerable petrogenetic significance (Green and Ringwood, 1973). Its composition testifies to the existence of an extremely primitive class of basaltic or picritic magma at the lunar surface. This magma possessed an extremely regular "primordial" REE pattern, closely parallel to the chondritic pattern, with only a very small Eu anomaly. The REE and other involatile incompatible elements (including titanium) are present at levels of only 4–5 times the chondritic abundances. Whereas the least fractionated Apollo 12 and Apollo 15 magmas have liquidus olivines Fo_{74-76}, the liquidus olivine of Green Glass is Fo_{84}. Green Glass is believed to represent a magma formed by a much higher degree of partial melting (40–60 percent) of a source region similar to that of 15555, 12002, and 12009. [The latter may represent 10–20 percent partial melts (Green and Ringwood, 1973; Ringwood, 1977a)].

By experimentally determining the nature and compositions of liquidus and near-liquidus phases of mare basalts over a wide range of pressures, powerful constraints may often be placed upon the nature of their source regions (see, for example, Ringwood and Essene, 1970; Green et al., 1971b; Ringwood, 1977a). In applying this method, we seek to choose basalts (using criteria previously discussed) which have undergone the minimum amount of fractionation en route to the surface. The method is capable, in principle, of characterizing the residual phases remaining in the source region after magma segregation. These residual phases are similar in composition to the liquidus and near-liquidus phases of the basalt at the appropriate pressure. It is not, however, directly capable of yielding the proportions of mineral phases remaining in the residuum, yet this knowledge is essential if the bulk composition of the source region is to be estimated. There are, nevertheless, additional constraints imposed by phase equilibria and chemical compositions which can be applied to facilitate a solution to the latter problem. These are discussed in greater detail by Ringwood (1977a).

Applying these methods, it is possible to construct a reasonably self-consistent compositional model for mare basalt source regions. Examples based on high-pressure phase equilibria displayed by 12009 and Green Glass are given in Tables 10.3 and 10.4. It is found that the source region of low-Ti mare basalts consisted of a mineral assemblage of orthopyroxene + clinopyroxene + olivine. Plagioclase was absent. The abundance of pyroxenes exceeded that of olivine, contrary to the situation in the dunitic lunar upper mantle. The $MgO/(MgO + FeO)$ ratio in the mare basalt source region was 0.75–0.80, as compared to an $MgO/(MgO + FeO)$ ratio of 0.88 in the upper mantle. The contents of CaO and Al_2O_3 in the mare basalt source region were smaller than 5 percent and probably in the vicinity of 3.5–4 percent, or about twice the (ordinary) chondritic abundances. The involatile, incompatible elements (e.g., REE, Ti, Zr, and U) in the mare basalt source region were also present at approximately twice the ordinary chondritic levels. A detailed discussion of the basis for these conclusions is given by Ringwood (1977a).

Depth of Origin

The following lines of evidence indicate that the source region of mare basalt partial melts was situated deep within the lunar interior.

(a) Mare basalts were generated over a period extending at least to 1.4×10^9 years after the formation of the Moon. Thermal history considerations employing a wide range of boundary conditions (see, for example, Urey, 1962b; Toksöz et al., 1976) show that deep-seated cooling would occur to a depth beyond 200 km over a period of 10^9 years. It is extremely difficult to understand how mare basalt magmas might be formed by partial melting within this outer cool shell some 3.2×10^9 years ago. Crater counts indeed imply that mare volcanism extended to even younger ages (Boyce, et al., 1974). The thermal problem has been compounded by the recent downward revision of lunar heat flow (Langseth et al., 1976), implying lower abundances of U, Th, and K than were postulated in many earlier studies of lunar thermal history. It follows that the source regions lay at depths exceeding 200 km.

(b) The mascons were presumably formed before or during the flooding of the mare basins. Their continued existence for up to 3.8×10^9 years implies the existence of a strong, cool, and thick (>150 km)

Table 10.3 Construction of model lunar olivine-pyroxenite source composition from 12009 plus near-liquidus phases at 15 kbar. Compositional data from Green et al. (1971b)

	$12009 + 10\% \; Fo_{75}$	Liquidus Opx^a	Liquidus Cpx^b	Liquidus olivine[c]	Refractory residue 40% cpx 30% opx 30% ol	Model lunar source 10% modified 12009 90% residua
SiO_2	44.6	54.0	50.3	38.5	47.9	47.6
TiO_2	2.6	0.3	0.7	—	0.4	0.6
Al_2O_3	7.8	2.3	5.0	—	2.7	3.2
Cr_2O_3	0.5	0.8	0.9	0.4	0.5	0.5
FeO	21.2	13.0	15.9	22.2	16.9	17.3
MnO	0.3	0.3	0.3	0.2	0.3	0.3
MgO	14.1	27.7	21.2	38.0	28.2	26.8
CaO	8.6	2.0	6.1	0.2	3.0	3.6
Na_2O	0.2	0.06	0.1	—	0.06	0.07
$\dfrac{MgO}{(MgO + FeO)}$	0.54	0.79	0.71	0.75	0.75	0.73

[a] Orthopyroxene was crystallized from modified 12009 composition containing 10 percent additional olivine (Fo_{75}).

[b] Clinopyroxene represents average of near-liquidus phases in 12009 and 12009 + 2 percent enstatite (En_{80}) at 1390°C and 15 kbar.

[c] Olivine analysis represents liquidus phase at atmospheric pressure but would be unchanged at higher pressure.

Table 10.4 Construction of model lunar olivine-pyroxenite source composition from Green Glass plus near-liquidus phases at 15 kbar. Compositional data from Green and Ringwood (1973)

	Green glass	Liquidus olivine 15 kbar	Liquidus orthopyroxene 15 kbar	Refractory residue 60% ol 40% opx	Model lunar source composition 50% GG 50% Refractory residue
SiO_2	45.2	40.6	56.1	46.8	46.0
TiO_2	0.4	–	0.1	–	0.2
Al_2O_3	7.6	–	2.1	0.8	4.2
Cr_2O_3	0.4	0.4	0.7	0.5	0.5
FeO	19.7	14.8	8.8	12.4	16.1
MnO	0.2	0.2	0.1	0.2	0.2
MgO	17.9	43.5	30.2	38.2	28.1
CaO	8.1	0.4	1.8	1.0	4.5
Na_2O	0.1	–	–	–	0.05
$\dfrac{MgO}{(MgO + FeO)}$	0.62	0.84	0.86	0.85	0.76

lithosphere at the time of mare volcanism (see, for example, Kaula, 1969). An origin for mare basalts by partial melting in this region implies loss of strength and destruction of the lithosphere. Preservation of mascons would be inexplicable unless mare basalts had been derived from deeper regions.

(c) Primitive low-Ti mare basalt magmas crystallize olivine as a liquidus phase up to moderate pressures, where olivine is joined by sub-calcic clinopyroxene or orthopyroxene. The olivine–pyroxene cotectics occur at the following pressures. 12009—7 kbar; 12002—14 kbar; 15555—12 kbar; Green Glass—20 kbar. It is possible that all of these magmas have crystallized some olivine during their ascent to the surface, so that these are *minimum* pressures. Constraints on the degree of partial melting imposed by trace element abundances (Ringwood, 1977a) require that pyroxenes remained in the residuum after partial melting and magma segregation. Accordingly, the pyroxene–olivine cotectic pressures are likely to represent the *minimum* depths at which the magmas were generated. The above pressures correspond to depths of 140—280 km. It is possible that the ultimate sources were deeper, below 400 km, owing to varying degrees of olivine crystallization during ascent. It is also quite likely that partial melting occurred on release of pressure, as diapirs ascended from deeper source regions (Green and Ringwood, 1967). The experimental cotectic pressures correspond to the depths of magma segregation rather than the ultimate depth of origin of the source diapirs (Fig. 10.5).

(d) The density, seismic velocities, and MgO/(MgO + FeO) ratios of the mare basalt region characterized in Tables 10.4 and 10.5 on petrological grounds are in good agreement with the inferred properties of the lunar lower mantle, but not with those of the upper mantle (Figs. 10.2 and 10.5). In particular, the density of the upper mantle is lower and its intrinsic seismic velocities are higher than the olivine pyroxenite source region of mare basalts.

The above evidence strongly implies that mare basalts were derived *ultimately* from the lower mantle, probably as indicated in Figure 10.5. However, the petrogenesis of mare basalts has been complex, involving at least a two-stage process, in which differentiated dense cumulates from the upper mantle are believed to have sunk into the lower mantle, contaminating this region, and giving rise to the hybrid characteristics of mare basalt source regions.

Zonal Structure

The mean composition of the outer few hundred kilometers of the Moon (crust and mantle), as estimated according to the discussion in Section 10.2, is given in Table 10.5. Except for lower amounts of Na_2O and SiO_2, it is similar to the pyrolite model composition for the Earth's upper mantle. The source region for low-Ti basalts resembles that of pyrolite in some aspects—e.g., the dominant minerals are orthopyroxene and olivine and it contains

Table 10.5 Comparison of estimated lunar bulk compositions with model pyrolite composition of the terrestrial mantle

	I[a] Mare basalt source region composition	II[b] Bulk composition of outer 400 km of moon	III[c] Bulk composition of entire moon	IV[d] "Primary matter"	V[e] Pyrolite
SiO_2	46.8	42.8	44.8	45.6	45.1
TiO_2	0.4	0.2	0.3	—	0.2
Al_2O_3	3.7	4.6	4.2	4.6	3.9
Cr_2O_3	0.5	0.3	0.4	0.4	0.3
FeO	16.7	11.0	13.9	13.0	7.9
MgO	27.5	37.8	32.7	32.3	38.1
CaO	4.1	3.2	3.7	3.8	3.1
Na_2O	0.06	0.05	0.05	0.06	0.4
$\dfrac{MgO}{(MgO + FeO)}$	0.75	0.86	0.81	0.82	0.89
$\dfrac{\sum REE}{Chondrites}$	~2	~2	~2	~2	~2
Ab	0.5	0.4	0.4	0.5	3.4
An	9.9	12.3	11.2	12.2	8.9
Di	8.5	2.8	5.7	5.5	5.2
Hy	41.8	11.9	26.0	29.2	18.4
Ol	37.9	71.8	55.5	51.7	63.3
Chr	0.7	0.4	0.6	0.6	0.5
Ilm	0.8	0.4	0.6	—	0.4

[a] I From Tables 10.4 and 10.5 (average of both).
[b] II 25 percent PLC magma + 75 percent olivine (Fo_{88})
[c] III Average of Columns I and II.
[d] IV "Primary matter" representing bulk composition of Moon (Wänke et al., 1977).
[e] V From Table 1.6

187

similar amounts of CaO, Al_2O_3, and other involatile elements (REE, U, Th). However, there are differences which are believed to be real. The most important difference is that the low-Ti basalt source region seems to have a lower $MgO/(MgO + FeO)$ ratio (~ 0.75–0.80) than pyrolite (0.89). Chromium (and manganese) are also significantly higher in the mare basalt source (see, for example, Huebner et al., 1976). Finally, silica is higher, causing a higher pyroxene/olivine ratio in the mare basalt source region (Ringwood, 1977a).

It seems likely, therefore, that the chemical zoning within the Moon implied by petrological arguments is a real feature (Fig. 10.5). Moreover, this zoning is also implied by physical observations of seismic velocities, mass, and moment of inertia (Section 10.1, Fig. 10.2). Indeed, the quantitative agreement between the two models derived independently from physical observations and petrological arguments is most impressive.

There is an interesting complementarity between CaO/Al_2O_3 ratios and silica contents in the two major petrologic provinces of the lunar interior. While neither individual region is identical to the inferred pyrolite composition for the Earth's mantle, it is a remarkable fact that if these compositions are combined in similar proportions, the pyrolite composition (minus elements more volatile than sodium) is closely approached (Table 10.5). The principal differences between the mean lunar bulk composition and pyrolite (apart from volatile elements) seems to be that the Moon is significantly richer in FeO.

A plausible explanation for this inferred primary chemical zoning in the Moon is difficult to find. A tentative suggestion relating to this problem is advanced in Chapter 12.

10.4 Further Limits on the Composition of the Lunar Interior

Several authors have proposed models requiring that the bulk composition of the Moon contains much more Al_2O_3 (and CaO) than was deduced in the previous section. Models in this category which have been widely discussed include those of D. L. Anderson (1973) (27.2 percent Al_2O_3); Wänke et al. (1974) (17.4 percent Al_2O_3); Ganapathy and Anders (1974) (11.6 percent Al_2O_3); S. R. Taylor and Jakeš (1974) (8.1 percent Al_2O_3); and S. R. Taylor and Bence (1975) (6.0 percent Al_2O_3). According to most models of this type, the outer regions of the Moon melted and differentiated about 4.6–4.4×10^9 years ago to form a plagioclase-rich lunar crust underlain by a sequence of complementary olivine + pyroxene cumulates, some hundreds of kilometers thick. In some models, melting and differentiation extended throughout the entire Moon. Mare basalts were interpreted as being formed by subsequent partial melting of the cumulates (see, for example, D. Walker et al., 1975a; Taylor and Jakeš, 1974) or by assimilative interactions between the cumulates and the primordial lunar interior (see, for example, Hubbard and Minear, 1975).

Ringwood (1976) carried out a detailed experimental investigation of melting equilibria displayed by the first four of the above compositions over a wide range of pressures and temperatures. A corresponding study of the Taylor–Bence composition was carried out by Kesson and Ringwood (1977). These investigations made it possible to determine the chemical and mineralogical structure of the Moon for each of these bulk compositions under conditions of (a) melting and differentiation of the entire Moon, and (b) melting and differentiation of an outer layer, a few hundred kilometers thick. As an example, the phase equilibria displayed by the Taylor–Jakeš model composition are shown in Figure 10.8. These phase relationships

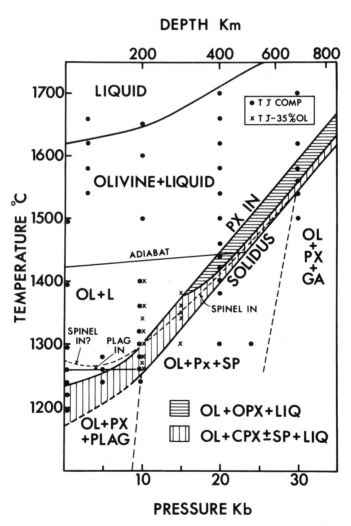

Figure 10.8 High-pressure phase equilibria for the Taylor–Jakeš model lunar bulk composition (from Ringwood, 1976; Fig. 1, with permission).

demonstrate that crystallization would occur from the base upwards, until the stage of plagioclase crystallization, when a surface crust enriched in plagioclase might form. The resultant structures of lunar models, as determined by these phase relationships, are shown in Figure 10.9. The results of these investigations showed unequivocally that mare basalts could not have formed by partial melting of ferromagnesian cumulate layers appropriate to these compositional models.

Several distinct difficulties were encountered by each of these compositional models and the reader is referred to the detailed papers for a full discussion (Ringwood, 1976; Kesson and Ringwood, 1977). One of the most fundamental difficulties was that each of these models necessarily produced mare basalt magmas containing much higher contents of alumina (12–18 percent Al_2O_3) than observed in the more primitive natural samples (~ 8.5 percent Al_2O_3). The only way in which this problem could be alleviated was by reducing the Al_2O_3 content of the bulk composition well below the range investigated. It was concluded that an acceptable compositional model would contain less than 5 percent Al_2O_3.

An analogous difficulty appears in explaining the composition of the pyroxene component of the lunar crust in terms of fractional crystallization of these high Ca–Al compositions. The bulk pyroxenes from highland anorthositic gabbros and gabbroic anorthosites tend to be relatively poor in CaO. The mean normative pyroxenes (diopside + hypersthene) in the Wänke et al. (1975) and S. R. Taylor (1973a) model compositions for the highlands contain only 6 and 3 mole percent, respectively, of CaO.

For the Taylor–Jakeš composition (6.6 percent CaO, 8.1 percent Al_2O_3) the pyroxene(s) crystallizing at the stage of plagioclase saturation contains a bulk average of at least 11 percent CaO, and this increases with further fractionation. It does not seem that models of this kind which postulate extended fractional crystallization prior to, and during plagioclase precipitation, can account for the low mean CaO content of pyroxenes in the lunar highlands. This problem becomes increasingly acute for other bulk composition models even richer in calcium, such as those for Ganapathy and Anders (1974), Wänke et al. (1974), and D. L. Anderson (1973).

All of these models, which invoke extensive fractional crystallization and differentiation of the lunar mantle, encounter another basic problem. The effect of differentiation is to cause increases of MgO/(FeO + MgO) and olivine/pyroxene ratios with depth. This, in turn, leads to seismic velocities that increase substantially with depth. Observations show, however, that P- and S-wave velocities are either constant in the 70–400 km region, or decrease significantly. Below about 400 km there is either a marked decrease in P- and S-wave velocities or continuing strong negative velocity gradients with depth (Fig. 10.1). If the lower mantle was primordial and contained Al_2O_3 contents as high as postulated by the models under discussion, garnet would become an important phase (Ringwood, 1977a, Fig. 3). This would cause a substantial *increase* in seismic velocities, contrary to observation.

Nobody has yet successfully explained how cumulates formed during the

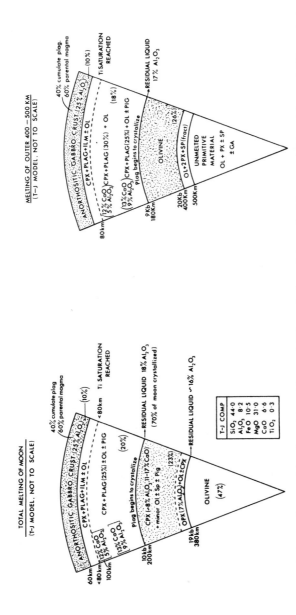

Figure 10.9 Mineralogical zonation resulting from (a) melting and differentiation of the entire Moon, and (b) melting and differentiation of only the outer 400 km. In both cases, the model lunar bulk composition is that of S. R. Taylor and Jakeš (1974). (From Ringwood, 1976; Figs. 3 and 6, with permission).

4.4–4.6 billion-year lunar differentiation could become remelted up to 1.5×10^9 years later, to provide mare basalts. This thermal problem has become greatly exacerbated since the recent downward revision of the lunar heat flow (Langseth et al., 1976).

It should be noted that Wänke and coworkers have recognized some difficulties in their earlier model which postulated a high content of Al_2O_3 in the bulk Moon. They have constructed a new model (Dreibus et al., 1977) which now contains 4.6 percent Al_2O_3, in good agreement with the conclusions of this chapter. The bulk lunar composition according to this new model of Dreibus et al. is given in Table 10.5.

Kaula (1977) has stated that if the lunar crust were to become differentiated from the upper mantle, the bulk system should contain about 7.9 percent Al_2O_3. A related argument has been made by S. R. Taylor and Bence (1975). Kaula's conclusion, however, is dependent upon his assumptions concerning the amount of Al_2O_3 remaining behind in the upper mantle after the crust had differentiated and the total thickness of the system subject to differentiation. The latter was assumed to be 200 km, on the basis of a thermal expansion argument of S. E. Solomon and Chaiken (1976), discussed in the next section. These authors, however, have increased their estimate of the thickness of differentiated regions to 300 ± 100 km (Solomon, personal communication), while the amount of Al_2O_3 in the residual dunitic upper mantle could well be smaller than assumed by Kaula. The differentiation of a 70-km-thick crust of acceptable composition from an outer primordial zone 400 km thick containing about 4 percent of Al_2O_3 is entirely feasible in principle.

The basic conclusion arising from these discussions is that the bulk composition of the Moon contains abundances of CaO and Al_2O_3 that are comparable to those in the Earth's mantle and probably in the range 3.5–5 percent.

10.5 Lunar Thermal Regime

Heat Flow and Radioactivity

The early lunar heat flux measurements obtained at the Apollo 15 and 17 sites (Langseth et al., 1972, 1973) were surprisingly high, and seemed to imply that the Moon possessed more than five times the chondritic abundances of uranium and thorium. Despite the cautious interpretations urged by the above authors, numerous geochemists adopted these values as fundamental boundary conditions and concluded that the bulk Moon was enriched in Ca, Al, and involatile elements to comparable degrees. Although in conflict with the evidence of experimental petrology (Ringwood, 1970c, 1977a), which required lower contents of involatile elements, these models received a wide degree of credence.

A major reduction in estimates of lunar heat flow has since been made on the basis of a much more reliable estimate of the bulk thermal conductivity of the lunar regolith (Langseth et al., 1976). The revised heat flow value at the Apollo 17 site is 1.4 μw/cm^2 and at the Apollo 15 site, 2.1 μw/cm^2. The reduction of these values from earlier estimates, combined with their large difference, emphasizes the need for considerable caution in using arguments based on "mean" lunar heat fluxes. Any such estimates, based on only two measurements in a body possessing the demonstrably complex history and diverse surface chemical heterogeneity of the Moon are necessarily highly model dependent, and likely to possess large uncertainties. The practice of many authors in using these estimates as the key boundary conditions in defining the bulk lunar chemical composition is difficult to justify. Perhaps the most that can be hoped for is to demonstrate that, within their limits of uncertainty, the measurements are *consistent* with given models of lunar bulk composition.

The higher heat flow at the Apollo 15 site was also correlated with much higher than average concentrations of thorium (and, by inference, U and K) than in the Apollo 17 region. A possible interpretation is that magmatic fractionation processes have resulted in strong upward concentration of radioactive elements in the near-surface layer, as has happened in the ter-restrial crust. Thus, the higher heat flow at the Apollo 15 site may be inter-preted as being due to a layer of rocks abnormally rich in radioactive elements (Langseth et al., 1976). This would justify an analysis similar to that which is commonly applied to measurements of terrestrial heat flow and radioactivity, in which heat flow is plotted against surface radioactivity. This is done in Figure 10.10. The slope of the line defined by the two data points defines the thickness of the surface layer of variable radioactivity. On the basis of this model, the mean lunar heat flow could be estimated if the average crustal content of thorium were known, in addition to the areal distribution of thorium in the surface-enriched layer.

Langseth et al. (1976) estimated the above thorium abundances on the basis of the orbital gamma-ray measurements of Metzger et al. (1974). Using these estimates, together with the observed heat flow measurements, they obtained a mean lunar heat flux of 1.8 μw/cm^2 on the basis of a plot similar to Figure 10.10. For a Moon in thermal equilibrium, and taking Th/U = 3.7, K/U = 2500, a mean bulk lunar concentration of 46 ppb U would be required.

This estimate, however, is almost certainly high in the light of the subse-quent, more complete study of thorium distribution by Metzger et al. (1977), who obtained substantially lower mean crustal and surface abundances of thorium. If we take Metzger et al.'s (1977) estimate of 1.3 ppm for the mean *surface* Th abundance, then, on the basis of Figure 10.10, the mean lunar heat flux is 1.3 μw/cm^2, equivalent to 33 ppb of U, for the bulk Moon. (A minimum estimate is provided by the intercept of the line on Figure 10.10 with the mean crustal value of 0.63 ppb. This yields 1.1 μw/cm^2, equivalent to 28 ppb U.) Corrections for the higher heat production of radioactive elements

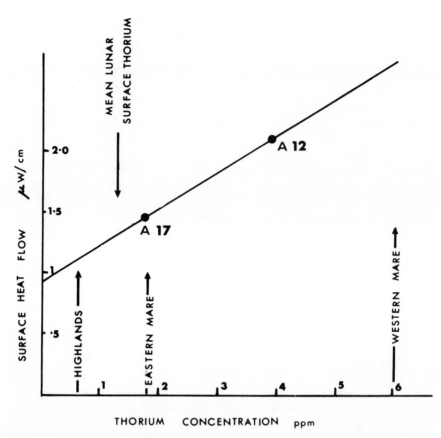

Figure 10.10 Plot of heat flow observed at Apollo 15 and 17 sites versus mean thorium contents estimated for the soil and rocks of these regions based on orbital gamma-ray measurements and laboratory measurement of samples. Based on Langseth et al. (1976). Estimates of Th content in major lunar petrologic provinces (Metzger et al., 1974, 1977) are included.

in the past in relation to lunar thermal inertia (Wänke et al., 1974) could easily reduce the previous "best" estimate of 33 ppb to about 25 ppb U.

In comparison, we have estimated that the bulk Al_2O_3 content of the Moon lies within the limits 3.5–5.0 percent, and is most probably close to 4.2 percent Al_2O_3 (Table 10.5), which is 2.3 times the average Al_2O_3 content of ordinary chondrites. Assuming that the Moon has a similar U/Al ratio to ordinary chondrites and using the abundance data of Mason (1971), the uranium content of the Moon is found to be 28 ppb. Considering all the uncertainties, the agreement is satisfactory. Using different assumptions, Toksöz et al. (1976) and Solomon and Chaiken assume mean lunar U concentrations of 37 ppb and 30 ppb, respectively.

For a bulk uranium content of 25 ppb, a Th/U ratio of 3.8, and a K/U ratio of 2500 applicable to the entire Moon, the radioactive heating would be about 500°C in the first 1.5×10^9 years and ~ 1000°C over 4.5×10^9 years.

Thus, radioactivity alone cannot have been responsible for the early (~ 4.5–4.4×10^9 years) differentiation of the outer Moon, or for subsequent (3.8–3.2×10^9 years) mare volcanism at depth. Additional sources of heat must have been present.

Further Thermal Boundary Conditions

The early ($\sim 4.5 \times 10^9$ years) melting and differentiation of the outer several hundred kilometers of the Moon provides a fundamental boundary condition. It appears that a large source of energy was present, capable of heating the outer regions, but not the deep interior.

There are strong reasons for believing that formation of the mascons was associated either with the major circular basin-forming events on the Moon around 4×10^9 years ago, or with the subsequent filling of the basins by mare basalts, mainly between 3.8 and 3.2×10^9 years ago. It is clear that at the time of mare basalt volcanism, a rigid and thick lithosphere had developed, which was capable of supporting the stresses imposed by the mascons over the past 3×10^9 years. This implies extensive cooling of the outer regions of the Moon over the time interval 4.5–3.0×10^9 years. Kaula (1969) estimated the thickness of the rigid lithosphere at the time of mare volcanism to be about 150 km. On the basis of a detailed study of stresses imposed by the mare Serenitatis mascon, Arkani-Hamed (1974) estimated that the thickness of the lunar lithosphere was about 400 km at the time of formation of the mascon. The stress differences associated with the Serenitatis mascon were estimated to be about 70 bar at the present time.

This is an important constraint, pointing to an ultimate origin of mare basalts at considerable depth within the Moon and below the lithosphere. This conclusion is also in harmony with the results of experimental petrology (see, for example, Ringwood, 1977a).

Although the total timespan was probably greater, it seems that eruption of the bulk of the mare basalts occurred between 3.8 and 3.2×10^9 years ago. This would seem to require that those source regions below the lithosphere were being heated by radioactivity during this time. The sharp decline of volcanic activity around 3.0×10^9 years is also an important constraint on lunar thermal evolution. At present, the high Q values to about 1000 km imply a solid lower mantle, so this region has presumably been subjected to a subsequent cooling regime. However, below about 1000 km, the absence of transmission of S waves suggests the possibility that this region might be partially molten. In view of the inferred chemical zoning of the Moon (Fig. 10.5), the occurrence of moonquakes to a depth of about 1000 km, and the relatively high Q values to this depth, it seems doubtful whether thermal convection is the dominant mode of heat transfer in this region. Below 1000 km, however, it might well be an important process.

Solomon and Chaiken (1976) have established another important boundary condition for lunar thermal evolution. They point out that since the time of emplacement of the maria, physiographic data show that there has been a negligible change in the Moon's radius (≤ 1 km). Thus, the combination of thermal contraction of the outer regions by conductive cooling plus thermal expansion of the deep interior by radiogenic heating have effectively cancelled each other out. Solomon and Chaiken find that this behavior can be explained only if the Moon was first formed with a hot, melted outer zone and a cool, deep interior ($\sim 500°C$). Initially, they placed the thickness of the heated outer zone as 200 ± 100 km, but Solomon (personal communication) now believes that uncertainties in the estimates could permit the thickness of the early melted region to be about 400 km, which is in better agreement with petrologic models (Fig. 10.5). Moreover, Solomon and Chaiken overlooked the contribution of phase changes (from garnet–olivine–pyroxenite to olivine–pyroxenite) on the thermal expansion of the deep interior during heating. This would favor extension of the initially melted region to greater depths than in their original estimate.

Electrical Conductivity

Early estimates of internal temperature distributions in the Moon (see, for example, Sonett et al., 1971) were inverted to obtain a lunar temperature profile using existing experimental data on the relationship between temperature and thermal conductivity in olivine. The temperatures they obtained were remarkably low, reaching only 800°C at a depth of 900 km.

Subsequently, Duba and Ringwood (1973) demonstrated that the electrical conductivities of olivines were extremely sensitive to ambient oxygen fugacity and that this effect required considerable changes in earlier estimates of internal temperature distribution in the Moon. Assuming that the oxygen fugacities of the lunar interior were controlled by the equilibrium of metallic iron with FeO-bearing olivines and pyroxenes, they concluded that temperatures approached the solidus at depths below 500 km. This evidence favored the occurrence of thermal convection as the dominant mode of heat transport below this depth.

Unfortunately, the picture has become confused by later data. In Section 11.6 it is shown that metallic iron is probably not present as a distinct phase in the lower mantle and that oxygen fugacities are probably substantially higher than formerly estimated. This would imply the existence of lower internal temperatures than estimated by Duba and Ringwood. Also, recent experimental work by Huebner et al. (1978) has shown that the electrical conductivity of orthopyroxenes is considerably increased by the presence of species such as Cr^{3+} that are believed to be present. This will also have the effect of reducing the calculated temperatures in the lunar interior, though not to the 1971 values. The lower temperatures now inferred for the lunar lithosphere seem to be more readily in accord with other physical properties, such as its long-term strength (Arkani–Hamed, 1974), high Q values, and occurrence of moonquakes to depths of about 1000 km (Fig. 10.2).

Thermal Evolution

In a remarkable pioneering study of the rheology and thermal state of the Moon, based largely on the profiles of highland impact craters and their estimated relative ages, R. B. Baldwin (1963) showed that the outer layers of the Moon were hot when formed, thereby permitting isostatic rebound of the floors of the oldest generation of craters. However, this was followed by a period of deep-seated cooling during which the strength of the lithosphere increased markedly, thereby preventing isostatic rebound of the later highland craters. He also concluded that the lunar maria were comprised of basaltic-type magmas which had been derived from deep within the lunar interior and flooded preexisting lunar basins long after the latter had been formed during the early lunar highland bombardment. Moreoever, he concluded that the period of mare basaltic volcanism had occurred approximately 10^9 years after the early highland cratering episode. Since these magmas were derived from considerable depths within the lunar interior, below the cooling lithosphere, it followed that this deeper source region must have been heated, presumably by radioactivity, while the lithosphere was cooling.

The thermal implications of Baldwin's conclusions were explored by Ringwood (1966b, 1970c). He pointed out that they could be most readily explained if it were assumed that the Moon had accreted sufficiently quickly to be heated by accretional energy, which would cause high temperatures in the outer regions, whereas the interior would remain relatively cool (Ringwood, 1966b; Fig. 6). The early near-surface heat pulse would then decay by conduction, leading to the formation of a lithosphere which thickened and became stronger with time. Superimposed on this early temperature distribution governed by the energy of accretion was long-lived radioactive heating by U, Th, and K, this being most effective at greater depths within the Moon where the heat could not escape by conduction. Thus, radioactive heating at depths of 300–500 km was believed to have been responsible for the episode of mare volcanism occurring, according to Baldwin, about 10^9 years after formation of the Moon.

The thermal history evidence reviewed earlier in this section is closely consistent with these pre-Apollo interpretations.

CHAPTER 11

Geochemistry of the Moon

11.1 Introduction

In this chapter, discussion of the very broad field of lunar geochemistry is limited to two topics, each of which has a particularly important bearing on the origin of the Moon. The first topic concerns the occurrence of siderophile elements within the Moon. Previously (in Sections 2.4 and 8.4) we described some remarkable aspects of the distribution of siderophile elements in the Earth's mantle and showed that this distribution is uniquely terrestrial, being caused primarily by the process of core formation within the Earth. The low density of the Moon shows that it is strongly depleted in iron as compared to the Earth and other terrestrial planets. It will be important to investigate the effect of this fractionation on the distribution of other siderophile elements. Such an investigation might provide clues as to whether the fractionation of metallic iron occurred within the solar nebula, prior to accretion of the Moon, or within a preexisting planet which was parental to the Moon.

The second topic concerns the abundances and distribution of volatile elements in the Moon. Lunar mare basalts show remarkable depletion patterns for many volatile elements when compared to meteoritic and terrestrial abundance patterns for these elements. The conditions under which these highly specialized depletions occurred clearly provide clues to the chemical environment in which the source material of mare basalts was formed. The volatile element geochemistry of the lunar highlands is found to be yet more complex, requiring the operation of additional endogenic and exogenic processes.

11.2 Siderophile Elements in Lunar-Mare and Terrestrial Basalts

When Apollo 11 basalts were first analyzed, they were found to be strongly depleted in many siderophile elements, compared to terrestrial basalts. This characteristic was generally interpreted as being due to crystallization and segregation of a metallic iron phase, either in the source region or during ascent of the magma. Many siderophile elements were also found to be strongly depleted in Apollo 12, 15, and 17 basalts, and most workers adopted the same explanation.

Ringwood and Kesson (1976, 1977) emphasized the significance to lunar geochemistry of some basic concepts which had long been familiar in terrestrial petrology. In order to establish the chemical characteristics of the sources for lunar and terrestrial basalts, it is necessary to select classes of basalts which carry the most direct chemical "memory" of their respective primordial source regions and which also have experienced the most "simple" petrogenetic evolution. These authors pointed out that lunar high-Ti basalts had experienced highly complex petrogenetic histories and were unsatisfactory for this purpose. It would be more appropriate to employ low-Ti basalts, since their petrogenetic evolution had been comparatively simpler (Section 10.3). In the case of the Earth, oceanic tholeiites were chosen for comparison, in view of their wide distribution and the evidence that they had formed by comparable degrees of partial melting to low-Ti basalts (Ringwood and Kesson, 1977).

Nickel and Cobalt

These two key siderophile elements merit a detailed discussion. In comparing their distributions in low-Ti mare basalts and terrestrial oceanic tholeiites, two factors must be born in mind. The first is that both classes of magmas have been subjected to varying degrees of fractional crystallization of olivine (and other minerals) since leaving their source regions. This process, particularly olivine separation, has a drastic effect on the abundances of nickel and, to a lesser extent, cobalt, because of high olivine/magma partition coefficients for these elements. The second factor is that the data base for low-Ti mare basalts is very much smaller than for oceanic tholeiites. The sampling of the latter on the Earth has been reasonably representative. This is not the case for low-Ti basalts on the Moon.

The relationships between Ni contents of terrestrial ocean floor tholeiites and low-Ti mare basalts and their corresponding $MgO/(MgO + FeO)$ ratios are shown in Figure 11.1. The solid lines delineate the field occupied by oceanic tholeiites. The observed relationship is caused by near-surface fractional crystallization, particularly of olivine, and the curvature is caused by the fact that the crystal/liquid partition coefficients for Ni in olivine and pyroxenes are much greater than for Mg. It is seen from Figure 11.1 that the corresponding points for Apollo 12 basalts fall squarely in the middle of the

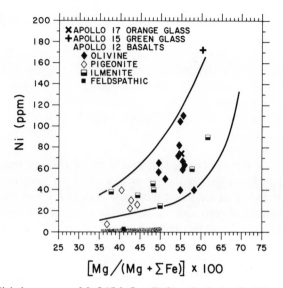

Figure 11.1 Nickel contents vs MgO/(MgO + FeO) ratios in Apollo 12 basalts compared with corresponding field observed for terrestrial ocean-floor basalts. The boundaries of the latter are shown by the solid lines. The corresponding field for high-Ti basalts is shown in the stippled region. (From Delano and Ringwood, 1978b, with permission.)

terrestrial field and display a similar overall trend for Ni to fall as the Mg number decreases. (Some of the scatter in the lunar rocks probably results from near-surface reduction and limited degrees of metal segregation, as discussed later in this chapter). The vast majority of Apollo 15 analyses would also plot within the terrestrial field, although they tend to have higher Ni contents than Apollo 12 mare basalts for the same MgO/(MgO + FeO) ratio (Ringwood and Kesson, 1977; Delano and Ringwood, 1978b).

Many Apollo 12 and 15 basalts are known to have experienced substantial degrees of near-surface olivine fractionation (see, for example, Compston et al., 1971; Chappell and Green, 1973). There can be little doubt that the Ni vs MgO/(MgO + FeO) trend for Apollo 12 basalts shown in Figure 11.1 is caused by this factor. *The correspondence in Ni fractionation behavior in terrestrial and lunar basalt systems (Fig. 11.1) strongly suggests that both possess generally similar amounts of nickel at equivalent stages of differentiation.* This conclusion is supported by comparison of the *mean* Ni contents of oceanic tholeiites and low-Ti mare basalts. Delano and Ringwood (1978b) reviewed available data and obtained a mean value of 75 ppm, Ni in ocean-floor basalts corresponding to a mean MgO/(MgO + FeO) ratio of 0.55, as compared to a mean of 59 ppm Ni in low-Ti mare basalts (mean MgO/(MgO + FeO) ratio of 0.48).

It is important to compare the Ni content of parental, primary low-Ti mare basalts (prior to olivine fractionation) with that of primary oceanic tholeiite magma. Terrestrial ocean-floor tholeiites containing more than 250 ppm Ni are comparatively rare. The abundances of Ni in ocean-floor tholeiites tend to level off at about 200 ppm, corresponding to MgO/(MgO + FeO) ratios of 0.68–0.70 (Kay et al., 1970). The most primitive crystalline low-Ti basalts contain about 110 ppm Ni, while Apollo 15 Green Glass contains 180 ppm Ni. However, the total number of low-Ti basalts sampled is very small. The low viscosities of mare basalt magmas and the long distances travelled by individual flows are very conducive to olivine fractionation, and it seems quite likely that all of the small population so far recovered may have been subjected in some degree to this process. Assuming that the most primitive mare basalts had lost only 5 percent of olivine, the parental magmas must have contained about 150 ppm Ni. This is the basis of the estimate in Table 11.1.

In the light of the preceding discussion, it seems reasonable to conclude that the Ni abundances in the least-fractionated, *parental* low-Ti basalts and primitive terrestrial ocean-floor basalts were generally similar (within a factor of about 2), implying correspondingly similar relative abundances of Ni in the olivine and pyroxene components of their respective source regions.

Note also, in Figure 11.1, the very low Ni abundances of high-Ti basalts caused by their more complex fractionation history, which involved saturation with an iron-rich metallic phase[1] (unlike low-Ti basalts—see Ringwood and Kesson, 1977).

Finally, the relationship between Ni contents and MgO/(MgO + FeO) ratios of Apollo 12 basalts (Fig. 11.1) strongly implies that nickel was present in the parental magma as an *oxidized species*, which permitted it to be fractionated by olivine. The high nickel content of Apollo 15 Green Glass has also been shown to be present as NiO (Ringwood and Kesson, unpublished measurements). The metallic phase now present in Apollo 12 basalts was evidently produced by post-eruption reduction processes (Reid et al., 1970; Brett et al., 1971; Hewins and Goldstein, 1974). We will return to this topic in Section 11.6.

In Figure 11.2, the cobalt abundances of Apollo 12 basalts are plotted against MgO/(MgO + FeO) ratios. The decrease of cobalt content with decreasing MgO/(MgO + FeO) ratio is caused dominantly by the effects of olivine crystallization. The linearity of the relationship arises from the similar crystal/liquid partition coefficients of cobalt and magnesium. The existence of this relationship also demonstrates that cobalt was present in the parental magma prior to eruption as an oxidized species, Co^{2+}, capable of entering the olivine structure.

The mean cobalt abundances in low-Ti mare basalts (41 ppm) and in terrestrial oceanic tholeiites (45 ppm) are practically identical (Table 11.1) and this is probably true of their respective source regions in the lunar and terrestrial mantles.

---

[1] Probably a molten FeS–Fe alloy.

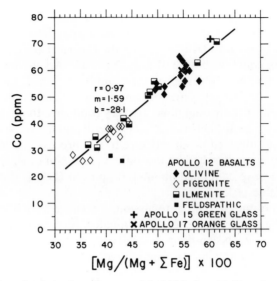

Figure 11.2 Plot of cobalt abundances vs MgO/(MgO + FeO) ratios for Apollo 12 basalts; also for Apollo 15 Green Glass and Apollo 17 Orange Glass. (From Delano and Ringwood, 1978b, with permission.)

Siderophile Elements: General Discussion

We have discussed the cases of Ni and Co in some detail because they are key siderophile elements and because any interpretation of their abundances in basalts may be greatly complicated by the effects of near-surface olivine fractionation. Fortunately, this process does not have a comparable effect on the abundances of most other siderophile elements. Available data on the abundances of many siderophile elements in low-Ti mare basalts and in ocean-floor tholeiites have been compiled by Ringwood and Kesson (1977), and are set out in Table 11.1.

It is seen that the siderophile elements fall into two clearly defined groups. The first group, including Fe, Ni, Co, W, Os, Ir, P, S, and Se have similar (i.e., within about × 2) abundances in lunar and terrestrial basalts, whereas the second group—Cu, Ga, Ge, As, Re, Ag, Sb, and Au—are greatly depleted by factors of 5–500 in lunar basalts. All the members of the second group, except Re,[2] are relatively volatile elements and can be demonstrated to have

---

[2] Rhenium is usually regarded as a highly involatile, siderophile element. However, the free energy of formation of ReO_2 implies that the siderophile behavior of Re more closely resembles that of Ni than Ir. Since it is demonstrated in this chapter that nickel occurs as an oxidized species in parental mare basalts, it seems possible that the rhenium may also be present as Re^{4+}. In this state, rhenium forms a highly stable oxide, ReO_2, which sublimes at 1363°C and 1 atm., and is, therefore, much more volatile than gold, which boils at 2660°C. The large depletion of Re in mare basalts compared to terrestrial basalts may accordingly be due to the volatility of ReO_2. Thus, the depletion of Re may be a particular example of the general depletion of volatile elements known to have affected the Moon.

Table 11.1 Estimated abundances of some siderophile elements in low-Ti mare basalts and ocean-floor tholeiites. (From Ringwood and Kesson, 1977; with modifications by Delano and Ringwood, 1978a). (Values are in ppm except where indicated otherwise.)

	Ocean-floor tholeiite	Low-Ti mare basalt	Terrestrial basalt / Lunar basalt
Group I			
Fe	6.3 (%)	15.5 (%)	0.4
Ni	250	150	1.7
Co	41	45	0.9
W	0.08	0.14	0.6
Os	0.3 ppb	0.5 ppb	0.6
Ir	0.04 ppb	0.07 ppb	0.6
P	300	380	0.8
S	900	1150	0.8
Se	0.17	0.14	1.2
Group II			
Cu	70	12	6
Ga	20	4	5
Ge	1.5	0.007	214
As	1	0.006	167
Re[a]	0.8 ppb	0.01 ppb	80
Ag	30 ppb	1 ppb	30
Sb	29 ppb	0.06 ppb	500
Au	1 ppb	0.03 ppb	33

[a] See footnote on page 202 for discussion of the behavior of rhenium.

displayed this property in the lunar environment (Section 11.7, Table 11.4). It is seen in Figures 11.12 and 11.13 that volatile elements in the Moon have been generally depleted, often by very large factors when compared to the Earth and Cl chondrites. Accordingly, Ringwood and Kesson concluded that the depletions of Group II siderophile elements were probably caused by their volatility and were not related to their siderophile characteristics.

The siderophile elements of Group I, on the other hand, are relatively *involatile*. This characteristic is shared by S and Se in an environment characterized by a low P_{H_2}/P_{H_2S} ratio (Ringwood, 1977a). Since the overall partial melting equilibria and other processes by which both terrestrial and lunar basalts were formed are believed to have been generally similar, it seems likely that the similarities in Group I siderophile element abundances extends also to the *source regions* of the basalt classes in the respective terrestrial and lunar mantles.

We have already noted that the abundances of Group I siderophile elements in the Earth's mantle were established by processes *unique to the*

Earth and connected with core formation (Sections 2.4 and 8.4). These processes could not possibly have operated within the Moon, a body in which the core, *if present*, amounts to less than 2 percent of the lunar mass, and which, if present, must have formed under very different conditions from the Earth's core, and in a pressure field extending to a maximum of only 47 kilobars as compared to 3.6 megabars for the Earth. Moreover, it will be shown in Section 11.6 that low-Ti mare basalts were not saturated with an Fe-rich metallic phase in their source region, so that the observed siderophile element abundances cannot be connected with the separation of such a phase. Ringwood and Kesson concluded, therefore, that the similarity in Group I siderophile element abundances between the Moon and the Earth's mantle implies that *the Moon was derived from the Earth's mantle after the Earth's core had segregated.*

11.3 Tungsten and Phosphorus

These elements are discussed separately, since data on their distribution in the Moon are unusually complete and, moreover, relate not only to mare basalts, but also to the highland crust system. Wänke et al. (1977) have demonstrated a strong covariance between La and W which applies to mare and highland samples (Fig. 11.3). The covariance over a large (× 30) range in absolute abundances results from the fact that both La and W are behaving as incompatible elements during lunar igneous fractionation processes. The data establish a well-defined W/La ratio of 0.019, which seems to be applicable

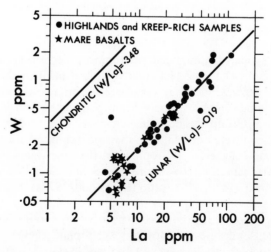

Figure 11.3 Tungsten vs lanthanum systematics in mare and highland samples. Data from Wänke et al. (1977). This paper gives full references to earlier work by this group. Data have been replotted by Delano and Ringwood (1978a); from which this diagram was taken (with permission).

to the entire Moon. This may be compared with the Cl chondritic W/La ratio of 0.35, which is higher by a factor of 19. It is a remarkable fact that the W/La ratio in terrestrial rocks is similar to the lunar ratio (Wänke et al. 1977). Moreover, the absolute abundances of W are similar in terrestrial oceanic tholeiites, low-Ti mare basalts, and the parental magma of the lunar highlands (Table 11-3).

Tungsten is one of the "high-temperature condensate" group (Table 6.2), so that the La/W ratio in the bulk Moon, bulk Earth, and Cl chondrites is expected to have been similar. The depletion of W relative to La in the Earth's mantle is clearly due to its siderophile properties, which have caused its preferential entry into the Earth's core (Wänke et al., 1977).

A detailed experimental investigation of the partition of W between basalt magmas and Fe–Ni alloys as a function of temperature, metal composition, and oxygen fugacity was carried out by Rammensee and Wänke (1977). They showed that the metal–silicate partition coefficient was very sensitive to changes in these variables, increasing strongly with temperature (Fig. 11.4) and decreasing drastically with increasing oxygen fugacity and nickel content (Fig. 11.5) of metal phase (the latter variables are correlated). In the light of these observations, Rammensee and Wänke pointed out that the similarity of W/La ratios and of absolute W abundances in the Moon and Earth's mantle was most remarkable.

The argument was extended still further. From the measured partition coefficients at an appropriate temperature ($\sim 1300°C$), Rammensee and Wänke demonstrated that equilibration with 26 percent of metallic iron

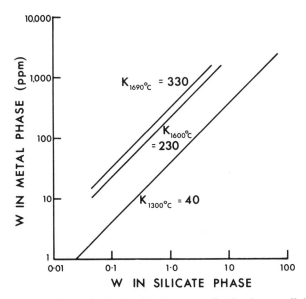

Figure 11.4 Experimental relationship between distribution coefficients (K) for W between metallic iron and a basaltic silicate melt as a function of temperature. Based on Rammensee and Wänke 1977.

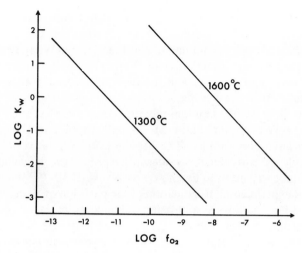

Figure 11.5 Relationships between log distribution coefficients for W between metal and basaltic silicate melt as a function of oxygen fugacity at 1300 and 1600°C. (Based on Rammensee and Wänke, 1977.)

would be needed to reduce the W abundance and W/La ratios by the observed amounts. This is a proportion of metal comparable to the Earth's core. However, the *maximum* size of a lunar core is only 2 percent of the mass of the Moon (Section 10.1). A small core of this size could not possibly account for the observed depletion. The discrepancy exceeds a factor of 10.

Indeed, the real discrepancy is much greater than this. The covariance of La and W in Figure 11.3 shows that W was present in the Moon as an *oxidized species*, not as a metal. We will find in Section 11.6 that the source regions of low-Ti basalts and the parental lunar crust magma were not saturated with an iron-rich phase. These systems evolved under conditions of higher oxygen fugacity than would be maintained by equilibrium with an iron-rich metal phase. There is no conclusive evidence for the existence of *any* kind of metallic core (Section 10.1). However, *if* present, it would have to consist of a nickel-rich alloy, which would be consistent with the observed oxygen fugacities of the parental mare and highland systems (Section 11.6). Rammensee and Wänke's data (Fig. 11.5) show that the metal–silicate partition coefficient for W under these conditions would be reduced by a factor of over 10. Thus, a lunar metallic core of the maximum permissible mass would fail to explain the depletion of W by a factor exceeding 100.

Some workers (e.g., Brett, 1973) have suggested that a lunar core (if present) may consist of FeS rather than Ni–Fe metal. The partition coefficient for W into FeS is even smaller than for metallic Ni–Fe (Rammensee and Wänke, 1977), so this proposal is not viable.

The conclusion seems inescapable that the depletion of W in the Moon was not caused by any process of metal segregation within the Moon, but must have been characteristic of the material from which the Moon accreted. It does not seem possible that this depletion occurred within the parental solar nebula prior to accretion of the Moon. Rammensee and Wänke point out that tungsten condenses completely with the high-temperature condensate fraction (Table 6.2) and this indeed is demonstrated by its occurrence in primordial proportions in the Ca–Al–rich inclusions of the Allende meteorite. Metallic nickel–iron condensing at lower temperatures contains practically no tungsten. Because of its exceedingly low vapor pressure, it does not seem possible for tungsten to migrate to dispersed grains of the metallic Ni–Fe phase at the lower temperatures ($< 1100°C$—Table 6.2), at which this latter phase condenses from the nebula. Moreover, because of the decrease of partition coefficient with falling temperature (Fig. 11.4), the observed W depletion would require transport to, and equilibration with, a much larger mass of Ni–Fe metal relative to silicates than is present in a condensed medium of solar composition.

Rammensee and Wänke conclude that the observed lunar depletion of W must have occurred via the separation of a large (~ 26 percent) iron-rich core within another planetary body, and that the material now comprising the Moon must then have been derived from the mantle of this differentiated planetary body. The similarity of W/La ratios in the Earth's mantle and in the Moon strongly suggests that the parental mantle was, in fact, that of the Earth. An alternative possibility, that the differentiation occurred within an earlier generation of parent bodies similar to those from which eucritic meteorites are derived, is considered subsequently in this section and rejected.

Phosphorus

As pointed out by Dreibus et al. (1976), the behavior of phosphorus in lunar and terrestrial systems resembles that of tungsten. Figure 11.6 illustrates the systematic variation of P and La in mare and highland samples. The La/P ratio varies only between 0.01 and 0.03 through a large (factor of 100) range in absolute abundance of La. Thus P, like La, is behaving essentially as an incompatible element throughout the petrological fractionations that occurred in the Moon. Incidentally, this implies that P was present dominantly as an oxidized species throughout these differentiations (Section 11.6). Figure 11.6 shows that P has been depleted (relatively to La) by a factor of about 90, as compared to Cl chondrites. It shows also that terrestrial oceanic tholeiites and alkali basalts possess La/P ratios within a factor of 2 of the lunar values. The similarity in absolute abundances of P in terrestrial basalts, low-Ti mare basalts, and the parental magma of the lunar crust is shown in Table 11.3.

Phosphorus is a relatively involatile element in the solar nebula, condensing as Fe_3P at a temperature of $1200°C$ compared to $1290°C$ for metallic Ni–Fe alloy (Table 6.2). Indeed, it seems that substantial condensation may occur at even higher temperatures when solution of P in metal is considered (Olsen et al., 1973). It is most improbable, therefore, that the gross (but similar) depletions of P in the terrestrial mantle and Moon are due to volatility.

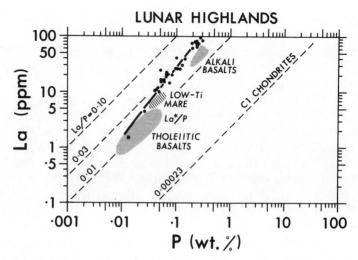

Figure 11.6 Phosphorus versus lanthanum systematics for lunar highland samples, low-Ti mare basalts, terrestrial tholeiitic basalts, and terrestrial alkali basalts. (From Delano and Ringwood, 1978a, with permission.)

Almost certainly, the depletion of P in the Earth's mantle is caused by its siderophile nature at high temperatures. Ringwood and Kesson (unpublished experiments) have shown that the partition coefficient for P between metallic iron and a basaltic melt at 1300°C exceeds 20.

As in the case of tungsten, phosphorus becomes more lithophile as temperature decreases and is dominantly lithophile below 1100°C (Friel et al., 1976). Although the relevant partition coefficient data are not as complete as in the case of tungsten, it seems impossible to account for the similarities in P abundances in the Earth's mantle and the Moon in terms of differentiation of metallic cores separately within each of these bodies. This follows from the very different relative sizes of possible cores (a factor exceeding 15), the sensitivity of the partition equilibria to oxygen fugacity, and the evidence that the mare basalt and highland crust systems of the Moon have never been in equilibrium with an iron-rich metallic phase (Section 11.6). Note also that P is not a particularly chalcophile element, so that the deficiency in the Moon cannot be ascribed to entry into an FeS core.

11.4 Eucritic Parent Body

We conclude, with Dreibus et al. (1977), Rammensee and Wänke (1977), and Delano and Ringwood (1978a,b) that the lunar P and W abundances were not established by metal/silicate fractionation within the Moon, but were apparently produced within another body that had segregated a proportionally large, Fe-rich core amounting to about 26 weight-percent of the planet. It has been suggested above that the Earth was the parent body. However, it has also been suggested (see, for example, Kaula, 1974) that the Moon was derived from differentiated planetesimals analogous to the bodies on which some of the achondritic, stony–iron, and iron meteorites were formed. In this model, metal/silicate differentiation occurred within the bodies, causing extraction of siderophile elements from the silicate-rich portions. Subsequently, in a complex scenario, the silicate-rich portions of these parent bodies became physically separated from the metallic cores and the former accreted in geocentric orbit to form the Moon.

There are some intriguing chemical similarities between low-Ti mare basalts and eucrites, as well as some important differences. Eucrites possess generally similar abundances of the major elements and some siderophile elements, including W and P, relative to low-Ti mare basalts, and were evidently formed by melting and differentiation of a small "chondritic-type" parent body which was strongly depleted in volatile elements (see, for example, Duke and Silver, 1967; Stolper, 1977). While these similarities provide some support to models relating low-Ti mare basalts to eucrites, the profound chemical differences discussed below show that the Moon cannot have formed from such material.

The crucial difference between the eucrites and the low-Ti mare basalts is that the eucritic melts were saturated with respect to an Fe-rich, metallic phase (Stolper, 1977), whereas the low-Ti basalts (and PLC magma) were not (Section 11.6). This important distinction between these two systems is amply demonstrated through comparison of their Ni, Co, Cu, Ga, Ir, and Os abundances (Table 11.2). The strong depletions of these elements in eucrites relative to low-Ti mare basalts is clear evidence that these basaltic melts were derived from fundamentally different systems. The eucrites equilibrated with about 20 weight-percent (Rammensee and Wänke, 1977; Delano and Ringwood, 1978a) of an Fe-rich, Ni-poor metallic phase (Stolper, 1977) during

Table 11.2 Comparison of some siderophile element abundances in the parental lunar crust (PLC) magma, low-Ti mare basalts, and eucrites. (From Delano and Ringwood, 1978a,b.)

	PLC magma	Low-Ti mare basalt	Eucrites
Ni (ppm)	200	150[a]	3.5
Co (ppm)	30	45	8
Cu (ppm)	10	12	4.7
Ga (ppm)	4.6	4	1.4
Ir (ppb)	unknown	0.07	0.004
Os (ppb)	unknown	0.5	0.02

[a] This is the estimated nickel content of a primary magma which has not undergone olivine fractionation en route to surface. The mean nickel content of existing fractionated low-Ti mare basalts is 59 ppm (Section 11.2).

differentiation of their parent body. In contrast with the eucritic parent body; this process did not operate within the Moon.

A second important difference between the eucrites and the Moon is their oxygen isotopic composition. Although both systems lie on the terrestrial mass fractionation line, thereby indicating that the Earth, Moon, and eucrites came from a single, well-mixed reservoir with respect to ^{16}O content, it is important to appreciate that the Moon and eucrites do not have the same oxygen isotopic composition. Whereas the Moon has $\delta^{18}O = 5.5 \pm 0.2^0/_{00}$ (Clayton and Mayeda, 1975), the eucrites have values typically from $+3.0$ to $+3.5^0/_{00}$ (Clayton et al., 1976). Since this difference is too large to have resulted from igneous processes in either the Moon or the eucritic parent body, Clayton and Mayeda (1975) concluded that the eucrites observed today cannot be directly related to the Moon. For comparison, the Earth has a value between $+4.5$ and $+6.0$ (H. P. Taylor, 1968; A. T. Anderson et al., 1971; Onuma et al., 1972, which is indistinguishable from the lunar value.

The differences, therefore, between the eucrites and low-Ti mare basalts in siderophile element abundances, oxygen fugacity, and oxygen isotopic composition indicate that the Moon's distinctive chemistry is not a consequence of an earlier evolution within a generation of planetesimals similar to the eucritic parent body.

Note added in press

Stolper (1979) has pointed out that the abundances of a group of 9 siderophile elements in shergottite meteorites (a very rare class consisting of two falls) are similar (within a factor of ten) with those in terrestrial basalts. It should be noted, however, that some of the differences are real. Thus, the P/Sm and P/Ti ratios of shergottites are 8 to 9 times higher than those of terrestrial oceanic tholeiites. It will be interesting to see whether W is also comparatively more abundant in shergottites. The oxidation state of these meteorites (quartz-fayalite-magnettite buffer) is also similar to, but somewhat higher than that of the Earth's mantle.

Stolper's observation is interesting since it demonstrates that another independent body in the solar system evolved geochemically in a manner approximately similar (but not identical) to the Earth's mantle. To that extent, it weakens the conclusion expressed elsewhere in this book that the siderophile element abundance pattern of the Earth's mantle was established by processes *unique to the Earth*. Contrary to Stolper's interpretation, however, it does not weaken the conclusion expressed herein that the similarity in siderophile element abundance patterns (combined with other geochemical similarities) implies a genetic relationship between Earth and Moon.

The composition of the shergottites can be explained if they differentiated within a small parent body containing the cosmic abundances of Fe, Mg, Si, Ca and Al, somewhat similar to the parent body of eucrites but richer in volatiles and more highly oxidized. In fact, the oxidation state must have been such that metallic iron was not present as a separate phase, all the iron occurring as Fe_3O_4 plus lesser FeS, as proposed for the bulk compositional model of Mars (Fig. 9.1). Thus the siderophile element abundance pattern of the mantle was established by equilibrium between an Fe–O–S melt and silicates under a relatively high oxygen fugacity. The Fe–O–S liquid presumably segregated to form a core (e.g. Fig, 9.1).

In the case of the Earth, it is also believed that the siderophile element abundance pattern in the mantle is a consequence of separation of an Fe–O–(S) alloy to form the core under conditions of high oxygen fugacity. However, in this case, entry of oxygen into the metal was a consequence of solubility of FeO in Fe at high pressures.

The study of lunar siderophile element abundances and original oxidation state as described in this chapter shows that the Moon could have been derived from the mantle of a large parent body which had differentiated under high oxygen fugacity. As discussed in this book the only available parent body is the Earth. The Moon *could not* have been derived by processes of disintegrative capture from the mantles of a hypothetical population of former small differentiated shergottite parent bodies because of their divergent oxygen isotope compositions, their grossly different volatile element abundance patterns and also because of the compelling arguments against disintegrative capture given on pages 241 and 250 (Chapter 12).

A more speculative possibility which cannot, however, be dismissed is that the shergottites were derived from a former moon of Venus, which had originated in a manner analogous to Earth's Moon (Chapter 12).

11.5 Indigenous Siderophile Component of the Lunar Highlands

In Table 10.2, a strong resemblance between the major element compositions of the parental lunar highland (PLC) magma and terrestrial ocean floor tholeiites was demonstrated. This similarity extended to minor element abundances such as the rare earths. In view of these observations, it seemed tempting to explore the relationships between siderophile element abundances in the PLC magma and terrestrial tholeiites.

Interpretation of siderophile element abundance patterns in the lunar highlands is greatly complicated by the widespread occurrence of contamination resulting from the intense meteoritic bombardment experienced by these regions early in lunar history. For some years, the prevailing view (see, for example, Anders et al., 1973; Hertogen et al., 1977), was that the vast bulk of siderophiles in the lunar highlands were of meteoritic origin and that any indigenous component was very, very small. On the other hand, Wänke et al. (1975, 1976, 1977) maintained that most of the Ni in the lunar highlands was accreted as part of a "primary component" and is therefore indigenous. Although the presence of an important meteoritic component is recognized, this model implies that significant indigenous abundances of other siderophile elements may exist in the lunar highlands. Similarly, Jovanovic and

Reed (1976, 1977) argued that significant indigenous abundances of Os and Ru may exist in the lunar highlands.

A detailed study of this topic was carried out by Delano and Ringwood (1978a,b), who concluded that the highlands indeed contained important indigenous concentrations of many siderophile elements. Their results are reviewed in this and the following sections. At the Ninth Lunar Science Meeting in March 1978, these conclusions were contested by Anders on behalf of the Chicago School. I believe that the Chicago position on this issue is incorrect. A comprehensive account of the reasons for this opinion is given in Delano and Ringwood (1978b).

On the basis of extensive trace-element data on highland rocks, Baedecker et al. (1973) and Gros et al. (1976) concluded that, *to a first approximation*, the meteoritic contamination component has a generally "ordinary chondritic" nature in the sense that it contains roughly primordial abundances of many cosmochemically involatile elements, but is systematically depleted to varying degrees for a wide range of volatile elements. In an attempt to test a very simple model, Delano and Ringwood (1978a,b) assumed that the composition of the meteoritic component in the lunar highlands corresponded specifically to that of H-group ordinary chondrites. It should be noted, however, that their overall conclusions do not depend strictly upon this assumption. Similar results would have been obtained had it been assumed that the contamination had a terrestrial (i.e., whole Earth) abundance pattern, which is more strongly depleted in many volatile elements than H-group chondrites.

It was assumed, furthermore, that all of the iridium in the lunar highlands was supplied by meteoritic contamination. Justification for this assumption is discussed by Delano and Ringwood (1978a,b). The chondritic contamination components for the elements studied were then simply subtracted from the observed concentrations by using the H-chondritic ratios of the given elements to iridium. Note that if there were any indigenous lunar iridium present in addition to the meteoritic contamination, this model would yield *minimum* values for abundances of indigenous lunar siderophiles. A further correction (generally small) was made for the presence of cumulate plagioclase, so that any siderophile residuals found could then be attributed to the PLC[3] magma (Table 10.2).

In view of persuasive evidence that individual (~ 1 g) lunar samples cannot be regarded as having been closed chemical systems during the energetic interactions of hypervelocity meteorites with lunar surface materials, Delano and Ringwood selected abundance data on as many highlands rock and soil samples as possible from the literature in order to maximise the size of the chemical system. The reported abundances in the PLC magma were thus based on *averages* of residuals that are believed to have comprised a sufficiently large chemical system to have remained effectively closed during meteoritic impact processes.

---

[3] The magma parental to the lunar crust.

Comparison of PLC Siderophile Residuals with Mare Basalts

Results of the above exercise are given in Table 11.3. Note that the residuals were in all cases positive and of substantial magnitude and are interpreted as representing indigenous lunar siderophiles. Data are also given on a few lithophile elements possessing varying volatilities for purposes of comparison.

The highland siderophile (and lithophile) data of Table 11.3 are plotted against corresponding abundances in low-Ti mare basalts in Figure 11.7. It is seen that the abundances of the siderophile elements W, Ni, Co, P, S, Se, Cu, and Ga are very similar in low-Ti mare basalts and in the PLC magma. Since the abundances of these elements in low-Ti mare basalts are undoubtedly indigenous to the Moon, the agreement provides powerful support for

Table 11.3 Abundances of siderophile and selected lithophile elements in the PLC magma, low-Ti mare basalts, and terrestrial oceanic tholeiites. The elements in each set have been arranged in a sequence of relative volatility. (From Delano and Ringwood, 1978b.)

	PLC magma	Low-Ti mare basalts	Terrestrial oceanic tholeiites
Siderophile Elements			
W	63 ppb	140 ppb	80 ppb
Ni	200 ppm	150 ppm	250 ppm
Co	30 ppm	45 ppm	41 ppm
P	220 ppm	380 ppm	300 ppm
Au	~0.7 ppb[a]	0.03 ppb	~1 ppb
S	700 ppm	1150 ppm	900 ppm
Se	180 ppb	140 ppb	170 ppb
As	200 ppb[a]	6 ppb	1000 ppb
Cu	10 ppm	12 ppm	70 ppm
Ga	4.6 ppm	4 ppm	20 ppm
Ag	4 ppb[a]	1 ppb	30 ppb
Sb	2 ppb[a]	0.06 ppb	29 ppb
Ge	400 ppb[a]	7 ppb	1500 ppb
Cd	110 ppb[a]	2 ppb	140 ppb
Lithophile Elements			
Li	8 ppm	6 ppm	19 ppm
Mn	730 ppm	2090 ppm	1360 ppm
K	230 ppm	250 ppm	1500 ppm
Na	3100 ppm	1800 ppm	19440 ppm
Zn	11 ppm[a]	1 ppm	100 ppm

[a] See Section 11.8 for a detailed interpretation of these abundances. It is not believed that they are necessarily representative of the PLC magma, since these elements have been concentrated at the Apollo 16 landing site by volatile-transport processes (e.g., fumarolic activity).

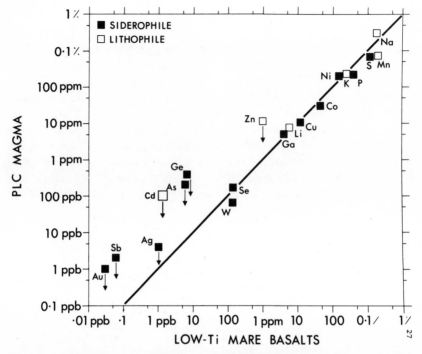

Figure 11.7 Comparison of siderophile and lithophile element abundances in the PLC magma and low-Ti mare basalts. Elements falling on the diagonal line have similar abundances in both systems. Siderophile and lithophile elements are plotted as closed and open symbols, respectively. The downward-pointing arrows associated with Zn, Ge, Ag, Sb, Au, and As signify that these elements have been found to be associated with a fumarolic component on the surface of Apollo 15 Green Glass and Apollo 17 Orange Glass. The presence of this fumarolic component in the Apollo 16 highlands may be the cause of this disparity between the PLC magma and low-Ti mare basalts. (From Delano and Ringwood, 1978b, with permission.)

Delano and Ringwood's conclusion that the siderophile element abundances so obtained for the PLC magma are also genuinely indigenous. Moreover, the consistency supports the general validity of the very simple model employed to deduce these abundances.

Table 11.3 and Figure 11.7 seem to imply that Au, As, Sb, Ge, Cd, and Zn are strongly enriched in the PLC magma as compared to low-Ti mare basalts, while Ag is moderately enriched ($\times 4$). The key factor relating these elements is their volatility under lunar conditions as demonstrated by their exceptionally high abundances on the surfaces of Green and Orange Glasses (Table 11.4), and concentration in the finest fractions of lunar soil. This behavior is discussed in greater detail in Section 11.8.

Krähenbühl et al. (1973) have shown that many Apollo 16 rocks contain anomalously high abundances of volatile elements that are believed to have been introduced by volcanic and/or fumarolic processes. It seems very likely that the "overabundance" of Zn, Cd, Ge, As, Ag, Sb, and Au in the PLC magma is due to these processes, perhaps supplemented by upward concentration caused by meteoritic impact remobilization (Wänke, personal communication). There do not appear to be any compelling reasons for suggesting that the PLC magma was actually enriched in Zn, Cd, Ge, As, Ag, Sb, and Au relative to low-Ti mare basalts, although the possibility cannot be totally eliminated (see Section 11.8).

Ni/Co Systematics

When plotted on a Ni/Co vs Ni diagram, the Ni and Co abundances in highland rocks and soils from Apollo 16 (Fig. 11.8) and from Apollo 14, 15, and 17 samples provide strong evidence for a mixing model involving two sources for these elements (Delano and Ringwood, 1978a,b). One component is characterized by high Ni (≥ 800 ppm) and a primordial Ni/Co ratio of 19. Since all major varieties of meteorites possess high Ni abundances and near-primordial Ni/Co ratios, this component in Figure 11.8 could well be of meteoritic origin. The second component is characterized by much lower Ni abundances and much lower Ni/Co ratios. Note that the Ni/Co vs Ni systematics of the highlands samples project directly towards and intersect the corresponding Ni/Co vs Ni field occupied by low-Ti mare basalts (Fig. 11.8).

This effect is seen even more clearly in Figures 11.9 and 11.10, in which the Ni and Co components due to meteoritic contamination have been subtracted out via the Ir-ratioing procedure. The trends of the residuals are approximately parallel to, and intersect, the mare basalt trend. Since the

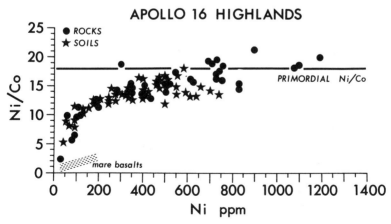

Figure 11.8 Diagram showing relationship between Ni contents and Ni/Co ratios for Apollo 16 samples (uncorrected raw data). Diagram provides clear evidence for a mixing relationship between a component possessing a primordial Ni/Co ratio, present at high Ni levels, and an indigenous lunar component possessing a much lower Ni/Co ratio. (From Delano and Ringwood, 1978b, with permission.)

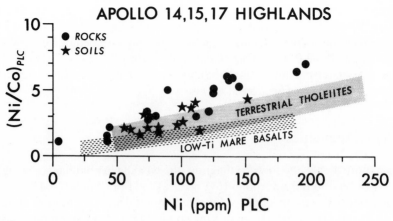

Figure 11.9 Ni/Co vs Ni diagram for the PLC magma after removal of meteoritic contamination and compensation for dilution effect by cumulus plagioclase in Apollo 14, 15, and 17 highland samples. The PLC Ni/Co vs Ni data is compared with corresponding fields for low-Ti mare basalt and terrestrial oceanic tholeiites. (From Delano and Ringwood, 1978b, with permission.)

latter is demonstrably of indigenous origin, we conclude that the second Ni and Co component is also indigenous. Figures 11.9 and 11.10 also demonstrate that the Ni/Co vs Ni trends for the PLC magma are similar to that produced in terrestrial oceanic tholeiites by olivine fractionation. This suggests that olivine fractionation was also responsible for the Ni/Co vs Ni highland trends shown in Figures 11.9 and 11.10. The terrestrial Ni/Co vs Ni trend is seen to be generally parallel and intermediate to the corresponding trends for the PLC magma and for low-Ti mare basalts, providing further

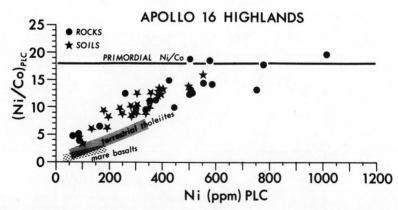

Figure 11.10 Ni/Co vs Ni diagram for the PLC magma after removal of meteoritic contamination and compensation for dilution effect by cumulus plagioclase in Apollo 16 highland samples. The PLC Ni/Co vs Ni data is compared with corresponding fields for low-Ti mare basalts and terrestrial oceanic tholeiites. (From Delano and Ringwood, 1978b, with permission.)

evidence of a close relationship between siderophile element systematics in the Earth and Moon.

An alternative explanation of the trend in Figure 11.10 is possible in terms of the model of lunar highland chemistry advanced by Wänke et al. (1976, 1977, 1978b) and discussed in Section 10.2. According to this model, the Ni/Co vs Ni residual trends may be caused by the mixing of a nonmeteoritic "primary component" representing the bulk composition of the Moon and having a near-primordial Ni/Co ratio of 15–18 and ≥ 1000 ppm of Ni, with a differentiated indigenous crustal component possessing Ni/Co ~ 3, Ni ~ 50 ppm (similar to low-Ti mare basalts). The implications for the origin of the Moon of Wänke's et al.'s model and the model favored in this book are similar (Wänke et al., 1978b).

Comparison of PLC Siderophile Residuals with Terrestrial Oceanic Tholeiites

We have already commented on the similarity in major element and REE abundances between the PLC magma and terrestrial tholeiites. It is seen from Table 11.3 that the PLC magma and terrestrial oceanic tholeiites also contain similar abundances of the relatively involatile siderophile elements W, Ni, Co, P, S,[4] and Se.[4] Moreover, Ga and Cu are depleted in the PLC magma compared to oceanic tholeiites only by factors of 5–7. They have volatilities intermediate between those of K, Na, and Li, but closer to K and Na (Table 6.2). These elements are depleted in the PLC magma compared to oceanic tholeiites by factors of 7 (K), 6 (Na), and 2.4 (Li)—see Table 11.2. The depletions of the alkali metals are clearly due to volatility, and it would therefore be expected that this factor would cause similar depletions in Cu and Ga. Thus, the abundances of Ga and Cu in the PLC magma (and also in mare basalts) correspond closely to expectations for terrestrial oceanic tholeiite magma depleted in volatiles to the observed degree.

We have previously commented on the significance of the similarity between siderophile element abundances in low-Ti mare basalts and terrestrial oceanic tholeiites. The data in Figure 11.7 and Table 11.3 show clearly that this similarity extends to the PLC magma, which is by far the most voluminous magma to have been formed in the Moon, representing the melting and differentiation of about half of its volume. We conclude, therefore, that this feature is probably characteristic of the entire Moon. The significance of this similarity in providing strong evidence that the material of the Moon was ultimately derived from the Earth's mantle subsequent to core formation has been commented on in Sections 11.2 and 11.3.

---

[4] S and Se have been listed among the relatively involatile elements because their condensation temperatures are sensitive to ambient hydrogen fugacity. They are relatively involatile under conditions of low hydrogen fugacity that may have prevailed during the formation of the Moon (Ringwood and Kesson, 1977).

We see from Table 11.2 that the abundances of gold in the PLC magma and oceanic tholeiites are also similar. Originally, Delano and Ringwood (1978a) considered this agreement to be significant. Subsequently, however, they (1978b) concluded that it was probably fortuitous, resulting from the self-compensating effects of an initial depletion due to volatility (as in mare basalts) combined with subsequent addition and enrichment of gold in the lunar upper crust via volcanic and fumarolic activity. The subject is further discussed in Section 11.8. The abundances of the remaining volatile sidero-phile elements, As, Ag, Sb, Cd, and Ge, are probably also the result of these competing processes.

11.6 Geochemical Constraints on the Existence of a Lunar Core

Small quantities of iron-rich metal phase occur widely in lunar highland rocks and in mare basalts. It has often been assumed, therefore, that the mafic cumulates underlying the crust and the source regions of mare basalts might likewise contain significant amounts of an iron-rich metallic phase. Several workers (e.g., Ganapathy et al., 1970) have ascribed the depletion of indi-genous siderophile elements in mare basalts and lunar crustal rocks to separa-tion of a metal phase in the lunar interior prior to formation of the lunar crust and/or mare basalts. This evidence, believed to favor the widespread occur-rence of metal phase within the Moon, combined with global-scale lunar differentiation, has been used to support the proposal that the Moon might possess a small, iron-rich metallic core.

Detailed studies of olivine-metal relationships in low-Ti mare basalts (see, for example, Brett et al., 1971: Hewins and Goldstein, 1974) have demon-strated that (a) liquidus olivines contain as much as 400 ppm Ni and 250 ppm Co, (b) the very earliest liquidus olivines to crystallize do not co-precipitate with a metal phase, and (c) subsequent liquidus olivines may co-precipitate with a metal alloy of unusual composition (up to 55 percent Ni). These observations show that the most primitive low-Ti liquids were far from being saturated with Fe-rich metal prior to eruption, and it thus follows that the source regions of these basalts likewise were not saturated with Fe-rich metal. Moreover, in Green Glass the high nickel content (180 ppm) is not attribut-able to a dispersed metal phase, but in fact occurs as NiO uniformly distri-buted throughout the glass (Kesson and Ringwood, unpublished results). A nickel content of around 180 ppm is much too high for Green Glass ever to have been in equilibrium with metallic iron at near-liquidus tempera-tures. Finally, Figure 11.1 demonstrates a correlation between total Ni and MgO/(MgO + FeO) ratios in low-Ti basalts which implies that nickel was present dominantly as the oxidized species prior to eruption and that nickel abundances were controlled by olivine fractionation. It appears that the most primitive mare basalts, essentially unaffected by olivine fractionation, probably contained over 150 ppm Ni and 40 ppm Co as oxidized species and were similar in this respect to terrestrial oceanic tholeiites. The enormous depletion of some siderophile elements such as germanium (Fig. 11.12) cannot, therefore, have been due to prior separation of a metallic iron phase. Moreover, since the source regions have never been in equilibrium with metallic iron, the chemical justification for proposing the existence of an iron core on these grounds appears decidedly weak. The present occurrence of metallic Fe–Ni alloys in mare basalts is ascribed to near-surface reduction processes during eruption (Reid et al., 1970; M. H. Sato et al., 1973).

In lunar highland and mare rocks, a strong covariance of W with La, and of P with La, is observed (Figures 11.3 and 11.6). This covariance was clearly caused by incompatible crystal chemical behavior of W and P during magmatic differentiation processes and requires that W and P were dominantly present as oxidized species in the original lunar highland rock system. The partition coefficient for W between metallic iron and silicate melts is shown in Figure 11.4, while Kesson and Ringwood (unpublished results) have measured the partition coefficient for P at 1300°C between metallic iron and PLC magma. Both W and P are found to be strongly siderophile under these conditions with metal–silicate partition coefficients exceeding 20. The constancy of the W/La ratio precludes the possibility of equilibration and substantial quantities of an Fe-rich metallic phase at any time during the petrogenesis of the mare and highland systems because even minute (e.g., 0.5 percent) amounts of metallic Fe would have buffered the W abundance causing the W/La ratio to vary strongly as a function of La abundance (Fig. 11.11). A similar argument can be applied to the P–La correlation of Figure 11.6 (Delano and Ringwood, 1978a,b).

Figures 11.8–10 illustrate that highland samples possess not only substantial indigenous abundances of Ni and Co, but also a prominent fractionation trend that is similar to the one produced in terrestrial oceanic tholeiites by olivine fractionation. This suggests strongly that the indigenous Ni and Co were present as oxidized species during the petrogenesis of the lunar highlands. A similar conclusion is implicit in Wänke et al.'s (1976) observation of a correlation between Ni and Mg in lunar highland samples. The occurrence of Ni and Co mainly as oxidized species precludes the occurrences of iron-rich metal. The fact that most Ni and Co in highland samples today occur in metallic form is due to the presence of meteoritic metal and to reduction during impact metamorphism/melting, subsequent to the 4.4 billion-year differentiation event. The high W/Ni ratio (i.e., a factor of about 80 greater than the primordial value) in metal grains in the highlands is strong evidence for this *in situ* reduction process (Wlotzka et al., 1973). Nevertheless, it should be noted that lunar olivines containing as much as 700 ppm NiO have been found (Steele and Smith, 1975).

FeS Hypothesis

The eutectic in the Fe-FeS system has been proposed as the composition of hypothetical pods at a depth of 250 km (Murthy et al., 1971) and of a hypothetical lunar core (Brett, 1973). Murthy et al. (1971) suggest that the

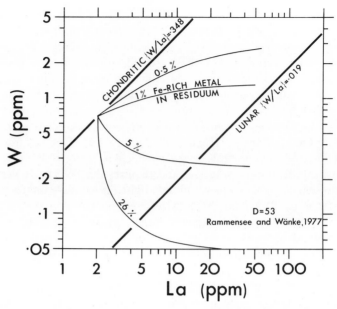

Figure 11.11 Calculated W and La abundances in silicate liquids produced by various degrees of partial melting of the hypothetical source regions containing constant abundances of W and La (i.e., 0.7 ppm W and 2 ppm La), but different quantities of residual metallic Fe (i.e., 0.5, 1.0, 5.0, and 26 weight-percent). The metal/silicate partition coefficient used in these calculations was 53 (Rammensee and Wänke, 1977). The diagram shows that liquids equilibrating with only 0.5 percent of residual metallic Fe during different degrees of partial melting would acquire more dispersion in their W/La ratios than is present in highlands and mare rocks (Fig. 11.2). (From Delano and Ringwood, 1978b, with permission.)

depletions in siderophile elements may be at least partially attributable to FeS segregation. However, this hypothesis is contradicted by the depletions of W and P, which are not chalcophile (see, for example, Imamura and Honda, 1976). Moreover, the fact that the PLC magma and primitive low-Ti mare basalts were not saturated with FeS (Kesson and Ringwood, 1976b) suggests that segregations of FeS or a core of FeS are either nonexistent or never equilibrated with the mare and highland systems.

Further arguments against the existence of an FeS-rich lunar core are provided by cosmochemistry. If a small lunar core of FeS corresponding to the upper bound set by seismic observations were assumed, the Moon would contain about 0.4 percent of sulphur. Compared to the primordial abundances, this would imply that the Moon successfully accreted about 3 percent of the primordial complement of sulphur, an extremely volatile element in the solar nebula. This may be compared with the corresponding proportions of other elements believed to have condensed in the Moon, on the basis of their abundances in low-Ti mare basalts (Fig. 11.12: Cs—1.5 percent; F—1.4 percent; Zn—0.5 percent; Ag—0.2 percent; As—0.1 percent; Au—0.01 percent; Ge—0.006 percent). All of these elements are much less

volatile in the hydrogen-rich solar nebula than sulphur (Table 6.2). Therefore, it is very difficult to conceive of conditions under which the Moon
formed that would permit sulphur to have been accreted so much more
effectively than these other, much less volatile elements. Several of these (e.g.,
Zn, Ge, and F) are not chalcophile, so that their depletions cannot be ascribed
to preferential partition into a hypothetical FeS core.

Geophysical evidence relating to the occurrence of a lunar metallic core is
generally regarded as being highly equivocal (Section 10.1). The geochemical
evidence discussed above does not favor the existence of a core comprised
predominantly of metallic iron or iron sulphide. This evidence would not
necessarily be inconsistent with the presence of a core composed dominantly
(>60 percent) of metallic nickel which, theoretically, could separate under
conditions of much higher oxygen fugacities than would be possible for an
iron core. However, a nickel-rich core amounting to 1–2 percent of the lunar
mass would raise many more geochemical problems than it would solve.
These include the constraint arising from the relative vapor pressures of
metallic iron and nickel—that metals condensing from the solar nebula
cannot contain more than 15 atomic percent of Ni (Grossman and Olsen,
1974).

While the geochemical evidence discussed above does not favor the existence of a lunar metallic core, it should be recognized that this may raise
additional problems in explaining the magnetizations of lunar rocks. One of
the hypotheses advanced to explain these magnetizations has been dynamo
action within a lunar metallic core. However, specialists in this field are far
from being unanimous on this question.

11.7 Volatile Elements in Mare Basalts and Their Source Regions

One of the most spectacular chemical characteristics of lunar rocks is the
degree of depletion of volatile elements that they have experienced (Figs.
11.12 and 11.13). In the cases of mare basalts and their source regions, it can
be demonstrated from isotopic arguments that the considerable depletion of
rubidium compared to strontium occurred near the time of formation of the
Moon, about 4.5×10^9 years ago, and *not* during the extrusion of the basalts
at the lunar surface which occurred mostly 3.2–3.9×10^9 years ago (see, for
example, Compston et al., 1971; Papanastassiou and Wasserburg, 1973).
Likewise, the bulk of the drastic depletion of lead relative to uranium occurred
around 4.4–4.5×10^9 years ago. In this case, however, there was also a
limited degree of subsequent fractionation of Pb from U (e.g., $\times 2$), most
probably during and soon after extrusion of the lavas (Tatsumoto et al.,
1971, 1972; Tera and Wasserburg, 1974; Unruh and Tatsumoto, 1977). The
key point is that by far the largest proportion of depletion of Pb and Rb
relative to C1 chondrites and terrestrial rocks occurred near the time of
formation of the Moon. Almost certainly, this is true of the other volatile
elements found to have been strongly depleted, including Ge, As, and Sb,

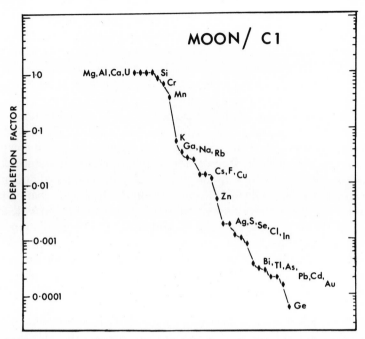

Figure 11.12 Comparison of abundances of (mainly) volatile elements in the source regions of low-Ti mare basalts with corresponding abundances in Cl chondrites. (From Ringwood and Kesson, 1977, with permission.)

which are relevant to subsequent discussions in this chapter (Figs. 11.12 and 11.13). The simplest and most satisfactory interpretation of these observations is that the volatile element depletions were characteristic of the material from which the Moon accreted.

Abundances of volatile elements in the source regions of mare basalts are compared with the primordial Cl abundances in Figure 11.12. It is notable that the overall pattern differs dramatically from that displayed by ordinary chondrites (Fig. 9.2) and implies that the Moon formed in a chemical and thermal environment very different from that which prevailed during the formation of ordinary chondrites. Moreover, the depletion patterns cannot readily be interpreted in terms of the solar nebula condensation sequence of Figure 6.6 and Table 6.2, or of various nebula fractionation models based on this sequence (see, for example, Ganapathy and Anders, 1974). The following contrasts and inconsistencies should be noted.

(a) For elements more depleted in the Moon than potassium (Fig. 11.12), there is little or no correlation between the magnitude of the depletion factor and condensation temperature in the solar nebula. For example, the condensation temperature of germanium in a nebula of solar composition is higher than those of Zn, S, Cd, Pb, Bi, In, and Tl. Yet Ge is depleted in the Moon by large factors (up to 100-fold) compared to these elements.

(b) In a cooling nebula of solar composition, significant quantities of FeO begin to enter Mg_2SiO_4 and $MgSiO_3$ at about 500°C and the FeO content of the source region of mare basalt $[MgO/(MgO + FeO) \sim 0.8]$ is reached only when the temperature has fallen to 250°C (Fig. 6.5). At this temperature, all elements in Table 6.2, excepting Bi, Pb, Tl, and In, would be fully condensed. In models wherein the Moon is formed from material ultimately condensed directly from the solar nebula (e.g., that of Ganapathy and Anders, 1974), more than 20 percent of a primordial, oxidized component should be present (Ringwood and Kesson, 1977). As shown in Table 6.2, this low-temp-erature component should contain the primordial abundances of Na, Rb, Cs, Ge, As, Zn, Cd, Se, and S. The maximum depletion factor of these elements in the Moon should thus be about 0.2 as compared to the observed depletion factors, which range between 0.03 and 0.00006 (Fig. 11.11).

(c) The elements As, Ga, Cu, F, Ag, Ge, S, and Zn are depleted in ordinary chondrites relative to Cl chondrites by small factors in the range 0.2–0.4. Anders (1968, 1971) attributes the similarity among these small depletion factors to the small range of temperatures over which these elements condense in the solar nebula (Table 6.2). However, these elements are depleted in the Moon by much greater factors (from 0.04 for Ga to 0.00006 for Ge). The enormous *range* of these depletion factors should be emphasized in view of Anders' conclusion that condensation of the above elements in a nebula of solar com-position would result in a high degree of coherence for the entire group.

The above considerations strongly imply that the Moon was not formed from material which condensed directly from the solar nebula and furthermore, that the processes responsible for the present composition of the Moon did not resemble those involved in the formation of ordinary chondrites.

Comparison with Terrestrial Mantle Abundances

Abundances of (mainly) volatile elements in the Moon and in the Earth's mantle are compared in Figure 11.13. This shows that the Moon is depleted in most volatile elements compared to the Earth's mantle and, although important exceptions exist, there is a trend for the depletion factors to cor-respond to differing degrees of volatility. For example, fluorine (depletion factor 0.13) is less depleted than the more volatile chlorine (depletion factor 0.02). Likewise, the alkali metals and Ga (depletion factors 0.09–0.25) are less depleted than Zn, Ag, Cd, In, Bi, and Tl (depletion factors 0.01–0.03), as would be expected from the relative volatilities shown in Table 6.2.

On the other hand, Ge, As, and Sb (depletion factors 0.001–0.003) are more depleted in the Moon than the above group of elements, a characteristic which is contrary to the predictions of Table 6.2. Calculations by Wai and Wasson (1977) show that the condensation of As, Sb, and Ge in the solar nebula is largely controlled by compound formation and solid solution with

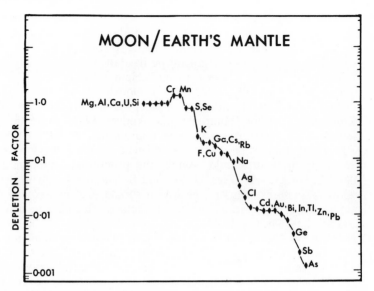

Figure 11.13 Comparison of abundances of (mainly) volatile elements in the source regions of low-Ti mare basalts with corresponding abundances in the Earth's mantle. (From Ringwood and Kesson, 1977, with permission.)

previously condensed metallic iron, and that these elements would condense at lower temperatures in the absence of metallic iron. The strong depletions of these elements might thus be explained if the material now in the Moon had formed in an environment characterized by the absence of metallic iron.

Note that although As, Sb, and Ge are also siderophile elements, their remarkable depletions in the source regions of low-Ti mare basalts cannot be ascribed to the prior crystallization of a metallic iron phase *within the Moon*. We have already seen that primitive low-Ti mare basalts were not saturated with an Fe-rich metallic phase in their source regions, so that their siderophile element abundance patterns were not established by metal fractionation within the Moon. Moreover, Ge and Sb are less siderophile than nickel, and are generally intermediate between Ni and Co with respect to this characteristic (see, for example, Wai et al., 1968). Yet it was shown earlier (Sections 11.2, 11.5, and 11.6) that these latter elements were present dominantly as *oxidized* species in their source regions.

It should be mentioned that the abundances of Ge, As, and Sb in eucrites are generally similar to those in low-Ti mare basalts. However, the mechanisms by which these strong depletions were caused differ in important respects. The extreme depletions of these elements in eucrites are apparently caused by a combination of two processes dependent upon their volatility and siderophile nature: (a) the general pattern of depletion of all volatile elements in eucrites, and (b) preferential partition into the iron core of the differentiated eucritic parent body. We have already seen that this latter process could not have operated within the Moon. A corresponding process may, however, have operated at an earlier stage within the Earth.

Two elements which are normally relatively volatile (S and Se) display similar abundances in the Earth's mantle and in the Moon (Figs. 11.7 and 11.13). This is most remarkable in view of the strong relative depletions of all other volatile elements in the Moon. Most of the elements so strongly depleted in the Moon are much less volatile in the solar nebula than S or Se (Table 6.2). The volatilities of S and Se are strongly affected by the abundance of hydrogen, since the volatile species are H_2S and H_2Se. Volatilization of alkali metals and Ga, while S and Se remain condensed, can only occur in an environment in which the H_2/S and H_2/Se ratios are much smaller than in the solar nebula (Ringwood, 1977b). It seems, then, that the extremely strong fractionation of most volatile elements between the Moon and the Earth may have occurred in an environment in which hydrogen was strongly depleted compared to the solar nebula.

11.8 Volatile Elements in the Lunar Highlands

The geochemistry of volatile elements in the lunar highlands is much more complicated than in mare basalts. It is seen in Figure 11.7 and Table 11.3 that several elements (Zn, Cd, Ge, As, Sb, Ag, and Au) appear to be much more abundant, by factors ranging up to 60, in the parental magma of the lunar highlands (PLC magma) than in mare basalts. Krähenbühl et al. (1973) and others have shown that, on the ,average, many elements are much more abundant in highlands rocks than in mare basalts. These include Ag, Au, Bi, Br, Cd, Ge, Sb, Te, Tl, and Zn. The property shared by all of these elements is their comparatively high *volatility* under lunar conditions.

It is important to appreciate that relative volatilities of elements in the lunar thermochemical environment may differ considerably from the corresponding relative volatilities in a nebula of solar composition (Table 6.2). The degree of volatility of an element under lunar conditions can be estimated only from empirical evidence obtained by the study of lunar materials. Thus, many elements—including Zn, Cd, Ge, Ag, Sb, and Au—are found to be strongly enriched on the *surfaces* of Green and Orange glasses, as compared to their interiors (see, for example, Chou et al., 1975, Table 11.4). These glasses are believed to have formed by volcanic fire-fountaining. Because the enrichments are exclusively associated with the surfaces of the spherules, it has been concluded that they were condensed from the vapor phase associated with the eruptions (see, for example, Chou et al., 1975). It is notable that some of the elements most strongly concentrated in this manner (e.g., Sb and Au) are not particularly volatile in the solar nebula (Table 6.2) when compared to others that have not been concentrated by such large factors (e.g., Ga and Cu).

Meyer et al. (1975) and Chou et al. (1975) have suggested that the volatilities of some elements might be enhanced by the formation of Cl-, F-, and S-bearing complexes. The high volatility of Ge, Sb, As, and Au might also be attributed to the absence of metallic iron in the original petrologic system in

Table 11.4 Comparison between element abundances on the surface of, and the interior of, Apollo 15 Green Glass and Apollo 17 Orange Glass. Data from Chou et al. (1975). All of these elements have been found to be substantially depleted in low-Ti mare basalts relative to the PLC magma (Fig. 11.7).

Element	Surface/Interior
Cd	420
Zn	71
Ge	54
As	(no data)[a]
Ag	> 30
Sb	100
Au	> 110
Cu	~ 1.5
Ga	4

[a] However, Wänke et al. (1978a,b) have concluded that the high enrichment of As in Apollo 16 rocks is due to concentration processes controlled by its volatility.

which the highlands were formed (Section 11.6). It has been shown by Larimer (1967) and Wai and Wasson (1977) that these elements are much more volatile in the absence of free metallic iron.

Important evidence pointing towards extensive volatile transport of certain elements on the lunar surface is provided by analyses of mature lunar highland soils as a function of grain size. It was found by Krähenbuhl et al. (1977), Boynton and Wasson (1977), and von Gunten et al. (1978) that Ge, Cd, Zn, Te, In, Au, Hg, Sb, and In were markedly to strongly enriched in the *finest* fractions of the lunar soil. Their interpretation was that these elements are surface correlated as a result of redistribution as volatile species, probably ultimately of volcanic origin.

Krähenbuhl et al. (1973) also interpreted the very high abundances of volatile elements that they observed in Apollo 16 highlands rocks to be the result of widespread fumarolic volcanic processes. Reed et al. (1977) reached an analogous conclusion based on investigations of volatile elements in highland rocks from several sites. Boynton et al. (1976) studied the compositions of mature Apollo 16 soils in relation to their development from several components of underlying parental rocks. They were unable to obtain a mass balance for In and Cd, and concluded that > 80 percent of these elements had been added to the soils via volcanic activity. Silver (1972) showed that a substantial proportion of the lead in highland rocks was highly labile and present in excess quantities relative to the amount of uranium present. Wänke et al. (1978a,b) concluded that arsenic had been greatly enriched in Apollo 16 rocks by volatile transport connected with magmatism. Wänke (personal communication) also suggested that strong upward enrichment of volatile elements in the highlands crust may have occurred via a process of mobilization connected with extensive meteorite impact, accompanied by selective volatilization and recondensation.

In addition to the very high *average* abundances of many volatile elements in highlands rocks as compared to mare basalts, another feature is that the abundances of these elements show very wide *dispersions* in the former. The data of Krähenbuhl et al. (1973) show that many volatile elements in Apollo 16 rocks vary in abundance by a factor of more than 1000. Perhaps the most dramatic evidence of such dispersions is provided by the lead isotopic studies of Tera and Wasserburg (1974), who show that the ratio of lead to uranium in highlands rocks varies by a factor of over 10,000! They conclude that this dispersion was caused essentially by large-scale volatilization and remobilization as a result of the "late-heavy" meteoritic bombardment of the highlands between 3.8×10^9 and 4.0×10^9 years ago. Boynton et al. (1976) also obtained evidence for impact-induced remobilization of germanium. In the light of the lead isotopic evidence, it seems likely that the process has been of general occurrence and is largely responsible for the wide dispersions in abundances of volatile elements in highland rocks.

In the previous section, we found that the gross depletions of volatile elements in the source regions of mare basalts occurred around 4.5×10^9 years ago, close to the time of origin of the Moon, and *not* during the subsequent times when the mare basalts were extruded on the lunar surface. Rubidium–strontium isotopic systematics likewise imply that the depletion of Rb with respect to Sr in the highlands petrogenetic system occurred around 4.5×10^9 years ago, and that the *initial* Rb/Sr ratio of this system at that time was comparable to that of the source regions of mare basalts (Papanastassiou and Wasserburg, 1973, 1975). An analogous situation prevails in the case of lead–uranium isotopic systematics (Tatsumoto et al., 1971, 1972; Nunes et al., 1975; Unruh and Tatsumoto, 1977; Tera and Wasserburg, 1974). The gross depletion of lead in the highlands system occurred around 4.5×10^9 years ago, and the initial Pb/U ratio of the highlands petrogenetic system was similar (within a small factor) to that of the source regions of mare basalts. In view of the evidence that the original abundances of Rb and Pb were approximately similar, both in the mare basalt and highland petrogenetic systems, it seems reasonable to conclude that the original abundances of many other elements possessing volatilities intermediate between those of Rb and Pb were also similar in both petrogenetic systems. The presently observed "overabundances" of these elements in the highlands system can thus reasonably be attributed to subsequent additions to near-surface rocks as a result of fumarolic volcanism and upward concentration via meteoritic impact remobilization.

There is one factor which complicates this basic interpretation. The isotopic composition of lead found on the surfaces of Green Glass and Orange Glass spherules implies derivation from a source region which possessed a much higher Pb/U ratio than generally found for other highland and mare rock systems (Tatsumoto et al., 1973; Meyer et al., 1975; Tera and Wasserburg, 1976). These studies clearly imply the existence of source regions within the Moon that, at an early stage of lunar evolution, were less depleted in volatile elements than the petrogenetic systems from which the vast majority of mare and highland samples were derived. Nevertheless, it should be noted that these source regions were depleted in Pb relative to U by factors of about four, as compared to the Earth's mantle.

Recent studies (Unruh and Tatsomoto, 1978) also indicate a substantially lower Pb/U ratio in the source region of Luna 24 basalt (Mare Crisium) than in other mare basalts.

CHAPTER 12

Towards a Theory of Lunar Origin

12.1 Introduction

The nature and origin of the Moon is a topic which has interested not merely scientists, but mankind generally, since the dawn of civilization. Our literature contains a most diverse range of speculations about this subject. A hypothesis which appears not infrequently in children's tales regards the Moon as being constructed from a variety of green cheese. Some of the early measurements on Apollo samples did not entirely dispel this hypothesis. It is seen from Table 12.1 that the seismic velocities of lunar surface materials resemble those of cheese and are dissimilar from those of terrestrial rocky materials. Of course, we now understand that this is merely a coincidence. The absence of water in lunar rocks, combined with porosity, greatly reduces their velocities in comparison with terrestrial rocks.

There is yet a second equality which is relevant to the nature of the Moon. The mean density of the Moon (3.34 g/cm^3) is very similar to that of the Earth's upper mantle (3.30–3.40 g/cm^3). Is this simply a coincidence, or is it a clue of profound genetic significance? The example of green cheese suggests the need for caution in interpreting its meaning. Indeed, many scientists have adopted hypotheses of lunar origin which imply that the similarity is coincidental. On the other hand, Sir George Darwin (1880, 1908) believed that the similar densities of Earth's mantle and Moon provided the fundamental clue to the origin of the Moon. It will already be obvious from the discussions of the previous chapter that I am in agreement with Darwin.

There are three basic classes of models for the origin of the Moon, together with many variants and combinations. According to the *capture* hypothesis,

Table 12.1 Seismic velocities in lunar rocks, terrestrial rocks, and selected cheeses. (From Schreiber and Anderson, 1970).

	V_p (km/sec)
Cheeses	
Sapsego (Swiss)	2.12
Romano (Italy)	1.74
Cheddar (Vermont)	1.72
Muenster (Wisconsin)	1.57
Lunar rocks	
Basalt 10017	1.84
Basalt 10046	1.25
Near-surface layer	1.2
Terrestrial rocks	
Granite	5.9
Gneiss	4.9
Basalt	5.8
Sandstone	4.9

the Moon formed as an independent planet in a heliocentric orbit elsewhere in the solar system. It made a close approach to the Earth, permitting a large amount of energy to be dissipated via tidal interactions, thereby causing the Moon to be captured into a geocentric orbit. The *binary planet* hypothesis regards the Earth and Moon as having accreted nearby from the same cloud of parental material. The accreting bodies were gravitationally bound together from the outset, and orbited about each other, as in a binary star system. According to the *fission* hypothesis, the material now constituting the Moon was derived from the Earth's mantle, after the core had formed.

The sources of evidence bearing on the origin of the Moon are complex and highly diverse. They are derived from several scientific fields covering a broader spectrum than can be adequately comprehended by individual scientists. Faced with this situation, it is inevitable that there will be an element of subjectivity in the selection and use of evidence by individual scientists when they attempt to address the problem of lunar origin. Almost invariably, the scientist will give greater weight to evidence from his own field, which he understands and with which he is familiar. Thus, the synthesizer usually espouses a particular "point of view" which he would incline to call a "perspective," whereas others would call it a "prejudice." I see no escape from this situation. The most that can be hoped for is that the specialist will at least make a serious attempt to comprehend boundary conditions arising from fields other than his own and to incorporate these boundary conditions into his thinking. If this attempt is not made, the interpretation is likely to be superficial.

Two sources of evidence which have proven to have a particularly important bearing on lunar origin are derived from dynamics and geochemistry, fields normally regarded as being far apart. It is obvious that I shall be approaching this topic with the "perspective" of a geochemist. It should be clearly understood, however, that there is no prospect of a final solution without a complete integration of the relevant evidence from dynamics, geochemistry, and other relevant fields.

Geochemists, too, have their prejudices, and perhaps it is time that I stated mine. Ringwood (1959) concluded that existing evidence was consistent with the hypothesis that Mars, the Earth, and Venus possessed generally similar abundances of the major elements Fe, Mg, Si, Ca, and Al, and moreover that these abundances were generally similar to those in the Sun and in Cl carbonaceous chondrites. Differences in intrinsic density between these planets were ascribed to differing redox states. As discussed in Chapters 4, 6, and 9, a large amount of evidence obtained subsequent to 1959 has remained consistent with this conclusion. Carbonaceous chondrites are now believed to represent the predominant material of the asteroid belt. Thus, it appears that during the formation of the solar system, there was no substantial *large-scale* fractionation of iron from silicates between the asteroid belt, the zones of accretion of Mars, the Earth, and Venus, and the Sun. Only in the case of Mercury is such a fractionation proven. This may be connected with Mercury's singular position as the planet closest to the Sun.

As we shall see subsequently, the Moon *must* have been born from material which originated between the orbits of Mars, the Earth, and Venus, whatever its mode of origin—capture, binary planet, or fission. The low density of the Moon, implying a gross depletion in iron relative to silicates as compared with these planets, strongly suggests that the Moon has a unique origin, that it is not, in fact, a true terrestrial planet. It was these considerations that led the author in 1960 to revive Darwin's hypothesis that the Moon had ultimately been derived from the Earth's mantle. Since then I have deliberately searched for evidence which might be relevant to this hypothesis. The results of this search were detailed in Chapters 10 and 11.

12.2 The Pre-Apollo Scene

Prior to the Apollo program, our knowledge of the physical and chemical properties of the Moon was so meagre that the field for speculation on these topics was virtually unlimited. There was, however, a considerable amount of accurate observational data concerning the dynamical properties of the Earth–Moon system, and it is little wonder that pre-Apollo theories were dominated by dynamical considerations. Excellent reviews of this subject have been published by Kaula (1971); Kaula and Harris (1975), and Öpik

(1972). The detailed investigations of the morphology of the lunar surface by Baldwin (1949, 1963) and its implications for the rheological development and thermal evolution of the Moon were preeminent among lunar studies during this period. Many of Baldwin's interpretations and predictions have proved to be remarkably accurate.

Capture

As the Moon revolves around the Earth, it raises tides on the latter body which dissipate energy, dominantly in shallow seas. As a result, there is a transfer of angular momentum from the Earth's rotation to the Moon. In consequence, the rotation of Earth is slowing down, while the increased angular momentum requires the Moon to recede from the Earth. The current rate of slowing of Earth's rotation and of the consequent recession of the Moon is known from astronomical observations.

Classical studies of the tidal evolution of the Earth–Moon system were carried out by Sir George Darwin (1880, 1908), who calculated the position of the lunar orbit backwards in time. He showed that the Moon must have been much closer to Earth in the past. Indeed, he concluded that the Moon ultimately had been derived from within Roche's Limit [2.89 Earth Radii (ER)] and had originated by fission from the Earth's mantle (the present distance of the Moon from the Earth is 60 ER).

The tidal evolution of the system was further studied by Gerstenkorn (1955), who followed the orbit further back in time than Darwin. Assuming that the Moon did not break up when it passed within Roche's Limit, he found that the inclination of the orbit increased greatly as it made its closest approach to the Earth (2.7 ER, later found to be 1.5 ER—Gerstenkorn, 1969), after which the orbit flipped over the Earth's pole and the Moon receded from the Earth on a retrograde orbit. This would have occurred only about 2×10^9 years ago.

Gersternkorn thereby concluded that the Moon had been captured by the Earth, approaching it initially on the retrograde path. This explanation appealed to many, since it permitted the Moon to have originated elsewhere in the solar system, far away from the Earth, where its great depletion in iron relative to the Earth might more readily be ascribed to unknown processes. (Incidentally, the Earth's initial rotation period was found to be 2.6 hours, very close to the period at which fission occurs). A further apparent attraction of the capture hypothesis was a possible explanation of the embarrassingly short time scale for tidal evolution, which places the Moon very close to the Earth only 2×10^9 years ago.

Since Gerstenkorn's (1955) pioneering study, the capture hypothesis has been investigated by many authors, and numerous variants proposed. These are reviewed in detail by Kaula (1971) and Kaula and Harris (1975). Three important later studies were by MacDonald (1964), Goldreich (1966), and Gerstenkorn (1969). These show that the inclination of the Moon to the

Earth's equator at closest approach is much smaller than originally obtained by Gerstenkorn (1955).

There seems to be general agreement among all concerned that capture is an event of extremely low probability, even when the most favorable configuration of the Moon's initial orbit is assumed. This would have the Moon initially in an orbit very close to the Earth's, so that the relative approach velocity was very low. For a more distant origin, e.g., in an orbit extending beyond Mars or Venus, the approach velocity becomes unacceptably high and the probability of capture astronomically low.

Because of the strong heating of the Earth and large tidal deformations, which would be caused by a close approach to the Earth as recently as 2×10^9 years ago, it seems that the assumption of constant tidal dissipation (phase lag) within the Earth must be incorrect. The theory of plate tectonics teaches us that the configuration of shallow seas on the Earth has changed greatly during geological time, so that there is no *a priori* reason to assume a constant phase lag. Thus the time scale problem for tidal evolution may be only apparent.

Acknowledging the low intrinsic probability of capture, Urey (1962b) proposed that it could be increased if it were assumed that a very large number of Moon-sized bodies were originally present in the solar system. Most of these accreted into planets. Earth's Moon was one of a very small proportion, which was captured. This concept is not without merit. However, it was largely based on the belief that the abundance of iron in the Moon was similar (relative to silicon) to the iron abundance in the Sun, and that the Moon was in other respects a geochemically "primitive" object. Subsequent studies have shown that this is not the case, and Urey has since abandoned this hypothesis in favor of a modified fission hypothesis (O'Keefe and Urey, 1977).

Singer (1968, 1970) and Alfvén and Arrhenius (1969) have presented modifications of the capture hypothesis aimed at relieving one of several dynamical problems it faces, namely the dissipation of energy and heating of the Earth and Moon during the assumed close approach. Singer is obliged to abandon Gerstenkorn's time scale for lunar tidal evolution, which was one of the primary reasons for turning to a capture-type hypothesis in the first place. Alfvén and Arrhenius, on the other hand, make a number of highly speculative assumptions about the possibility of spin-orbit coupling. These doubtful assumptions reduce still further the intrinsically low probability of the postulated capture mechanisms (Kaula, 1971).

The capture hypotheses considered above require that the Moon initially possessed a heliocentric orbit very similar to the Earth's and thus must have accreted originally in the Earth's feeding zone. The hypotheses do not, therefore, provide any explanation of the deficiency of iron in the Moon, or indeed of any of the other lunar compositional characteristics discovered during the Apollo project. When these serious defects are combined with the intrinsically low dynamical probability of proposed mechanisms, the case for an origin by capture of the Moon (as an integral planetary body) is seen to be very, very weak.

Binary Planet and Coagulation Hypotheses

Kuiper (1954, 1963) suggested that the Earth–Moon system is essentially analogous to a binary star system, where both bodies have formed in close proximity and in orbit around each other, from a common cloud of dust and gas. Actually, the analogy is not so close, since stars form by gravitational collapse processes mainly involving gases, while the Earth and Moon are believed to have formed by accretion from solid planetesimals. Urey (1963) pointed out that the dynamical conditions required for accretion of two bodies differing in size as much as the Earth and Moon, which finally end up in a closely bound orbit, are extremely restrictive. There has to be a critical balance between the rates of change of gravitational and centrifugal forces during accretion. By far the most likely result of such a process is collision between the two bodies or escape from each other. This variant of the binary planet hypothesis is thus also of low intrinsic plausibility.

A more plausible version of the binary planet hypothesis has been developed by Ruskol (1960, 1963, 1972a, 1977). During accretion of the Earth from heliocentric planetesimals, a substantial proportion of the planetisimals will undergo inelastic collisions in the Earth's neighborhood, thereby causing them to be captured into geocentric orbits. The swarm of geocentric planetesimals evolves into a circumterrestrial ring of matter estimated as 0.01 to 0.1 the mass of the Earth. The ring was maintained in a dispersed state by disruptive impacts from Earth-bound planetesimals. When the flux of these declined below some critical value, the ring coagulated rapidly to form the Moon. Ruskol's hypothesis has been examined by Kaula and Harris (1973, 1975) and found to be generally plausible from the dynamical point of view.

It should be noted that a Moon formed by this process would orbit initially in the Earth's equatorial plane (as would a Moon formed by fission from the Earth's mantle). The tidal calculations of Goldreich (1966) and others show, however, that the Moon's orbit must have possessed a substantial inclination to the Earth's equator when it was closer than 10 ER. This has been widely cited as an objection both to the coagulation and fission hypotheses (see, for example, MacDonald, 1964). The objection is easily countered, however, since it is based on the improbable assumption that Earth–Moon evolution has been exclusively of a two-body nature. The non-zero inclination might have been caused by one of several factors, e.g., by external angular momentum contributed by impacts on the Moon and/or Earth after their formation by one or more large planetesimals which had failed to accrete by that stage (Safronov and Zvjagina, 1969; Shoemaker, 1972; Kaula and Harris, 1973). Moreover, O'Keefe (1972a) and Rubincam (1975) have extended some earlier results of Darwin (1880, 1908) and shown that the details of Goldreich's calculations are sensitive to assumptions concerning the mean viscosity of the Earth. If this were substantially lower than at present, e.g., about 10^{15} poise, the dimensions of the problem would be greatly reduced.

The primary difficulty encountered by binary planet and coagulation hypotheses lies in explaining why, when both bodies formed from a common cloud of silicate and iron particles, the Moon was so strongly depleted in metallic iron. Physical processes so far invoked have been speculative and highly qualitative. Orowan (1969) suggested that the differential ductility and strength of iron compared to silicates would result in iron grains welding together to form relatively large planetesimals during collisions in the geocentric swarm, while silicates would be fragmented into fine dust. Orowan believed that the large iron planetesimals would be preferentially captured by the Earth. This suggestion has been followed up by Ruskol (1972b, 1977), Kaula (1975), and Kaula and Harris (1975). These authors argue that the finer silicate dust would be preferentially enriched in the circumterrestrial swarm from which the Moon ultimately formed. Harris and Tozer (1967) invoked ferromagnetic properties of metal grains to explain preferential aggregation and accretion in the Earth. However, Banerjee (1967) demonstrated that the mechanism was inadequate, by a factor of 10^4, to cause the desired effects.

Returning to the collisional fractionation mechanisms advocated by Orowan, Ruskol, and Kaula (above), one can agree with these authors that there may well be a *general trend* for the differentiation process to act in the desired direction. Indeed, this is as much as these authors seem to be claiming. The aspect which seems extraordinarily difficult to explain is the *remarkable efficiency* with which the process must be presumed to have operated. Anticipating the results of the next section, and citing the post-Apollo results discussed in Chapter 11, we concluded that the source regions of low-Ti mare basalts and the lunar highland petrologic system did not originally contain *any* iron-rich metal phase. Moreover, it was concluded that the Moon did not possess an iron-rich metallic core (Section 11.6).

It is scarcely credible that the highly complex collisional differentiation mechanisms advocated by Orowan, Ruskol, and Kaula could have achieved the *quantitative removal* of iron-rich metal phase implied by the geochemical observations. Additional difficulties confronting the binary planet hypothesis will be cited in the next section.

Fission and Precipitation Hypotheses

Darwin (1880, 1908, 1962) developed a theory according to which the Moon and Earth were originally combined in a single body which rotated with a period of about four hours. A resonance effect occurred between the free period of the parent body and the solar tides, leading to the development of an enormous tidal bulge. An instability developed and the tidal bulge was ejected from the Earth's mantle to form the Moon. This theory provided an elegant explanation of the Moon's density, as previously noted. However, serious objections to Darwin's hypothesis were raised by Jeffreys (1930), and it was generally discarded. Darwin's basic hypothesis was revived by Ringwood (1960), who suggested that the Earth might have been formed near the limit of rotational instability (period about two hours). Rapid segregation of the metallic core decreased the moment of inertia and caused a corresponding increase of rotational velocity, leading to fission of material from the mantle. According to Ringwood's model, the excess angular momentum of the Earth–Moon system was carried away by a massive primitive atmosphere which was also disrupted and escaped during the cataclysm. A similar hypothesis was subsequently suggested and further developed by Wise (1963) and Cameron (1963b).

By far the most complete investigations of the fission hypothesis have been carried out by O'Keefe (1966, 1969, 1970, 1972a). He showed that rotational destabilization by core segregation was a somewhat marginal mechanism and investigated the possibilities of other processes. Most recently, O'Keefe and Sullivan (1978) proposed that the Earth formed with the maximum amount of angular momentum consistent with dynamical stability, and at relatively high temperatures, implying a low viscosity. However, they show that a catastrophic instability may develop after core formation when convective cooling causes a large increase in mean viscosity of the mantle, leading to fission of the outer mantle. O'Keefe (1969) also showed that for conservation of energy and momentum, about 10–20 percent of the Earth's mass must be flung off. The Moon represents only a small part of this material. Intense generation of energy via tidal dissipation would have caused strong heating of the outer regions of the Moon and Earth, accompanied by devolatilization.

As emphasized by Kaula (1971), a major problem encountered by the fission hypothesis is the excessive amount of angular momentum with which, it must be assumed, the Earth was originally endowed, corresponding to a rotation period of about 2.5 hours. We have seen in Section 5.5 and Figure 5.4 that there are empirical grounds justifying an initial rotation period of about five hours for the proto-earth. Ringwood (1972) suggested that all planets had formed near the limits of rotational instability (periods of 1.4–3.4 hours) and that large amounts of angular momentum had subsequently been removed from all protoplanetary systems via the escape of large amounts of gases. It must be admitted, however, that the physical processes invoked were qualitative and highly speculative. Nevertheless, the possibility of extensive loss of angular momentum by the Earth–Moon system and by other planets cannot be finally dismissed.

A further point made by Ringwood (1972) concerned the widely held assumption that the distribution of mass in the Earth–Moon system is anomalous when compared to the corresponding distributions in other planet–satellite systems. In order to make valid comparisons, the light gases originally associated with the satellites of Jupiter, Saturn, Uranus, and Neptune, and which have since escaped, should be taken into account. He showed that the *original* mass ratios of protosatellite systems of Jupiter, Saturn, and Neptune to their respective primaries may well have been in the vicinity of 1/100. Likewise, the *minimum* mass of the primordial solar nebula was about one percent of the Sun's mass. These ratios are similar to that observed in the Earth–Moon system, which cannot, therefore, be held to be anomalous.

An essential element of the Darwin–O'Keefe hypothesis is that the Moon was fissioned from the Earth, initially in the solid or liquid state. Ringwood (1966b) proposed a hypothesis in which it was envisaged that the material now in the Moon had been evaporated from the Earth, spun out into a disk, and recondensed to form a sediment ring of planetesimals, which then coagulated to form the Moon.

This "precipitation" hypothesis was developed in conjunction with the "single-stage" model for the formation of the Earth, reviewed in Section 7.2. In an early version (Ringwood, 1966b), it was suggested that silicates were selectively evaporated from accreting metallic iron during the rapid, high-temperature accretion of the Earth that was envisaged (Fig. 7.1). In subsequent developments of the precipitation hypothesis (Ringwood, 1970a,b, 1972, 1975a), the emphasis was changed to direct evaporation of the outer mantle after core formation, the thermal energy being supplied primarily as a result of rapid core formation. In the "single-stage" model, formation of the Earth was accompanied by the loss of a massive, hot primitive atmosphere (Section 7.2), of which the evaporated silicates now forming the Moon constituted only a small part.

The removal of the primitive atmosphere was believed to have been accompanied by a combination of processes:

(1) high initial rotation rate of Earth;
(2) coupling of atmosphere to Earth's rotation via hydromagnetic torques or by turbulent viscosity;
(3) intense solar particle radiation during the T-Tauri phase of the sun;
(4) turbulent mixing of hydrogen from the solar nebula into the atmosphere, thereby lowering its mean molecular weight.

As a result of the combination of these processes, the massive primitive atmosphere was dissipated. On cooling, the silicate components were precipitated to form an assemblage of Earth-orbiting planetesimals. A further fractionation according to volatility occurred during the precipitation process, since the less volatile components were precipitated first at relatively high temperatures and close to the Earth, whereas the more volatile components were precipitated at lower temperatures and further from the Earth. The silicates precipitating at high temperatures would probably have grown into relatively large planetesimals that would tend to be left behind by the escaping terrestrial atmosphere. However, the more volatile components precipitating at relatively low temperatures were more likely to have formed fine micron-sized smoke particles that would be carried away with the escaping atmosphere by viscous drag, and hence lost from the Earth–Moon system. The Moon then accreted from the sediment ring of Earth-orbiting planetesimals (Fig. 12.1).

The precipitation hypothesis, as discussed above, is a derivation of Ringwood's "single-stage" model for the origin of the Earth. As discussed in Section 7.2, the latter has been criticized by several workers. The most serious problems are connected with removing the massive primitive

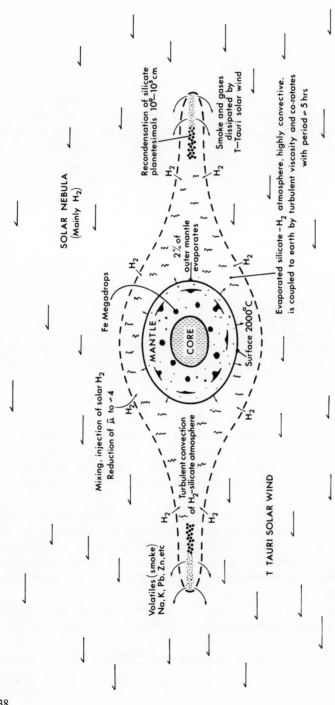

Figure 12.1 Structure of the Earth and its primitive atmosphere immediately after accretion according to the precipitation hypothesis as developed by Ringwood (1966b, 1970a,b, 1972, 1975a).

238

atmosphere after accretion and with the very rapid accretion time scale required to obtain the very high temperatures during the later stages of accretion. It was acknowledged in Section 7.2 that these difficulties reduced the plausibility of the model. The same objections apply, therefore, to the early versions of the precipitation hypothesis. We will investigate in Section 12.3 whether the latter can be revived in a modified form.

Before leaving this topic, it should be mentioned that the pre-Apollo fission and precipitation hypotheses successfully predicted characteristics of the Moon which were later discovered or inferred as a result of the Apollo project. For example, Ringwood (1966b), O'Keefe (1969), and Wise (1969) predicted that the Moon would be depleted in volatiles. The interpretation of lunar thermal history proposed by Ringwood (1966b; Fig. 6) has been followed by most subsequent investigators. This model, in conjunction with the conclusions of Baldwin (1963), interpreted the lunar maria as being composed of basaltic flows formed by partial melting within the Moon at depths of several hundred kilometers, about 10^9 years after the formation of the Moon.

Concluding Remarks

Of the three classes of pre-Apollo hypotheses which we have discussed, only the coagulation model of Ruskol (1963, 1977) appears to be free from serious dynamical difficulties. However, it seems to encounter severe problems explaining the low iron content of the Moon. The fission and precipitation hypotheses, on the other hand, encounter severe dynamical difficulties, but offer an attractive explanation of the composition of the Moon. Finally, the capture hypotheses appear seriously inadequate, both on dynamical grounds and in providing a basis for explaining the difference in iron content between Earth and Moon. In the following sections we will review the status of these hypotheses in the light of evidence obtained as a result of the Apollo project.

12.3 New Variations on Old Themes

If the lunar tidal evolution is extrapolated sufficiently far backwards in time, the Moon is found to approach within 1.5 Earth radii of the Earth, which is well within Roche's limit (2.89 ER). Thus, the Moon should be torn apart into small fragments at this stage. Proponents of whole-Moon capture have speculated that since the period during which the Moon was within Roche's limit was shorter than the Moon's free period of oscillation, disintegration would not necessarily have occurred. However, Öpik (1972) investigated the situation and concluded that the Moon would indeed disintegrate, forming a circumterrestrial ring of fragments.

Öpik (1955, 1961, 1967) had long maintained that the Moon formed ultimately from a ring of planetesimals orbiting the Earth within Roche's limit, something like the rings of Saturn, only relatively more massive. Tidal effects would cause planetesimals forming the ring to move outwards beyond Roche's limit,[1] forming first a small number of moonlets, which would coagulate on a time scale on the order of 100 years, causing strong heating of the outer regions of the Moon. In his earlier papers, Öpik did not consider the origin of this sediment ring in any detail. However, in 1972 he pointed out that formation of such a ring would be a highly probable result of both capture and fission hypotheses. He emphasized that capture of the Moon within Roche's limit, accompanied by total breakup was enormously more probable dynamically than capture of an intact Moon, because it offers a much more efficient means of energy dissipation. Öpik (1972) favored a capture–breakup hypothesis in which about half of the fragments were captured into the circumterrestrial ring while the remainder escaped.

It will be evident, at this stage, that there are indications of a certain unity developing among hypotheses concerning the dynamical aspects of lunar origin, with the probability that both capture and fission-type models would first form a circumterrestrial planetesimal ring, which could subsequently coagulate to form the Moon. This result also begins to approach the model of Ruskol (1963, 1977), although the dynamical basis of ring formation is different. According to Ruskol's model, the ring was formed initially outside Roche's limit, during the primary accretion process of the Earth (Section 12.2).

Formation of the Moon by coagulation of a circumterrestrial planetesimal ring provides an attractive explanation of the Moon's early thermal evolution. Ringwood (1966b) pointed out that Baldwin's (1963) interpretations of lunar rheology and of the origins of mare basalts implied early strong heating of the outer regions, while the deeper interior remained relatively cool. This temperature distribution, supported by the Apollo results discussed in Section 10.5, could be explained in the Moon had formed very rapidly, so that a significant proportion of accretional energy had been conserved, thereby heating the exterior but not the interior regions. Öpik (1961) had already shown that if the Moon had formed by coagulation of the circumterrestrial ring, accretion could have occurred on a time scale of about 100 years, leading to strong near-surface heating. A study by Mizutani et al. (1972) indicated that the early melting and differentiation of the outer Moon by accretional energy would result even if the Moon formed on a time scale of up to 1000 years. However Öpik (1961, 1969) had also shown that if the Moon had formed by accretion of planetesimals in *heliocentric* orbit, the time scale for accretion would be orders of magnitude longer. With such a long time scale, the possibility of significant accretional heating appeared remote. Ringwood (1972) concluded, therefore, that the thermal history of the Moon, implied by its petrological structure and early differentiation (Fig. 10.5), provided support for the view that the Moon originated from a

---

[1] Kaula and Harris (1975) argue that tidal interactions will be ineffective in causing planetesimals to move outwards from the Earth beyond Roche's limit because of comminution caused by collisions. An alternative mechanism for causing the ring to move away from Earth has been proposed by Cameron and Ward (1978).

circumterrestrial ring and was not formed as a separate body by accretion from sun-orbiting planetesimals.

Wetherill (1976a) attempted to show that if the accreting bodies were sufficiently large, the outer parts of the Moon might have been significantly heated, even if the accretion time scale was 10^7-10^8 years. However, the effect was marginal and incapable of producing the inferred early global magma ocean resulting from partial melting extending to depths of a few hundred kilometers (Section 10.2 and Fig. 10.5). Kaula (1978b) has recently completed a detailed study of impact heating processes and has showed that it is unlikely that the early melting and differentiation of the Moon could be explained by its accretion from heliocentric planetesimals on the required long time scale. He concluded, accordingly, that the thermal history of the Moon was best explained by accretion from Earth-orbiting material, probably derived ultimately from the Earth's mantle.

Disintegrative Capture

Öpik (1972) pointed out that disintegration of a large Earth-approaching body within Roche's limit was far more probable than capture as an intact body or direct impact with the Earth. Wood and Mitler (1974), Smith (1974), and Mitler (1975) proposed that prior to formation of the Earth, a population of large (>500 km radius), differentiated planetesimals possessing metallic iron cores and silicate mantles (possibly parental to meteoritic irons, stony–irons, and achondrites) existed in the region of the Earth's feeding zone. One or more of these bodies passed within Roche's limit and were disrupted. If rather stringent restrictions are placed upon the velocity and trajectory of approach, the Earth may have captured a large proportion of the mantles of such bodies, while most of the cores either escaped or impacted the Earth. In such a manner, the compositional difference between the Earth and Moon might be explained.

Objections to these hypotheses (Harris and Kaula, 1975) are that an enormous mass of such planetesimals (about one Earth-mass) must be passed through Roche's Limit, and that under such circumstances the process is intrinsically inefficient. Although there may be a *general tendency* for the Earth to capture more of the silicate mantle fraction, almost inevitably, a significant fraction of metallic iron would also be captured. This would be enhanced because the cores of the large (> 500 km) differentiated planetesimals would be molten at the time of disruption. (In bodies this size, there would have been insufficient time at the stage when the Moon was formed about 4.5×10^9 years ago for the cores to have frozen). The liquid cores would have been intensively disintegrated into small drops, a substantial proportion of which would have been captured with the silicates into geocentric orbits. Thus the *extreme deficiency* of metallic iron in the Moon, as shown by our conclusions (Section 11.6) that neither the source regions of mare basalts nor of the highlands system were originally saturated with iron, is unexplained. It seems clear that the process is much too inefficient to explain the deficiency of metallic iron in the Moon. Its plausibility is further reduced by the arbitrary nature of the initial assumptions.

Modified Precipitation Hypothesis

The author's early versions of the precipitation hypothesis were derived from a model according to which the Earth formed via single-stage accretion from Cl carbonaceous chondrite-like planetesimals. This latter model is now seen to have serious weaknesses (Section 7.2). However, these do not necessarily affect the viability of essential aspects of the precipitation hypothesis. In a modified version of the single-stage model, Ringwood (1975a) suggested that gases formed near the surface of the Earth during accretion (mainly H_2O produced by the reduction of FeO in the accreting planetesimals by H_2 from the solar nebula) escaped from the Earth continuously during accretion, and at no stage was a *massive* primitive atmosphere (mainly composed of CO in earlier models) amounting to about 0.2 M_E permitted to accumulate.

However, the revised model proposed by Ringwood (1975a) continued to maintain that the Moon had formed from material evaporated from the Earth's mantle. It was suggested that the process of core formation, once nucleated, occurred extremely rapidly, perhaps within 10^4 years, as suggested previously by Ringwood (1960). It was shown that under these circumstances, about half of the core-forming energy of about 600 cals/gm (for the entire Earth) would be utilized in causing *evaporation* of silicates from the mantle. The combination of a series of factors—rapid rotation of the primitive Earth ($\sim$ five-hour period), turbulent mixing of the silicate atmosphere with hydrogen from the solar nebula, and coupling of the atmosphere to the Earth's rotation via hydromagnetic torques or turbulent viscosity—caused the atmosphere to be spun out into a disk, from which the planetesimals parental to the Moon condensed (Fig. 12.1).

Impact Models of Lunar Origin

It is generally believed that tektites were formed by the effects of meteoritic impacts on the Earth's surface (see, for example, S. R. Taylor, 1973b). It is also known that at least one particular class, the australites, were accelerated by the impact to velocities of about 10 km/sec (Chapman and Larson, 1963). Thus it is empirically established that impact of meteorites or planetesimals after core formation has the capacity to remove material from the Earth's mantle and place it in geocentric orbit. Theoretical and experimental studies of high-velocity impact processes by Jakovsky and Ahrens (1978) have also demonstrated the capacity of high-velocity planetesimal collisions to evaporate material from the Earth's surface and place it into orbit.

It is now believed that accretion of the Earth proceeded via a hierarchy of planetesimals of varying sizes, some of which may have been quite large (see, for example, Safronov, 1972a; Wetherill, 1976a). The inclination of the Earth's equator to the plane of the ecliptic is best explained by impact with a planetesimal with a highly inclined orbit and amounting to at least 0.001 the mass of the Earth. Urey (1963) has long maintained that a population of lunar-sized planetesimals existed in the solar system prior to accretion of the planets. Studies of accretion processes by Hartmann and Davis (1975) and Wetherill (1976a) show that the second largest planetesimal to impact the Earth from its "feeding zone" may have been in the vicinity of 0.03–0.1 the mass of the Earth. These studies make it seem very likely that the Earth was impacted by many planetesimals in the mass range 0.001–0.01 Earth mass.

Hartmann and Davis (1975) point out that a planetesimal of about half a lunar mass impacting the Earth at 13 km/sec would have sufficient energy to eject two lunar masses of the Earth's mantle to near-escape speeds. Half of this energy may well have appeared as kinetic energy of the ejecta (Gault and Heitowit, 1963). A large proportion of this material could well have been placed in geocentric orbit. Hartmann and Davis hypothesize that, provided one or more of these collisions occurred subsequent to core formation, it would have been possible to produce a geocentric ring of material mainly derived from the Earth's mantle, from which the Moon might have formed.

A related model has been proposed by Cameron and Ward (1976), who suggest that, subsequent to core formation, the Earth was hit by a differentiated planetesimal of about 0.1 Earth masses at about 11 km/sec. Vaporization of extensive regions of the mantles both of Earth and of the planetesimal would have occurred, and a substantial proportion of the vapor would have expanded at a sufficient velocity to attain Earth orbit. Selective recondensation of the less volatile components of the vapor followed, as in Ringwood's model, forming a ring of devolatilized Earth-orbiting planetesimals, while the volatile-components condensed further away as micron-sized smoke particles which were lost.

Cameron and Ward apparently prefer impact by a Mars-sized planetesimal in order to provide the Earth–Moon system with additional angular momentum. Presumably, they feel that an initial rotational period of five hours for the proto-Earth would have been too short. However, we have previously considered the distribution of angular momentum density among the protoplanetary systems (Section 5.5 and Fig. 5.4) and concluded that an initial period of five hours for the proto-Earth is compatible with other evidence, including the *original* periods of other planets. If this conclusion is accepted, the requirement that the impacting planetesimal was as large as advocated by Cameron and Ward can be relaxed.

The attractive feature of these impact models as a means of placing material from the Earth into geocentric obit is that they rely on processes that must *inevitably occur* during the accretion of planets and can supply the enormous amounts of energy required. The models of Hartmann and Davis (1975) and Cameron and Ward (1976) can be regarded as special, but more plausible, cases of Ringwood's (1966b, 1970b) model, which invokes evaporation of the outer mantle by rapid accretion of the Earth to provide the parental circumterrestrial planetesimal ring. The essential difference is that in these later models, the required energy is supplied by one or more large impacts, thereby removing the requirement for an uncomfortably short accretion interval for the Earth.

The author now believes that the answer to the Moon's origin is to be sought in the general direction pointed by Hartmann and Davis, and by Cameron and Ward. But loose ends still remain. It is necessary that the impacts which provided material for the circumterrestrial swarm occurred only *after* core formation within the Earth. Earlier impacts (which must also have occurred) would have ejected material containing metallic iron into the ring. Moreover, it is desirable that the material placed into Earth orbit was derived from the Earth's mantle and not from the incoming projectile, which was unlikely to have been depleted in iron. Even if the latter were differentiated into a mantle and core, it is difficult to understand how some portion of the latter did not become mixed into the debris ultimately placed into orbit.

For these reasons, I believe that a more complicated scenario may be necessary. A highly speculative attempt in this direction was outlined by Ringwood (1978a). The proto-Earth was assumed to be spinning with a period of about five hours. As discussed in Section 3.4, core formation occurred quite rapidly towards the final stages of the Earth's accretion. The energy evolved would have caused melting of the outer mantle to depths of a few hundred kilometers (Ringwood, 1975a). Due to enhanced convection within the Earth's core at this early stage, the early terrestrial magnetic field may have been much stronger than at present. After core formation, the molten upper mantle was impacted by several relatively small (e.g., 0.2 lunar mass) and *cool* planetesimals at velocities in the vicinity of 13 km/sec. Because of the high initial temperature of the mantle target and the low temperature of the projectile, the most intense shock heating would have occurred in the mantle, and not in the projectile. This would have been enhanced by the higher compressibility of molten mantle material compared to the projectile. Thus, substantial amounts of the mantle magma ocean were totally *evaporated*, whereas the projectile was mainly shock melted. The gas evolved in the impact was partially ionized (mainly Na^+) and was, therefore, coupled to the Earth's rotation by the magnetic field. As it rapidly expanded, it was spun out by the field beyond the geosynchronous radius (2.4 ER), where the less volatile components condensed to form Earth-orbiting planetesimals, while the more volatile components condensed further away from the Earth as fine smoke particles and were blown away by the greatly enhanced solar wind. An essential element of the model was the transfer of angular momentum from the Earth to the gas cloud, permitting the latter to be driven outwards. The shock-melted projectile would not have been so coupled and hence would have been accreted by the Earth. Since the process requires (1) a molten outer mantle produced by the core-formation process, and (2) a powerful magnetic field generated by the early strongly convecting core, it is clear that the proposed mechanism could have operated only *after* core formation.

12.4 The Post-Apollo Scene: Geochemical Testimony

Bulk Composition

In Section 10.2, it was concluded that important resemblances exist between the outer regions of the Moon and the Earth's mantle. The parental magma (PLC) of the lunar crust was found to be very similar to that of primitive terrestrial oceanic tholeiites, providing that significant proportions of the two most volatile major components (Na_2O and SiO_2) were removed from the latter (Table 10.2). The rare-earth abundances of both magmas were similar. The physical properties of the upper mantle (Fig. 10.2), combined with evidence from experimental petrology, implied that the lunar upper mantle, to a depth of about 400 km, probably consisted dominantly of residual refractory dunite with an $MgO/(MgO + FeO)$ ratio similar to that of the Earth's upper mantle (Fig. 10.5).

Except for volatile elements (including Na_2O and SiO_2—Fig. 6.8) it was concluded in Section 10.2 that the bulk chemical composition of the combined lunar crust–upper mantle petrologic system to a depth of about 400 km was similar to that of pyrolite (Table 10.5). Moreover, the lunar heat flow data was consistent with similar mean abundances of U and Th existing in the entire Moon and in the Earth's mantle (Section 10.5). The principal difference between the outer Moon and the Earth's upper mantle was that the former region was depleted in elements more volatile than silicon by variable, and sometimes extreme, degrees. Ignoring this latter factor for the moment, the general similarity in abundances of the less volatile elements between both Earth's mantle and the outer regions of the Moon (amounting to half of its mass) are suggestive of some kind of genetic relationship between the two bodies. Similar arguments have been made by Binder (1974, 1976).

This is supported by a comparison of the isotopic composition of oxygen in the Moon (including mare basalt source regions in the lower lunar mantle) and the Earth's mantle. About half of the Moon and the Earth's mantle (by weight) is composed of oxygen, which possesses three isotopes—^{16}O, ^{17}O, and ^{18}O. Variations in oxygen isotope ratios between planets and various classes of meteorites are caused by two factors: (1) primary isotopic inhomogeneities in the oxygen in various regions of the solar nebula prior to accretion, due to the nonuniform distribution of a nuclear component rich in ^{16}O, possibly injected via a supernova explosion, and (2) chemical isotopic fractionations caused, for example, by differing temperatures at which the solid matter which accreted to form planets and meteoritic parent bodies equilibrated and became separated from the gases in the parental solar nebula.

The oxygen isotope compositions of both lunar and terrestrial basalts are identical (see, for example, Clayton et al., 1976). This implies that the material from which the Earth and Moon was formed was homogeneous with respect to the ^{16}O supernova component and was also separated from the nebula at the same mean temperature. In contrast, the oxygen in most classes of meteorites possesses distinctly different proportions of the ^{16}O-rich component. Some differentiated meteorites—the eucrites and howardites—possess oxygen which differs only very slightly in this respect from terrestrial and lunar oxygen, and may perhaps be identical (Clayton et al., 1976). However, the oxygen in these meteorites is distinctly different from terrestrial-lunar oxygen owing to a chemical fractionation effect (H. P. Taylor et al., 1965; Section 11.4). The only classes of meteorites which possess identical oxygen isotopic compositions to the Earth and Moon are the enstatite chondrites and achondrites. However, numerous other chemical characteristics of these latter objects rule out any genetic relationship with the Earth and Moon.

Meteorites provide clear evidence that the ^{16}O component of dust grains was not uniformly distributed within the solar nebula prior to accretion into meteorite parent bodies and Planets. Moreover, the existence of chemically derived oxygen isotopic fractionations indicates that the parental material of different kinds of meteorites and planets separated from the solar nebula at differing temperatures. In the light of these observations, the identity in oxygen isotopic compositions between the Moon and Earth acquires an added significance. It will be very important to determine the ^{16}O, ^{17}O, and ^{18}O ratios in the oxygen of Venus and Mars in the future. If these should differ from the Earth and Moon in their abundance of the ^{16}O supernova component, the case for a unique genetic relationship between the Earth and Moon would be greatly strengthened.

On the other hand, some of the chemical and physical properties of the lunar lower mantle (Figs. 10.2 and 10.5), which are believed to constitute the ultimate source regions of mare basalts (Section 10.3), pose some difficulties for this interpretation. Although the overall abundances of Ca, Al, U, Th, REE, and other involatile elements in this region are believed to be similar to those in the Earth's mantle, we concluded in Section 10.3 that the lunar lower mantle appeared to be richer in FeO ($MgO/MgO + FeO = 0.75$–0.80) and to have a higher pyroxene/olivine ratio than the pyrolite composition of the Earth's upper mantle. To some extent, these conclusions are model dependent. Nevertheless, in the author's opinion they represent the most acceptable interpretation of available petrologic, geochemical, and geophysical data pertaining to the Moon's lower mantle, and cannot reasonably be swept under the rug. An attempt to account for these properties in the context of a genetic relationship between the Moon and Earth is made later in this Section.

Volatile Elements

The remarkable depletion patterns of the Moon in relation to Cl chondrites, ordinary chondrites, and the Earth's mantle are displayed in Figures 9.2, 11.12, and 11.13. These patterns have a vital bearing on current hypotheses of lunar origin.

The dynamically attractive binary planet, or coagulation hypothesis of Ruskol (1963, 1972a,b), 1977) encounters considerable difficulties in explaining the lunar volatile depletion patterns. Ruskol (1977) and Kaula and Harris (1975) suggested that strong heating, caused by collisions between planetesimals within the circumterrestrial swarm, caused evaporation of volatile elements which were then swept away by the solar wind. A considerable amount of evidence now exists on the effect of hypervelocity impacts on lunar rocks. In most cases, elements of intermediate volatility such as sodium were not lost in these processes, while elements of greater volatility (e.g., Pb) were redistributed and recondensed.

The Moon appears to be depleted in sodium by a factor of 30 compared to Cl chondrites and by a factor of 11 compared to Earth (Figs. 11.12 and 11.13). Our knowledge of high-velocity impact phenomena provides very little support for the view that depletions of this magnitude could be caused by the relatively mild collisions in the circumterrestrial swarm as proposed by Ruskol. Such collisions may lead to a degree of impact melting but sodium and many other volatiles are not readily lost under these conditions. In the light of empirical evidence concerning impact phenomena, it is hard to believe that the spectacular depletions of elements such as Ge (Figs. 11.12 and 11.13) could be caused by this process. Depletions of this magnitude seem to require total evaporation of both projectile and target, followed by selective recondensation under conditions where the volatile components are recondensed in a separate environment where they can be removed by an appropriate process. Ruskol's model does not appear to envisage such conditions.

Depletions of volatile elements in the Moon are comparable to those in eucritic achondrites. In principle, therefore, the hypothesis of disintegrative capture of a series of previously differentiated planetesimals (Wood and Mitler, 1974; Smith, 1974) might appear to provide a plausible means of explaining the Moon's depletion in volatiles. This may be so, though it should be recalled that the abundance of volatile elements in eucrites are distinctly different from those in lunar basalts (Anders, 1977).

Ganapathy and Anders (1974) and Anders (1977) argue that lunar volatile element chemistry is favorable to their preferred binary planet hypothesis of lunar origin. This hypothesis maintains that the Moon, like the Earth, was formed by heterogeneous accretion in a cooling solar nebula, involving the mixing in arbitrary proportions of seven different components. The justification for this model is provided by their interpretation of the chemistry and origins of chondritic meteorites.

It was demonstrated in Section 7.3 that this hypothesis provides an unacceptable explanation of the origin and geochemical properties of the Earth. Reasons for regarding Anders' interpretation of the genesis of chondrites with considerable reserve were detailed in Section 9.6. Even if his chondrite-based model is accepted, it was shown in Section 11.7 that the processes responsible for the formation of the lunar volatile element depletion patterns were totally different from those which operated in chondrites. The fractionations responsible for the lunar volatile abundances appear to have been established in an environment strongly depleted in hydrogen compared to the solar nebula, and where metallic iron was probably not present as a free phase (Section 11.7).

O'Keefe (1969) and Wise (1969) correctly predicted, on the basis of the fission hypothesis, that the Moon would be depleted in volatiles compared to the Earth's mantle. According to their model, the proto-Moon separated in the condensed state as a large solid or liquid mass. The outer regions were strongly heated by tidal interactions (O'Keefe, 1969, 1970), leading to devolatilization. While conceding that some degree of depletion might have been caused in this manner, I am sceptical as to whether the model can account for the intense degrees of depletion observed for many elements (Fig. 11.13). The reason is that loss of volatiles from large volumes of either solids or liquids is a very slow process, being governed by diffusion to surfaces.

It is for the above reasons (and others besides) that I have long favored a hypothesis which required the material in the Moon to have been completely evaporated from the Earth's mantle and then selectively recondensed in circumstances under which volatiles were lost, as discussed in Section 12.3. I believe that such a process could be capable of explaining the relationship between volatile elements in the Moon and Earth's mantle, as shown in Figure 11.13. The spectacular depletions of Au, Ge, As, and Sb in the Moon are explained in this model as being partly due to the absence of metallic iron in the system during the volatilization–recondensation process. We noted earlier that this factor had an important effect on the condensation temperature of these elements. Finally, the similarity of S and Se in Earth's mantle and Moon suggest that any such separation occurred in an environment possessing a much lower relative abundance of hydrogen than the solar nebula (Section 11.7). These species are as highly volatile as H_2S and H_2Se in the solar nebula (Table 6.2).

Siderophile Elements

The important bearing of siderophile element geochemistry upon the problem of the Moon's origin was emphasized by O'Keefe (1972a,b) and O'Keefe and Urey (1977). The implications have been studied in detail by Ringwood and Kesson (1977), Rammensee and Wänke (1977), Delano and Ringwood (1978a,b) and Wänke et al. (1978b), and were discussed in Chapter 11, where it was shown that they provided key boundary conditions for the origin of the Moon. The principal conclusions reached were:

(a) The abundances of the siderophile elements Fe, Ni, Co, W, P, Os, Ir, S, and Se were similar (within a factor of about two) in low-Ti mare basalts and terrestrial oceanic tholeiites. This similarity probably extended to the respective source regions in the lunar and terrestrial mantles.

(b) The indigenous abundances of Fe, Ni, Co, W, P, S, and Se in the parental (PLC) magma of the lunar crust were also similar to those in terrestrial oceanic tholeiites.

(c) When the appropriate corrections were applied to depletions of Cu and Ga caused by volatility, the corrected abundances of these elements in low-Ti mare basalts and the PLC magma were similar to those in terrestrial oceanic tholeiites.

(d) The depletions of W and P in the Moon, Earth's mantle, and the parent bodies of eucritic meteorites require equilibration with, and separation of, about 20–30 percent of an iron-rich metallic phase.

(e) However, the maximum size of a lunar metallic core permitted by geophysical evidence amounts to only 2 percent of the mass of the Moon.

(f) Moreover, the source regions of low-Ti mare basalts and the PLC magma within the Moon have never been in equilibrium with an iron-rich metallic phase.

(g) The abundances of W, P, and perhaps S and Se in low-Ti mare basalts and the PLC magma are similar to those in eucritic meteorites. However, the abundances of Ni, Co, Ir, Os, Cu, and Ga are strongly depleted in eucrites as compared to low-Ti mare basalts and the PLC magma.

These relationships are truly remarkable. Siderophile elements have been strongly depleted in the Earth's mantle owing to their preferential entry into a metallic iron phase which segregated to form the core. However, we noted in Chapters 3 and 8 that the abundance patterns of siderophile elements in the Earth's mantle are complex. Several such elements—Ni, Co, Cu, Au, Ir, and Re—are much more abundant (factors of from 10 to > 100) in the mantle than would be expected if these elements had been partitioned under equilibrium conditions and low pressures into an Fe-rich metallic phase. According to the discussions in Chapters 3 and 8, these siderophile-element abundance patterns are the product of at least two major processes:

(a) the presence of large quantities of an element of low atomic weight, such as oxygen, within the segregating core, which caused large changes in metal/silicate partition coefficients;

(b) large changes in silicate/metal partition coefficients caused by high pressures deep within the Earth.

The important point is that each of these processes is intrinsically terrestrial. As shown by the discussion in Section 7.3 and Table 7.1, there is no apparent way by which these patterns could be caused by generalized cosmochemical fractionations within the solar nebula prior to accretion of the

Moon. *Thus, the relative abundances of siderophile elements in the Earth's upper mantle constitute a unique signature of terrestrial origin.*

These processes could not possibly have operated in the Moon, a body in which the pressure field extends to only 47 kbar as compared to 3900 kbar within the Earth, where the maximum size of the core is only 2 percent of the lunar mass as compared to a corresponding proportion of 32.5 percent in the case of the Earth, and where the source regions of low-Ti basalts and the PLC magma have never been in equilibrium with an iron-rich metallic phase. There is thus no way in which the siderophile-element abundance pattern of the Moon could have been established by differentiation processes occurring within the Moon.

We conclude that the siderophile element pattern of the Moon must have been generated within the mantle of the Earth, or in the mantle of an Earth-like planet. Economy in assumptions favor the former interpretation. It follows that the material in the Moon has been derived in some manner, ultimately, from the Earth's mantle.

The hypothesis of disintegrative capture (Wood and Mitler, 1974; Smith 1974) discussed in Section 12.3 derives its inspiration in part from the observation of differentiated meteoritic bodies such as eucrites, pallasites, and irons, and proposes that the Moon may have been derived from the mantles of such bodies. However, the large differences in Ni, Co, Ir, Os, Cu, and Ge abundances between low-Ti mare basalts and the PLC magma on the one hand, and eucrites on the other (Table 11.2), as well as oxygen isotope differences, show that the eucritic parent body could not have been involved. Likewise, the low Ni and Co contents of the olivines from pallasites show that their parent bodies were not involved. In fact, there is a general difficulty in that any small parent body which has melted and differentiated to segregate 20–30 percent of metal phase (necessary to account for the W and P depletions) will inevitably reduce the Ni, Co, Ir, and Os abundances to levels which are far below those inferred for the Moon.

Finally, we note that the disintegrative capture hypothesis is not likely to be highly efficient in separating the silicate mantle from the (molten) iron core (Section 12.3). There will inevitably be some iron mixed in with the silicates which would contradict the observation that low-Ti mare basalts and the PLC magma were not saturated with metallic iron. A similar dilemma confronts the coagulation hypothesis of Ruskol (1977), which invokes collisional processes to deplete the circumterrestrial planetesimal swarm in iron, prior to accretion of the Moon. We have already seen that the latter model cannot readily explain the Moon's depletion in volatile elements.

Further Evidence Relevant to a Possible Genetic Relationship Between the Earth and Moon

Lightner and Marti (1974) and Leich and Niemeyer (1975) found that some lunar highland rocks have trapped a xenon component possessing terrestrial isotopic composition. This component is released only at temperatures exceeding 1000°C. Since terrestrial xenon has a very characteristic and probably unique isotopic signature, the discovery of tightly bound terrestrial-type xenon in lunar rocks is of considerable interest. If proven indigenous to the Moon, rather than introduced by terrestrial contamination, a close genetic relationship between the Moon and Earth would be indicated.

Niemeyer and Leich (1976) have since demonstrated that fine grinding of lunar breccias in air can cause xenon to be tightly adsorbed, requiring temperatures above 1000°C to be released. Thus it is possible that the xenon in lunar rocks was introduced by terrestrial contamination. On the other hand, sample 60025, which was found by Lightner and Marti to contain terrestrial xenon, consisted of chips which were not ground. Moreover, this anorthosite contained anomalously high amounts of other volatile elements that are certainly of lunar origin (Krähenbuhl et al., 1973). Marti (personal communication) points out that the situation is not yet resolved and that the xenon may be indigenous to the Moon.

If the Moon was derived from the Earth's mantle subsequent to core formation, it is possible that there might be a measurable age difference between the Moon and smaller, less differentiated bodies such as the chondrites. Gancarz and Wasserburg (1977) and Tera and Wasserburg (1974) show that single-stage lead-isotope growth curves for the Moon and the Earth's mantle record essentially the same model age (4.4–4.45×10^9 years), which is significantly younger than the age of 4.55×10^9 years obtained for some meteorites. A time difference on the order of 10^8 years would be consistent with an acceptable time scale for accretion of the Earth and core formation (Sections 3.4 and 6.2).

Composition of the Moon's Lower Mantle

We concluded in Section 10.3 that the lower mantle of the Moon probably possessed higher FeO and SiO_2 contents than the Earth's mantle. The source regions of mare basalts may also have possessed significantly higher amounts of MnO and Cr_2O_3 than the Earth's mantle. How can these characteristics be explained in terms of a terrestrial origin for the Moon? The explanation proposed below is admittedly highly speculative.

Ringwood (1975a) pointed out that the heat produced during the early formation of the core would have caused strong convection in the mantle and melting of the outer mantle to a depth of 200–400 km (see also Section 3.4). This huge ultramafic magma ocean would then have been subjected to fractional crystallization, dominated by the separation of olivine. We have previously suggested that the Moon was formed from material evaporated from this magma ocean by several large planetesimal impacts. It is suggested that the material now constituting the highlands system was formed, perhaps by several impacts *soon after* core formation, before extensive fractional crystallization of the magma ocean had occurred. This material condensed outside Roche's limit to form several moonlets. Subsequently, olivine fractionation of the magma ocean led to a composition richer in FeO, SiO_2,

MnO, and Cr_2O_3 and significantly poorer ($\times 2$) in NiO, i.e., similar to the inferred composition of the source regions of mare basalts. A single large impact into the magma ocean then occurred at this stage, resulting in formation of a large moon nucleus with the composition of the mare basalt source regions. This large nucleus then rapidly swept up the earlier moonlets that accreted to form the lunar upper mantle. Accretion occurred on a time scale on the order of 100 years, causing the upper mantle to melt and differentiate, thereby forming a lunar crust.

There is one puzzling geochemical property of many lunar rocks which might also conceivably be explained by the above process. Mare basalts and many highlands rocks are usually depleted in the heavy rare earths as compared to the intermediate rare earths, as seen in Figure 10.6. Indeed, Nyquist et al. (1977) have suggested that this feature was characteristic of the Moon when it accreted, in contrast to the Earth, which has a "chondritic" REE pattern. The heavy REE depletion is suggestive of fractionation caused by the crystallization of garnet, but it is difficult to understand how this occurred within the Moon in view of the high pressures needed to bring garnet on to the liquidus. However, the pressures at depths of 200–300 km in the primitive terrestrial magma ocean (70–100 kbar) may well have caused garnet to have coprecipitated with olivine, thereby causing depletion in heavy rare earths. The lunar pattern could thus conceivably be of terrestrial origin.

12.5 Conclusion

There is, as yet, no general consensus concerning the origin of the Moon. Perhaps the most plausible theory on dynamical grounds is some variant of Ruskol's binary planet-coagulation hypothesis but the geochemical problems encountered are, in the author's view, insuperable. The same verdict applies to disintegrative capture and integral capture hypotheses. In addition, the latter is highly improbable on dynamical grounds.

In the final three chapters of this book, an array of geochemical evidence supporting the hypothesis that the Moon was derived from Earth's mantle subsequent to core formation has been marshalled. The key evidence, based on interpretations of siderophile and volatile element geochemistry, has been published only very recently, and there has not been sufficient time for a critical evaluation by scientific colleagues holding other views.[2] Also, the author is aware that he is too close to the issues at the present time to be capable of seeing them in a detached light. Under these circumstances, and until there has been an adequate degree of critical discussion in the literature, the reader would be well advised to approach the issues with a degree of cautious scepticism.

---

[2] See, however, Anders (1978), Warren and Wasson (1978), and the rejoinder by Delano and Ringwood (1978b).

Nevertheless, the evidence has been set out and must be considered by all who, in the future, wish to address the problem of lunar origin. It might be suggested that there is a valid analogy between the hypothesis of lunar origin favored herein and the hypothesis of continental drift as it stood during the mid 1950s. In that period, palaeomagnetic evidence, obtained particularly by Irving and Runcorn, strongly implied that in the past, continents had been displaced relative to one another by thousands of kilometers, as compared to their present positions. This evidence and its interpretation were rejected by most Earth scientists, primarily because they could not envisage a *mechanism* for causing continental drift. The rest is history. During the subsequent 10 years, further sources of evidence emerged together with new concepts, and the *theory* of plate tectonics involving a specific mechanism for causing continental displacements became generally accepted almost overnight.

In the case of the Moon, the author's opinion is that the geochemical evidence *uniquely requires* that is was derived, in some way, from the Earth's mantle subsequent to core formation. The problem of a suitable mechanism still exists, though recent developments in understanding the effects of impacts of large planetesimals on the Earth's mantle seem to provide some grounds for optimism. At least, the problem of mechanism does not look so intrinsically intractable today as continental drift seemed to most earth scientists in 1956.

An attractive feature of the present hypothesis of lunar origin is that it is capable in principle of being tested and evaluated through the acquisition of new data, using existing scientific methods and technology. The following investigations would be especially worthy in this respect. (The order does not imply any relative degrees of importance.)

(a) Comprehensive studies of the abundances and distribution of siderophile elements in lunar rocks on a much larger scale than has presently been attempted.

(b) Experimental measurements of partition coefficients for siderophile elements between metallic phases and silicate phases of relevant compositions over a wide range of T, P, and fO_2 conditions.

(c) Comprehensive studies of the abundances and distribution of siderophile elements in terrestrial basaltic and ultramafic rocks, particularly in the most primitive, least fractionated and least altered varieties. It is disquieting to find that, for many siderophile elements, much more is known about their distribution in lunar rocks than in terrestrial rocks.

(d) Experimental studies of the fractionation behavior of siderophile elements during the crystallization of ultramafic and mafic magmas and during partial melting processes leading to basalt formation.

(e) A search by Earth-based telescopes using reflection spectroscopy to identify regions on the Moon containing the most *primitive* mare basalts, high in Mg and low in Ti, that might be capable of providing the most direct chemical information about their source regions in the lunar interior. When such basalts have been identified, an attempt should be made to recover samples and return them to Earth via automated spacecraft of the type successfully employed by the USSR.

(f) Comprehensive chemical and petrological studies of such primitive basalts, combined with application of experimental petrology methods, to constrain mineralogical and chemical composition of source regions and depth of source regions.

(g) Development of automated spacecraft capable of landing on the Moon and measuring heat flow *in situ* at several different locations.

(h) Completion of the proposed lunar "polar-orbiter" project which would provide a much more accurate lunar moment-of-inertia coefficient, as well as vast amounts of key geochemical and other data about the lunar surface.

(i) Space flights to Mars and Venus, temporarily entering their atmospheres to "scoop up" some atmospheric gases and return them to Earth, where they can be accurately analyzed. In particular, it is essential to find out whether the oxygen isotope composition of the CO_2 in these atmospheres belongs to the lunar–terrestrial ^{16}O–^{17}O–^{18}O system or whether these planets have different amounts of the ^{16}O supernova component. Likewise, xenon isotopic analyses would show whether terrestrial xenon is, in fact, isotopically unique.

(j) Detailed studies of the tightly bound atmospheric xenon in lunar breccias to establish whether this xenon is indigenous to the Moon or results from terrestrial contamination.

(k) Higher resolution stable isotope studies of other elements, e.g., nitrogen, in lunar and terrestrial rocks, meteorites, and samples from other planets, to test whether the Moon and Earth are more closely related than are the Earth, meteorites, and other planets.

(l) Further detailed geochronological studies of lunar rocks, terrestrial rocks, and meteorites to determine whether the Moon is significantly younger than some classes of meteorites.

Several aspects of this program can be carried out using existing laboratory facilities available in many countries. Others, such as items (g) and (h), seem feasible in principle but would require further development involving international collaboration. It is to be hoped that as many as possible of these investigations can be carried out over the next decade. If so, there is a reasonable chance that by then we might achieve a widely acceptable *theory* of lunar origin, rather than a multitude of incomplete and competing hypotheses as is now the case. This would be a worthy sequel to the magnificently successful Apollo Program.

References

Adams, J. B. and T. B. McCord (1979). Mercury: Evidence for an anorthositic crust from reflection spectra. *Bull. Am. Astron. Soc.* 1979 (in press).

Adler, I., J. Trombka, R. Schmadebeck, P. Lowman, N. Blodget, L. Yin, and E. Eller (1973). Results of the Apollo 15 and 16 x-ray experiment. *Proc. Fourth Lunar Sci. Conf.* **3**, 2783–2791.

Ahrens, T. J. (1976). Shock wave data for pyrrhotite and constraints on the composition of the outer core (abstract). U.S.–Japan Seminar on *High-Pressure Research Applications in Geophysics*, 31p.

D. L. Anderson, and A. E. Ringwood (1969). Equations of state and crystal structures of high pressure phases of shocked silicates and oxides. *Rev. Geophys.* **7**, 667–701.

Akaogi, M. and S. Akimoto (1977). Pyroxene-garnet solid solution equilibria in the systems $Mg_4Si_4O_{12}$–$Mg_3Al_2Si_3O_{12}$ and $Fe_4Si_4O_{12}$–$Fe_3Al_2Si_3O_{12}$ at high pressures and temperatures. *Phys. Earth Planet Inter.* **15**, 90–106.

Akimoto, S. I. (1972). The system MgO–FeO–SiO_2 at high pressures and temperatures —phase equilibria and elastic properties. *Tectonophysics* **13**, 161–187.

and H. Fugisawa (1968). Olivine–spinel solid solution equilibria in the system Mg_2SiO_4–Fe_2SiO_4. *J. Geophys. Res.* **73**, 1467–1479.

Akimoto, S., M. Akaogi, K. Kawada and O. Nishizawa (1976a). Mineralogic distribution of iron in the upper half of the transition zone in the Earth's mantle. *Am. Geophys. U. Monogr.* **19**, 399–405.

Y. Matsui, and Y. Syono (1976b). High pressure crystal chemistry of orthosilicates and the formation of the mantle transition zone. In R. G. J. Strens (ed.): *The Physics and Chemistry of Minerals and Rocks.* Wiley, New York, pp. 327–363.

Alfvén, H. (1942). Remarks on the rotation of a magnetized sphere with application to solar rotation. *Avk. Mat. Astron. Fys. A*, **28**(6), 2–9.

(1954). *On the Origin of the Solar System*. Oxford Univ. Press, London.

(1964a). On the formation of celestial bodies. *Icarus* **3**, 57–62.

(1964b). On the origin of the asteroids. *Icarus* **3**, 52–56.

and G. Arrhenius (1969). Two alternatives for the history of the moon. *Science* **165**, 11–17.

Al'tshuler, L. V. and S. Kormer (1961). On the internal structure of the Earth. *Izv. Akad. Nauk. SSSR. Ser. Geofiz.* **1**, 33–37.

K. Krupnikov, K. Lebedev, B. Zhuchihin, and M. Brazhnik (1958) Dynamic compressibility and equation of state of iron at high pressure. *Zh. Eksp. Teor. Fiz.* **34**, 874–885.

A. Bakanova, and R. Trunin (1962). Shock adiabats and zero isotherms of seven metals at high pressures. *Zh. Eksp. Teor. Fiz.* **42**, 91–104.

R. Trunin, and G. Simakov (1965). Shock compression of periclase and of quartz and composition of the lower mantle of the Earth. *Izv. Akad. Nauk. SSSR, Fiz. Zemle* **10**, 1–6.

R. Trunin and G. Simakov (1968). On the question of chemical composition of the Earth's core. *Izv. Akad. Nauk. SSSR, Fiz. Zemle* **1**, 3–6.

Anders, E. (1964). Origin, age, and composition of meteorites. *Space Sci. Rev.* **3**, 583–714.

(1968). Chemical processes in the early solar system, as inferred from meteorites. *Accounts Chem. Res.* **1**, 289–298.

(1971). Meteorites and the early solar system. *Ann. Rev. Astron. Astrophys.* **9**, 1–34.

(1977). Chemical compositions of the Moon, Earth, and eucrite parent body. *Phil. Trans. Roy. Soc. London* **A295**, 23–40.

(1978). Procrustean Science: Indigenous siderophiles in the lunar highlands, according to Delano and Ringwood. *Proc. Ninth Lunar Plan. Sci. Conf.* **1**, 161–184.

R. Ganapathy, U. Krähenbühl, and J. W. Morgan (1973). Meteoritic material on the moon. *The Moon* **8**, 3–24.

Anderson, A. T., R. N. Clayton, and T. K. Mayeda (1971). Oxygen isotope thermometry of mafic igneous rocks. *J. Geol.* **79**, 715–729.

Anderson, D. L. (1967). Phase changes in the upper mantle. *Science* **157**, 1165–1173.

(1968). Chemical heterogeneity of the mantle. *Earth Planet. Sci. Lett.* **5**, 89–94.

(1970). Petrology of the mantle. *Miner. Soc. Am. Spec. Paper* **3**, 85–93.

(1972a). Internal constitution of Mars. *J. Geophys. Res.* **77**, 789–795.

(1972b). Implications of the inhomogeneous planetary accretion hypothesis. *Comments Earth Sci.: Geophys.* **2**, 93–98.

(1973). The composition and origin of the moon. *Earth Planet. Sci. Lett.* **18**, 301–316.

(1977). Composition of the mantle and core. *Ann. Rev. Earth Planet. Sci.* **5**, 179–202.

and T. Jordan (1970). The composition of the lower mantle. *Phys. Earth Planet. Inter.* **3**, 23–35.

C. Sammis, and T. Jordan (1971). Composition and evolution of the mantle and core. *Science* **171**, 1103–1113.

C. Sammis, and T. Jordan (1972). Composition of the mantle and core. In E. C. Robertson (ed.) *The Nature of the Solid Earth.* McGraw-Hill, New York, pp. 41–66.

Anderson, O. L. (1966). A proposed law of corresponding states for oxide compounds. *J. Geophys. Res.* **71**, 4963–4971.

Archbald, P. (1978). Unpublished results. Australian National University, Canberra.

Andre, C. G., R. W. Wolfe, and I. Adler (1977). Evidence for a high-magnesium sub-surface basalt in Mare Crisium from orbital x-ray fluorescence data. *Conference on Luna 24, Abstract,* Lunar Science Institute, Houston, Texas, pp. 8–11.

Argollo, R. M. (1974). Ga-Al and Ge-Si in volcanic rocks. M.Sc. Thesis (unpublished), Univ. Rhode Island.

Arkani-Hamed, J. (1974). Stress constraint on the thermal evolution of the Moon. *Proc. Fifth Lunar Sci. Conf.* **3**, 3127–3134.

Arndt, N. T. (1977). Ultra basic magmas and high degree melting of the mantle. *Contrib. Mineral. Petrol.* **64**, 205–221.

A. Naldrett, and D. Pyke (1977). Komatiitic and iron-rich tholeiitic lavas of Munro Township, Northeast Ontario. *J. Petrol.* **18**, 319–318.

Arnold, J. R., A. E Metzger, and R. Reedy (1977). Computer-generated maps of lunar composition from gamma-ray data. *Proc. Eighth Lunar Sci. Conf.* **1**, 945–948.

Baedecker, P. A., R. Shaudy, J. Elzie, J. Kimberlin, and J. Wasson (1971). Trace element studies of rocks and soils from Oceanus Procellarum and Mare Tran-quillitatas. *Proc. Second Lunar Sci. Conf.* **2**, 1037–1061.

C.-L. Chou, E. B. Grudewicz, and J. T. Wasson (1973). Volatile and siderophilic trace elements in Apollo 15 samples: Geochemical implications and characterization of the long-lived and short-lived extralunar materials. *Proc. Fourth Lunar Sci. Conf.* **2**, 1177–1195.

Baldwin, R. B. (1949). *The Face of the Moon.* Chicago University Press.

(1963). *The Measure of the Moon.* Chicago University Press.

Banerjee, S. K. (1967). Fractionation of iron in the solar system. *Nature* **216**, 781.

Barsukov, V. L., L. Tarasov, L. Dimitriev, G. Kolesov, I. Shevaleevsky, and A. Garinin (1977). The geochemical and petrological features of regolith and rocks from Mare Crisium. *Proc. Eighth Lunar Sci. Conf.* **3**, 3319–3332.

Bielefield, M. J., C. Andre, E. Eliason, P. Clarke, I. Adler, and J. Trombka (1977). Imaging of lunar surface chemistry from orbital x-ray data. *Proc. Eighth Lunar Sci. Conf.* **1**, 901–908.

Biemann, K., T. Owen, D. Rushneck, A. La Fleur, and D. Howarth (1976). The atmosphere of Mars near the surface: Isotope ratios and upper limits on noble gases. *Science* **194**, 76–77.

Bills, B. G. and A. Ferrari (1977). A lunar density model consistent with topographic, gravitational, librational, and seismic data. *J. Geophys. Res.* **82**, 1306–1314.

Binder, A. B. (1969). Internal structure of Mars. *J. Geophys. Res.* **74**, 3110–3118.

(1974). The origin of the Moon by rotational fission. *The Moon* **11**, 53–76.

(1976). On the petrology and early development of a moon of fission origin. *The Moon* **15**, 275–314.

and D. R. Davis (1973). Internal structure of Mars. *Phys. Earth Planet. Inter.* **7**, 477–485.

Birch, F. (1952). Elasticity and constitution of the Earth's interior. *J. Geophys. Res.* **57**, 227–286.

(1958). Differentiation of the mantle. *Bull. Geol. Soc. Am.* **69**, 483–486.

(1961). The velocity of compressional waves in rocks to 10 kilobars, 2. *J. Geophys. Res.* **66**, 2199–2224.

(1963). Some geophysical applications of high pressure research. In: W. Paul and D. Warschauer (eds.) *Solids Under Pressure*. McGraw-Hill, New York, pp. 137–162.

(1965). Speculations on the earth's thermal history. *Bull. Geol. Soc. Am.* **76**, 133–154.

(1969). Density and composition of the upper mantle: First approximation as an olivine layer. In *Geophysical Monograph* **13**, American Geophysical Union, Washington D.C., pp. 18–36.

Black, L. P., N. Gale, S. Moorbath, and R. Pankhurst (1971). Isotopic dating of very early Precambrian amphibolite facies gneisses from the Godthaab district, West Greenland. *Earth Planet Sci. Lett.* **12**, 245–259.

Blackshear, W. T. and J. P. Gapcynski (1977). An improved value of the lunar moment of inertia. *J. Geophys. Res.* **82**, 1699–1701.

Blanchard, D. P., J. Rhodes, M. Dungan, K. Rodgers, C. Donaldson, J. Brannon, J. Jacobs, and E. Gibson (1976). The chemistry and petrology of basalts from Leg 37 of the Deep Sea Drilling Project. *J. Geophys. Res.* **81**, 4232–4246.

L. A. Haskin, J. C. Brannon, and E. Aaboe (1977). Chemistry of soils and particles from Luna 24. *Conference on Luna 24* (Abstract). Lunar Science Institute, Houston, Texas, pp. 37–40.

Blerkom, D. V. and L. Auer (1976). The geometry of VY Canoris Majoris derived from SiO maser lines. *Astrophys. J.* **204**, 775–780.

Bodenheimer, P. (1972). Stellar evolution toward the main sequence. *Rep. Prog. Phys.* **35**, 1–54.

(1974). Calculations of the early evolution of Jupiter. *Icarus* **23**, 319–325.

(1977). Calculations on the effects of angular momentum on the early evolution of Jupiter. *Icarus* **31**, 356–368.

Boyce, J. M., A. L. Dial, and L. A. Soderblom (1974). Ages of the lunar nearside light plains and maria. *Proc. Fifth Lunar Sci. Conf.* **1**, 11–23.

Boynton, W. V. and J. T. Wasson (1977). Distribution of 28 elements in size fractions of lunar mare and highland soils. *Geochim. Cosmochim. Acta* **41**, 1073–1082.

C.-L. Chou, K. L. Robinson, P. H. Warren, and J. T. Wasson (1976). Lithophiles, siderophiles and volatiles in Apollo 16 soils and rocks. *Proc. Seventh Lunar Sci. Conf.* **1**, 727–742.

Brett, R. (1973). A lunar core of Fe-Ni-S. *Geochim. Cosmochim. Acta* **37**, 165–170.

(1971). The earth's core: Speculations on its chemical equilibrium with the mantle. *Geochim. Cosmochim. Acta* **35**, 203–221.

(1976). The current status of speculations on the composition of the core of the Earth. *Rev. Geophys. Space Phys.* **14**, 375–383.

P. Butler, C. Meyer, A. Reid, H. Takeda, and R. Williams (1971). Apollo 12 igneous rocks 12004, 12008, 12009, and 12022: A mineralogical and petrological study. *Proc. Second Lunar Sci. Conf.* **1**, 301–317.

Brey, G. and D. H. Green (1976). Solubility of CO_2 in olivine melilitite at high pressure and the role of CO_2 in the Earth's upper mantle. *Contrib. Mineral. Petrol.* **55**, 217–230.

Brown, H. (1952). Rare gases and the formation of the earth's atmosphere. In C. P. Kuiper (ed.): *The Atmospheres of the Earth and Planets*, 2nd Ed. Chicago University Press, pp. 258–266.

Brown, R. D. (1974). Organic matter in interstellar space. In J. P. Wild (ed.): "*In the Beginning . . .* " Copernicus 500th Birthday Symposium, Australian Academy of Science, Canberra, Chapter 1, pp. 1–14.

Bullen, K. E. (1936). The variation of density and the ellipticities of strata of equal density within the earth. *Mon. Not. Roy. Astron. Soc., Geophys. Suppl.* **3**, 395–401.

(1940). The problem of the earth's density variation. *Bull. Seismol. Soc. Am.* **30**, 235–250.

(1974). *An Introduction to the Theory of Seismology*. Cambridge Univ. Press, London.

(1952). Cores of the terrestrial planets. *Nature* **170**, 363–364.

(1963). *Introduction to the Theory of Seismology*. Cambridge Univ. Press, London.

(1973a). Cores of the terrestrial planets. *Nature* **243**, 68–70.

(1973b). On planetary cores. *The Moon* **7**, 384–395.

Cameron, A. G. W. (1963a). Formation of the solar nebula. *Icarus* **1**, 339–342.

(1963b). The origin of the atmospheres of Venus and the Earth. *Icarus* **2**, 249–257.

(1969). Physical conditions in the primitive solar nebula. In P. Millman (ed.): *Meteorite Research*. Reidel, Dordrecht, Holland, pp. 7–15.

(1973a). Accumulation processes in the primitive solar nebula. *Icarus* **18**, 407–450.

(1973b). Formation of the outer planets. *Space Sci. Rev.* **14**, 383–391.

(1975a). The origin and evolution of the solar system. *Sci. Am.* **233**, 32–41.

(1975b). Cosmogonical considerations regarding Uranus. *Icarus* **24**, 280–284.

(1977). Formation of the outer planets and satellites. In J. Burns (ed.): *Planetary Satellites*, Proceedings of the IAU Colloquium No. 28 (in press).

(1978). Physics of the primitive solar accretion disc. *The Moon and the Planets*, **18**, 5–40.

and M. R. Pine (1973). Numerical models of the primitive solar nebula. *Icarus* **18**, 377–406.

and J. B. Pollack (1976). On the origin of the solar system and of Jupiter and its satellites. In T. Gehrels (ed.): *Jupiter, the Giant Planet*. University of Arizona Press, Tucson.

and W. R. Ward (1976). The origin of the Moon. *Lunar Science* **7**, 120–122.

Carswell, D. A. and J. B. Dawson (1970). Garnet peridotite xenoliths in South African kimberlite pipes and their petrogenesis. *Contrib. Mineral. Petrol.* **25**, 163–184.

Carver, J. H., B. Horton, D. McCoy, R. O'Brien, and E. Sandercock (1975). Comparison of lunar ultraviolet reflectivity with that of terrestrial rock samples. *The Moon* **12**, 91–100.

Cassen, P. (1977). Planetary magnetism and the interior of the Moon and Mercury. *Phys. Earth Planet. Inter.* **15**, 113–120.

R. Young, G. Schubert, and R. Reynolds (1976). Implications of an internal dynamo for the thermal history of Mercury. *Icarus* **28**, 501–508.

Chapman, C. R. (1976). Asteroids as meteorite parent-bodies: The astronomical perspective. *Geochim. Cosmochim. Acta* **40**, 701–719.

and D. R. Davis (1975). Asteroid collisional evolution: Evidence for a much larger early population. *Science* **190**, 553–556.

Chapman, D. R. and H. K. Larson (1963). On the lunar origin of tektites. *J. Geophys. Res.* **68**, 4305–4358.

Chappell, B. W. and D. H. Green (1973). Chemical composition and petrogenetic relationships in Apollo 15 mare basalts. *Earth Planet. Sci. Lett.* **18**, 237–246.

Charette, M. P., S. R. Taylor, J. B. Adams, and T. B. McCord (1977). The detection of soils of Fra Mauro basalt and anorthosite gabbro composition in the lunar highlands by remote spectral reflectance techniques. *Proc. Eighth Lunar Sci. Conf.* **1**, 1049–1061.

Chou, C.-L., W. V. Boynton, L. L. Sundberg, and J. T. Wasson (1975). Volatiles on the surface of Apollo 15 green glass and trace-element distributions among Apollo 15 soils. *Proc. Sixth Lunar Sci. Conf.* **2**, 1701–1727.

Clark, S. P. and A. E. Ringwood (1964). Density distribution and constitution of the mantle. *Rev. Geophys.* **2**, 35–88.

K. Turekian, and L. Grossman (1972). Model for early history of the earth. In E. C. Robertson (ed.): *The Nature of the Solid Earth.* McGraw-Hill, New York, pp. 3–18.

Clarke, D. B. (1970). Tertiary basalts of Baffin Bay: Possible primary magma from the mantle. *Contrib. Mineral. Petrol.* **25**, 203–224.

Clayton, R. N. and T. Mayeda (1975). Genetic relations between the Moon and Meteorites. *Proc. Sixth Lunar Sci. Conf.* **2**, 1761–1769.

L. Grossman, and T. Mayeda (1973). A component of primitive nuclear composition in carbonaceous chondrites. *Science* **182**, 485–488.

N. Onuma, and T. K. Mayeda (1976). A classification of meteorites based on oxygen isotopes. *Earth Planet. Sci. Lett.* **30**, 10–18.

Cleary, J. R. (1974). The D″ region. *Phys. Earth Planet. Inter.* **9**, 13–27.

Coish, R. A. and L. A. Taylor (1977). Igneous rocks from Mare Crisium: *Mineralogy and Petrology Conference on Luna 24*, Abstracts. Lunar Science Institute, Houston, Texas, pp. 48–50.

Compston, W. H., H. Berry, M. Vernon, B. Chappell, and M. Kaye (1971). Rb–Sr chronology and chemistry of lunar material from the Ocean of Storms. *Proc. Second Lunar Sci. Conf.* **2**, 1471–1485.

Cook, A. H. (1977). The moment of inertia of Mars and the existence of a core. *Geophys. J.* **51**, 349–356.

Craig, H. and J. E. Lupton (1976). Primordial neon, helium, and hydrogen in oceanic basalts. *Earth Planet. Sci. Lett.* **31**, 369–385.

Crocket, J. H. (1971). Neutron activation analysis for noble metals in geochemistry. In A. Brunfelt and E. Steinnes (eds.): Proceedings N.A.T.O. Advanced Study Institute: *Activation Analysis in Geochemistry and Cosmochemistry*. Universitets-forlaget, Oslo, pp. 339–350.

and L. L. Chyi (1972). Abundances of Pd, Ir, Os, and Au in an ultramafic pluton. *Proceedings of the 24th International Geological Congress*, Montreal, *Section 10*, pp. 202–209.

and Y. Teruta (1977). Palladium, iridium and gold contents of mafic and ultra-mafic rocks drilled from the Mid-Atlantic ridge. *Can. J. Earth Sci.* **14**, 777–785.

Cumming, G. L. and J. R. Richards (1975). Ore lead ratios in a continuously changing Earth. *Earth Planet. Sci. Lett.* **28**, 155–171.

Dainty, A. M., M. N. Toksöz, S. Solomon, K. Anderson, and N. Goins (1974). Constraints on lunar structure. *Proc. Fifth Lunar Sci. Conf.* **3**, 3091–3114.

M. N. Toksöz, and S. Stein (1976). Seismic investigation of the lunar interior. *Proc. Seventh Lunar Sci. Conf.* **3**, 3057–3075.

Darwin, G. H. (1880). On the secular changes in the orbit of a satellite revolving around a tidally disturbed planet. *Phil. Trans. Roy. Soc. London* **171**, 713–891.

(1908). *Tidal Friction and Cosmogony. Scientific Papers 2*. Cambridge University Press, London.

(1962). *The Tides*. Friedman, San Francisco.

Davies, G. F. and A. M. Dziewonski (1975). Homogeneity and constitution of the Earth's lower mantle and core. *Phys. Earth Planet. Inter.* **10**, 330–343.

and E. S. Gaffney (1973). Identification of phases of rocks and minerals from Hugoniot data. *Geophys. J.* **33**, 165–183.

Dayhoff, M., R. Eck, E. Lippincott, and C. Sagan (1967). Venus: Atmospheric evolution. *Science* **155**, 556–558.

Delano, J. W. and A. E. Ringwood (1978a). Indigenous abundances of siderophile elements in the lunar highlands: Implications for the origin of the Moon. *The Moon and Planets* **18**, 385–425.

and A. E. Ringwood (1978b). Siderophile elements in the lunar highlands: Nature of the indigenous component and implications for origin of the Moon. *Proc. Ninth Lunar Planet. Sci. Conf.*

Dickinson, W. R. and W. C. Luth (1971). A model for plate tectonic evolution of mantle layers. *Science* **174**, 400–404.

Distin, P. A., S. Whiteway, and C. Masson (1971). Solubility of oxygen in liquid iron from 1785° to 1960°. A new technique for the study of slag-metal equilibria. *Can. Metall. Quart.* **10**, 13–18.

Dodd, R. T. (1969). Metamorphism of ordinary chondrites. *Geochim. Cosmochim. Acta* **33**, 161–203.

(1974). The metal phase in unequilibrated chondrites and its implications for calculated accretion temperature. *Geochim. Cosmochim. Acta* **36**, 485–494.

Dreibus, G., B. Spettel and H. Wänke (1976). Lithium as a correlated element, its condensation behavior, and its use to estimate the bulk composition of the moon and the eucrite parent body. *Proc. Seventh Lunar Sci. Conf.* **3**, 3383–3396.

    H. Kruse, B. Spettel, and H. Wänke (1977). The bulk composition of the moon and the eucrite parent body. *Proc. Eighth Lunar Sci. Conf.* **1**, 211–227.

Duba, A. and A. E. Ringwood (1973). Electrical conductivity, internal temperatures and thermal evolution of the Moon. *The Moon* **7**, 356–376.

Dubrovskiy, V. A. and V. Pan'kov (1972). On the composition of the earth's core. *Izvestiya, Physics of the Solid Earth*, **7**, 48–54.

Duke, M. B. and L. T. Silver (1967). Petrology of eucrites, howardites and mesosiderites. *Geochim. Cosmochim. Acta* **31**, 1637–1665.

Dziewonski, A., A. Hales and E. Lapwood (1975). Parametrically simple earth models consistent with geophysical data. *Phys. Earth Planet. Inter.* **10**, 12–48.

Edgeworth, K. E. (1949). The origin and evolution of the solar system. *Mon. Not. Roy. Astron. Soc.* **109**, 600–609.

Elsasser, W. M. (1963). Early history of the Earth. In J. Geiss and E. Goldberg (eds.): *Earth Science and Meteorites.* North Holland, Amsterdam, pp. 1–30.

Eucken, A. (1944). Physikalische-chemische Betrachtungen über der früeste Entwick-lungsgesichte der Erde. *Nachr. Akad. Wiss. Göttingen, Math-Phys. Kl.* **1**, 1–25.

Fish, F. (1967). Angular momenta of the planets. *Icarus* **7**, 251–256.

Fitch, T. J. (1975). Compressional velocity in source regions of deep earthquakes: An application of the master earthquake technique. *Earth Planet. Sci. Lett.* **26**, 156–165.

    (1977). *In situ* P-wave velocities in deep earthquake zones of the S. W. Pacific: Evidence for a phase boundary between the upper and lower mantle. In M. Talwani and W. C. Pitman III (eds.): *Island Arcs, Deep Sea Trenches and Back-Arc Basins* American Geophysical Union, Washington D. C., Maurice Ewing Series 1, pp. 123–136.

Flasar, F. M. and F. Birch (1973). Energetics of core formation: A correction. *J. Geophys. Res.* **78**, 6101–6103.

Florensky, C. P. (1965). On the initial stages of differentiation of earth materials. *Geokhimiya* **8**, 909–917.

    A. Basilevsky, G. Burba, O. Nicolaeva, A. Pronin, V. Volkov, and L. Ronca (1977). First panoramas of the Venusian surface. *Proc. Eighth Lunar Sci. Conf.* **3**, 2655–2664.

Fredriksson, K. (1969). The Sharps chondrite—new evidence on the origin of chon-drules and chondrites. In P. M. Millman (ed.): *Meteorite Research*, Reidel, Dordrecht, Holland, pp. 155–165.

    A. Noonan and J. Nelen (1973). Meteoritic, lunar and Lonar impact chondrules. *The Moon* **7**, 475–482.

Frey, F. A. and D. H. Green (1974). The mineralogy, geochemistry and origin of lherzolite inclusions in Victorian basanites. *Geochim. Cosmochim. Acta* **38**, 1023–1059.

    W. Bryan and G. Thompson (1974). Atlantic Ocean Floor: Geochemistry and petrology of basalts from Legs 2 and 3 of the Deep Sea Drilling Project. *J. Geophys. Res.* **79**, 5507–5528.

Friel, J. J. and J. Goldstein (1976). An experimental study of phosphate reduction and phosphorus-bearing lunar metal particles. *Proc. Seventh Lunar Sci. Conf.* **1**, 791–806.

Gaffey, M. J. and T. B. McCord (1977). Asteroid surface materials: Mineralogical characteristics and cosmological implications. *Proc. Eighth Lunar Sci. Conf.* **1**, 113–143.

Gahm, G. E., H. Nordh, and S. Olofsson (1975). The T-Tauri star RU Lupi and its circumstellar surrounding. *Icarus* **24**, 285–293.

Ganapathy, R. and E. Anders (1974). Bulk compositions of the moon and earth, estimated from meteorites. *Proc. Fifth Lunar Sci. Conf.* **2**, 1181–1206.

R. Keays, J. Laul, and E. Anders (1970). Trace elements in Apollo 11 lunar rocks: Implications for meteorite influx and origin of moon. *Proc. Apollo II Lunar Sci. Conf.* **2**, 1117–1142.

Gancarz, A. J. and G. J. Wasserburg (1977). Initial Pb of the Amitsoq gneiss, West Greenland and implications for the age of the Earth. *Geochim. Cosmochim. Acta* **41**, 1283–1301.

Garz, T., M. Kock, J. Richter, B. Baschwek, H. Holweger, and A. Ünsold (1969). Abundances of iron and some other elements in the sun and in meteorites. *Nature* **223**, 1254–1256.

Gast, P. W. (1960). Limitations on the composition of the upper mantle. *J. Geophys. Res.* **65**, 1287–1297.

(1968). Trace element fractionation and the origin of tholeiitic and alkaline magma types. *Geochim. Cosmochim. Acta* **32**, 1057–1086.

Gault, D. E. and E. Heitowit (1963). The partition of energy for hypervelocity impact craters formed in rock. *Sixth Hypervelocity Impact Symposium*, Cleveland, Ohio.

Gehrels, T., J. Gill and J. Haughey (1971). Introduction. In T. Gehrels (ed.): *Physical Studies of Minor Planets*. NASA, Washington, D.C.

Gerstenkorn, H. (1955). Über Gezeitenreibung beim Zweikörper problem. *Z. Astrophys.* **36**, 245–274.

(1969). The earliest past of the earth-moon system. *Icarus* **11**, 189–207.

Giuli, R. T. (1968a). On the rotation of the Earth produced by gravitational accretion of particles. *Icarus* **8**, 301–323.

(1968b). Gravitational accretion of small masses attracted from large distances as a mechanism for planetary rotation. *Icarus* **9**, 186–190.

Glikson, A. Y. (1970). Geosynclinal evolution and geochemical affinities of early Precambrian systems. *Tectonophysics* **9**, 397–433.

Goins, N. R., A. Dainty and M. N. Toksöz (1977). The deep seismic structure of the moon. *Proc. Eighth Lunar Sci. Conf.* **1**, 471–486.

Goldreich, P. (1966). History of the lunar orbit. *Rev. Geophys.* **4**, 411–439.

and S. Soter (1966). *Q* in the solar system. *Icarus* **5**, 375–389.

and W. R. Ward (1973). The formation of planetesimals. *Astrophys. J.* **183**, 1051–1061.

Goldschmidt, V. M. (1922). Über die Massenverteilung im Erdinneren, verglichen mit der Structur gewisser Meteoriten. *Naturwissenschaften* **42**, 1–3.

Goles, G. C. (1967). Trace elements in ultramafic rocks. In P. J. Wyllie (ed.): *Ultramafic and Related Rocks*, Wiley, New York, pp. 352–362.

Graboske, H. C., J. Pollack, A. Grossman, and R. Olness (1975). The structure and evolution of Jupiter: The fluid contraction stage. *Astrophys. J.* **199**, 265–281.

Graham, E. K. (1973). On the compression of stishovite. *Geophys. J.* **32**, 15–34.

Gray, C. M. and W. Compston (1974). Excess ^{26}Mg in the Allende meteorite. *Nature* **251**, 495–497.

Green, D. H. (1963). Alumina content of enstatite in a Venezuelen high-temperature peridotite. *Bull. Geol. Soc. Am.* **74**, 1397–1402.

(1970a). The origin of basaltic and nephelinitic magmas. *Trans. Leicester Lit. Phil. Soc.* **64**, 28–54.

(1970b). A review of experimental evidence on the origin of basaltic and nephelinitic magmas. *Phys. Earth Planet. Inter.* **3**, 221–235.

(1971). Composition of basaltic magmas as indicators of conditions of origin: Application to oceanic volcanism. *Phil. Trans. Roy. Soc. London* **A268**, 707–725.

(1972). Magmatic activity as the major process in the chemical evolution of the Earth's crust and mantle. *Tectonophysics* **13**, 47–71.

and A. E. Ringwood (1967). The genesis of basaltic magmas. *Contrib. Mineral. Petrol.* **15**, 103–190.

and A. E. Ringwood (1973). Significance of a primitive lunar basaltic composition present in Apollo 15 soils and breccias. *Earth Planet. Sci. Lett.* **19**, 1–8.

A. E. Ringwood, N. G. Ware, W. O. Hibberson, A. Major, and E. Kiss (1971a). Experimental petrology and petrogenesis of Apollo 12 basalts. *Proc. Second Lunar Sci. Conf.* **1**, 601–615.

N. G. Ware, W. O. Hibberson, and A. Major (1971b). Experimental petrology of Apollo 12 basalts. Part I, sample 12009. *Earth Planet. Sci. Lett.* **13**, 85–96.

I. Nicholls, M. Viljoen, and R. Viljoen (1975). Experimental demonstration of the existence of peridotitic liquids in earliest Archaean magmatism. *Geology* **3**, 11–14.

Gros, J., H. Takahashi, H. Hertogen, J. W. Morgan, and E. Anders (1976). Composition of the projectiles that bombarded the lunar highlands. *Proc. Seventh Lunar Sci. Conf.* **2**, 2403–2425.

Grossman, L. (1972). Condensation in the primitive solar nebula. *Geochim. Cosmochim. Acta* **36**, 597–619.

and J. Larimer (1974). Early chemical history of the solar system. *Rev. Geophys. Space Phys.* **12**, 71–101.

and E. Olson (1974). Origin of the high-temperature fraction of C2 chondrites. *Geochim. Cosmochim. Acta* **38**, 173–187.

Gunten, von H. P., U. Krähenbühl, G. Meyer, and F. Wegmüller (1978). On the partition of volatile trace elements between minerals and agglutinates in grain size fractions of Apollo 17 soils. *Lunar and Planetary Science* **9**, 436–438. Lunar and Planetary Institute, Houston.

Gurney, J. J. and L. H. Ahrens (1973). Zinc content of some ultramafic and basic rocks. *Trans. Geol. Soc. South Africa* **73**, 301–307.

Haddon, R. W. and J. R. Cleary (1974). Evidence for scattering of seismic PKP waves near the mantle–core boundary. *Phys. Earth Planet. Interiors* **8**, 211–234.

Hall, H. and V. R. Murthy (1971). Early chemical history of the earth: Some critical elemental fractionations. *Earth Planet. Sci. Lett.* **11**, 239–244.

Hanks, T. C. and D. L. Anderson (1969). The early thermal history of the earth. *Phys. Earth Planet. Inter.* **2**, 19–29.

Hansen, M. and K. Anderko (1958). *Constitution of Binary Alloys*, 2nd Ed. McGraw-Hill, New York.

Hapke, B. (1977). Interpretations of optical observations of Mercury and the Moon. *Phys. Earth Planet. Inter.* **15**, 264–274.

Hargraves, R. B., D. Collinson, R. Arvidson, and C. Spitzer (1977). The Viking magnetic properties experiment. *J. Geophys. Res.* **82**, 4547–4558.

Harris, P. and D. Tozer (1967). Fractionation of iron in the solar system. *Nature* **215**, 1449–1451.

Hartmann, W. K. (1969). Angular momentum of Icarus. *Icarus* **10**, 445–446.

    (1970). Growth of planetesimals in nebulae surrounding young stars. In: *Evolution Stellaire Avant La Sequence Principale*. Proc. Liege Symposium, June 30-July 2, 1969. *Liege Collec. in–8°, 5th Ser.* **19**, 215–227.

    (1972). *Moons and Planets*. Wadsworth, Belmont, California.

    (1976). Planet formation: Compositional mixing and lunar compositional anomalies. *Icarus* **27**, 553–559.

    and S. Larson (1967). Angular momenta of planetary bodies. *Icarus* **7**, 257–260.

    and D. Davis (1975). Satellite-sized planetesimals and lunar origin. *Icarus* **24**, 504–515.

    D. Davis, C. Chapman, S. Soter, and R. Greenberg (1975). Mars: Satellite origin and angular momentum. *Icarus* **25**, 588–594.

Haskin, L. A., R. O. Allen, P. A. Helmke, T. P. Paster, M. R. Anderson, R. L. Korotev, and K. A. Zweifel (1970). REE and other trace elements in Apollo 11 samples. *Proc. Apollo 11 Lunar Sci. Conf.* **2**, 1213–1231.

Hayashi, C. (1966). Evolution of protostars. *Ann. Rev. Astron. Astrophys.* **4**, 171–192.

Head, J. W. (1976). Lunar volcanism in space and time. *Rev. Geophys. Space Phys.* **14**, 265–300.

    M. Settle, and R. Stein (1975). Volume of material ejected from major lunar basins: Implications for the depth of excavation of lunar samples (abstract). In: *Lunar Science VI*. Lunar Science Institute, Houston, pp. 352–354.

Helmke, P. A., D. Blanchard, L. Haskin, K. Telander, C. Weiss, and J. Jacobs (1973). Major and trace elements in igneous rocks from Apollo 15. *The Moon* **8**, 129–148.

Herbig, G. H. (1966). On the interpretation of FU Orionis, In A. Beere (ed.): *Vistas in Astronomy*. Pergamon, Oxford, Vol. 8, pp. 109–125.

    (1970). Introductory Remarks to Symposium: "Evolution des Etoiles avant Sejour sur la Sequence Principale." *Mim. Soc. R. Sci. Liege 8°, 5th Ser.* **19**, 13–26.

Hertogen, J., M. J. Janssens, H. Takahashi, H. Palme, and E. Anders (1977). Lunar basins and craters: Evidence for systematic compositional changes of bombarding population. *Proc. Eighth Lunar Sci. Conf.* **1**, 17–45.

Hewins, R. H. and J. I. Goldstein (1974). Metal–olivine associations and Ni–Co contents in two Apollo 12 mare basalts. *Earth Planet. Sci. Lett.* **24**, 59–70.

Higgins, G. H. and G. C. Kennedy (1971). The adiabatic gradient and the melting point gradient in the core of the earth. *J. Geophys. Res.* **76**, 1870–1878.

Hill, M. N. (1957). Recent exploration of the ocean floor. *Phys. Chem. Earth* (eds. L. Ahrens, F. Press, K. Rankama and S. Runcorn), **2**, 129–163.

Hofman, A. W. and M. Magaritz (1977). Diffusion of Ca, Sr, Ba, and Co in a basalt melt: Implications for the geochemistry of the mantle. *J. Geophys. Res.* **82**, 5432–5440.

Holland, H. D. (1963). On the chemical evolution of the terrestrial and Cytherean atmospheres. In P. Brancazio and A. G. W. Cameron (eds.): *The Origin and Evolution of Atmospheres and Oceans*. Wiley, New York, pp. 86–101.

Hörz, F., R. Gibbons, K. Hill, and P. Gault (1976). Large scale cratering in the lunar highlands: Some Monte Carlo considerations. *Proc. Seventh Lunar Sci. Conf.* **3**, 2931–2945.

Hoyle, F. (1946). On the condensation of the planets. *Mon. Not. Roy. Astron. Soc.* **106**, 406–414.

(1960). On the origin of the solar nebula. *Q. J. Roy. Astron. Soc.* **1**, 28–55.

Huang, S. S. (1969). Occurrence of planetary systems in the universe as a problem in stellar astronomy. *Vistas in Astronomy* **11**, Pergamon, Oxford, 217–263.

(1973). Extrasolar planetary systems. *Icarus* **18**, 339–376.

Hubbard, N. L. and P. W. Gast (1971). Chemical composition and origin of nonmare lunar basalts. *Proc. Second Lunar Sci. Conf.* **2**, 999–1020.

C. Meyer, L. Nyquist and C. Shih (1971). Chemical composition of lunar anorthosites and their parent liquids. *Earth Planet. Sci. Lett.* **13**, 71–75.

and J. W. Minear (1975). A chemical and physical model for the genesis of lunar rocks: Part II. Mare basalts (abstract). *Lunar Science* **VI**, 405–407.

Hubbard, W. B. and R. Smoluchowski (1973). Structure of Jupiter and Saturn. *Space Sci. Rev.* **14**, 599–662.

Huebner, J. S., A. Duba, L. Wiggins, and H. Smith (1978). Electrical conductivity of orthopyroxene: Measurements and implications. *Lunar and Planetary Science* **9**, 561–563. Lunar and Planetary Institute, Houston.

B. Lipin and L. Wiggins (1976). Partitioning of chromium between silicate melts and crystals. *Proc. Seventh Lunar Sci. Conf.* **2**, 1195–1220.

Hutchinson, R., D. H. Paul, and P. G. Harris (1970). Chemical composition of the upper mantle. *Min. Mag.* **37**, 726–729.

Imamura, K. and M. Honda (1976). Distribution of tungsten and molybdenum between metal, silicate, and sulfide phases of meteorites. *Geochim. Cosmochim. Acta* **40**, 1073–1080.

Ingersoll, A. P. (1969). The runaway greenhouse: A history of water on Venus. *J. Atmos. Sci.* **26**, 1191–1198.

Ito, E., T. Matsumoto, K. Suito, and N. Kawai (1972). High-pressure breakdown of enstatite. *Proc. Jpn. Acad.* **48**, 412–415.

Y. Matsui, K. Suito, and N. Kawai (1974). Synthesis of γ-Mg$_2$SiO$_4$. *Phys. Earth Planet. Inter.* **8**, 342–344.

Jackson, E. D. (1967). In P. J. Wyllie (ed.): *Ultramafic and Related Rocks*, Wiley, New York, pp. 20–38.

Jackson, I. A., T. J. Ahrens, and H. Richeson (1977). Dynamic compression of forsterite to 150 GPa. *EOS* **6**, 518.

R. C. Liebermann, and A. E. Ringwood (1978). The elastic properties of (Mg$_x$Fe$_{1-x}$)O solid solutions. *Phys. Chem. Miner.* 11–31.

Jacobs, J. A. (1975). *The Earth's Core*. Academic, London.

Jacovsby, B. M. and T. J. Ahrens (1978). Constraints on lunar formation by impact volatilization. *Lunar and Planetary Science* **9**, 582–584. Lunar and Planetary Institute, Houston.

Jeans, J. H. (1917). The motion of tidally distorted bodies with special reference to theories of cosmogony. *Mem. Roy. Astron. Soc.* **62**, 1–48.

 (1919). *Problems of Cosmogony and Stellar Dynamics*. Cambridge Univ. Press, London.

Jeffreys, H. (1916). On certain possible distributions of meteoric bodies in the solar system. *Mon. Not. Astron. Soc.* **77**, 84–112.

 (1929). *The Earth*, 2nd Ed. Cambridge University Press, London.

 (1930). The resonant theory of the origin of the moon 2. *Mon. Not. R. Astron. Soc.* **91**, 169–173.

 (1937). The density distributions in the inner planets. *Mon. Not. R. Astron. Soc. Geophys. Suppl.* **4**, 62–71.

 (1939). The times of *P*, *S* and *SKS* and the velocities of *P* and *S*. *Mon. Not. R. Astron. Soc. Geophys. Suppl.* **4**, 498–533.

Johnson, L. R. (1967). Array measurements of *P* velocities in the upper mantle. *J. Geophys. Res.* **72**, 6309–6325.

 (1969). Array measurements of *P* velocities in the lower mantle. *Bull. Seism. Soc. Am.* **59**, 973–1008.

Johnston, D. H. and M. N. Toksöz (1977). Internal structure and properties of Mars. *Icarus* **32**, 73–84.

T. R. McGetchin, and M. N. Toksöz (1974). Thermal state and internal structure of Mars. *J. Geophys. Res.* **79**, 3959–3971.

Jordan, T. (1975). Lateral heterogeneity and mantle dynamics. *Nature* **257**, 745–750.

 (1976). Lateral heterogeneity in the lower mantle (abstract). *EOS* **57**, 326.

 (1977). Lithospheric slab penetration into the lower mantle beneath the Sea of Okhotsk. *J. Geophys.* **43**, 473–496.

Jovanovic S. and G. W. Reed, Jr. (1976). Chemical fractionation of Ru and Os in the moon. *Proc. Seventh Lunar Sci. Conf.* **3**, 3437–3446.

 and G. W. Reed, Jr. (1977). Is osmium chemically fractionated in the moon? A response. *Proc. Eighth Lunar Sci. Conf.* **1**, 53–56.

Kant, I. (1755). *Allegemeine Naturgeschichte und Theorie des Himmels*.

Kaula, W. M. (1968). *An Introduction to Planetary Physics*. Wiley, New York.

(1969). Interpretation of lunar mass concentrations. *Phys. Earth Planet. Interiors* **2**, 123–137.

(1971). Dynamical aspects of lunar origin. *Rev. Geophys. Space Phys.* **9**, 217–238.

(1974). Mechanical processes affecting differentiation of protolunar material. *Proceedings of the Soviet–American Conference on Cosmochemistry of the Moon and Planets*. (in press).

(1975). The seven ages of a planet. *Icarus* **26**, 1–15.

(1977). On the origin of the moon with emphasis on bulk composition. *Proc. Eighth Lunar Sci. Conf.* **1**, 321–331.

(1978a). Lecture given at *Ninth Lunar and Planetary Science Conference*, Houston, March 1978.

(1978b). Thermal evolution of Earth and Moon by planetesimal impacts. *J. Geophys. Res.* (in press).

(1979) The moment of inertia of Mars. *Geophys. Res. Lett* **6**, 194–196.

and P. E. Bigeleisen (1975). Early scattering by Jupiter and its collision effects in the terrestrial zone. *Icarus* **25**, 18–33.

and A. W. Harris (1973). Dynamically plausible hypotheses of lunar origin. *Nature* **245**, 367–369.

and A. W. Harris (1975). Dynamics of lunar origin and orbital evolution. *Rev. Geophys. Space Phys.* **13**, 363–371.

G. Schubert, R. Lingenfelter, W. Sjogren, and W. Wollenhaupt (1974). Apollo laser altimetry and inferences as to lunar structure. *Proc. Fifth Lunar Sci. Conf.* **3**, 3049–3058.

Kawai, N. and A. Nishiyama (1974a). Conductive SiO_2 under high pressure. *Proc. Jpn. Acad.* **50**, 72–75.

and A. Nishiyama (1974b). Conductive MgO under high pressure. *Proc. Jpn. Acad.* **50**, 634–635.

Kay, P. W., N. Hubbard, and P. Gast (1970). Chemical characteristics and origin of oceanic ridge volcanic rocks. *J. Geophys. Res.* **75**, 1585–1613.

Keihm, S. J. and M. G. Langseth (1977). Lunar thermal regime to 300 km. *Proc. Eighth Lunar Sci. Conf.* **1**, 499–514.

Kennedy, G. C. and G. H. Higgins (1973). Temperature gradients at the core–mantle interface. *The Moon* **7**, 14–21.

Kesson, S. E. (1975). Mare basalts: melting experiments and petrogenetic interpretations. *Proc. Sixth Lunar Sci. Conf.* **1**, 921–924.

and A. E. Ringwood (1976a). Mare basalt petrogenesis in a dynamic moon. *Earth Planet. Sci. Lett.* **30**, 155–163.

and A. E. Ringwood (1976b). Unpublished experimental data.

and A. E. Ringwood (1977). Further limits on the bulk composition of the moon. *Proc. Eighth Lunar Sci. Conf.* **1**, 411–431.

Kovach, R. L. and D. L. Anderson (1965). The interiors of the terrestrial planets. *J. Geophys. Res.* **70**, 2873–2882.

Krähenbühl, U., R. Ganapathy, J. W. Morgan, and E. Anders (1973). Volatile elements in Apollo 16 samples: Implications for highland volcanism and accretion history of the moon. *Proc. Fourth Lunar Sci. Conf.* **2**, 1325–1348.

A. Grütter, R. von Gunten, G. Meyer, F. Wegmüller, and A. Wyttenbach (1977). Volatile and non-volatile elements in grain-size fractions of Apollo 17 soils 75081, 72461 and 72501. *Proc. Eighth Lunar Sci. Conf.* **3**, 3901–3916.

Kuhi, K. V. (1964). Mass-loss from T-Tauri stars. *Astrophys. J.* **140**, 1409–1433.

Kuiper, G. P. (1951). On the origin of the solar system. In J. A. Hyneck (ed.): *Astrophysics*. McGraw-Hill, New York, pp. 357–424.

(ed.) (1952). *The Atmospheres of the Earth and Planets*, 2nd Ed. Chicago University Press, pp. 306–405.

(1954). On the origin of the lunar surface features. *Proc. Nat. Acad. Sci. Wash.* **40**, 1096–1112.

(1955). The formation of the planets, Part II. *J. Roy. Astron. Soc. Can.* **50**, 105–121.

(1963). The surface of the moon. In D. P. LeGalley (ed.): *Space Science*. John Wiley, New York, Chapter 15, pp. 630–649.

Kuno, H., K. Yamasaki, C. Iida, and K. Nagashima (1957). Differentiation of Hawaiian magmas. *Jpn. J. Geol. Geogr.* **28**, 179–218.

Kurat, G., K. Keil, M. Prinz, and C. Nehru (1972). Chondrules of lunar origin. *Proc. Third Lunar Sci. Conf.* **1**, 707–721.

Kusaka, T., T. Nakano, and C. Hayashi (1970). Growth of solid particles in the primordial solar nebula. *Prog. Theor. Phys.* **44**, 1580–1595.

Lamar, D. L. (1962). Optical ellipticity and internal structure of Mars. *Icarus* **1**, 258–265.

Lambeck, K. (1979). Comments on the gravity, topography and moment of inertia of Mars. *J. Geophys. Res.* (in press).

Langseth, M. G., S. P. Clark, J. L. Chute, S. J. Keihm, and A. E. Wecksler (1972). Heat-flow experiment. In *Apollo 15 Prelim. Sci. Rept.*, NASA Publication *SP 289*, 11–1 to 11–23.

S. J. Keihm, and J. L. Chute (1973). Heat-flow experiment. In *Apollo 17 Prelim. Sci. Rept.*, NASA publication *SP 330*, 9–1 to 9–24.

S. J. Keihm, and K. Peters (1976). Revised lunar heat-flow values. *Proc. Seventh Lunar Sci. Conf.* **3**, 3143–3171.

Laplace, P. S. (1976). *Exposition du Systeme du Monde*, Paris.

Larimer, J. W., (1967). Chemical fractionations in meteorites I. Condensation of the elements. *Geochim. Cosmochim. Acta* **31**, 1215–1238.

(1971). Composition of the earth. Chondritic or achondritic? *Geochim. Cosmochim Acta* **35**, 769–786.

and E. Anders (1967). Chemical fractionations in meteorites—II. Abundance patterns and their interpretation. *Geochim. Cosmochim. Acta* **31**, 1239–1270.

and E. Anders (1970). Chemical fractionations in meteorites—III. Major element fractionation in chondrites. *Geochim. Cosmochim Acta* **34**, 367–387.

Larson, R. B. (1969). The dynamics of a collapsing protostar. *Mon. Not. R. Astron. Soc.* **6**, 265–295.

(1972a). The evolution of spherical protostars with masses 0.25 M ⊙ to 10 M ⊙. *Mon. Not. R. Astron. Soc.* **157**, 121–145.

(1972b). Collapse calculations and their implications for the formation of the solar system. In H. Reeves (ed.): *The Origin of the Solar System.* Proceedings of a symposium in Nice. CNRS, Paris, pp. 142–149.

(1973). Processes in collapsing interstellar clouds. *Ann. Rev. Astron. Astrophys.* **11**, 219–238.

Latimer, W. M. (1950). Astrochemical problems in the formation of the earth. *Science* **112**, 101–104.

Laul, J. C., D. W. Hill, and R. A. Schmitt (1974). Chemical studies of Apollo 16 and 17 samples. *Proc. Fifth Lunar Sci. Conf.* **2**, 1047–1066.

Lee, T. and D. Papanastassiou (1974). Mg isotopic anomalies in the Allende meteorite and correlation with O and Sr effects. *Geophys. Res. Lett.* **1**, 225–228.

D. Papanastassiou, and G. J. Wasserburg (1977). Aluminium-26 in the early solar system: Fossil or fuel? *Astrophys. J.* **211**, L107–L110.

Leich, D. A. and S. Niemeyer (1975). Trapped xenon in anorthositic breccia 60015. *Proc. Sixth Lunar Sci. Conf.* **2**, 1953–1965.

Levin, B. J. (1949). Structure of the earth and planets and a meteoritic hypothesis of their origin. *Priroda* **10**, 3–14.

(1970). Internal constitution of terrestrial planets. In A. Dollfus (ed.): *Surfaces and Interiors of Planets and Satellites.* Academic Press, New York, pp. 462–510.

(1972a). Origin of the earth. *Tectonophysics* **13**, 7–29.

(1972b). Revision of initial size, mass and angular momentum of the solar nebula and the problem of its origin. In H. Reeves (ed.): *The Origin of the Solar System.* Proceedings of a symposium held in Nice. CNRS, Paris, pp. 341–357.

Lewis, J. S. (1971). Consequences of the presence of sulphur in the core of the earth. *Science* **112**, 101–104.

(1972). Metal/silicate fractionation in the solar system. *Earth Planet. Sci. Lett.* **15**, 286–290.

(1973). Chemistry of the planets. *Ann. Rev. Phys. Chem.* **24**, 339–352.

Liebermann, R. C. (1974). Elasticity of pyroxene–garnet and pyroxene–ilmenite phase transformations in germanates. *Phys. Earth Planet. Inter.* **8**. 361–374.

and A. E. Ringwood (1973). Birch's law and polymorphic phase transformations. *J. Geophys. Res.* **78**, 6926–6932.

and A. E. Ringwood (1976). Elastic properties of anorthite and the nature of the lunar crust. *Earth Planet. Sci. Lett.* **31**, 69–74.

A. E. Ringwood, and A. Major (1976). Elasticity of polycrystalline stishovite. *Earth Planet. Sci. Lett.* **32**, 127–140.

Lightner, B. and K. Marti (1974). Lunar trapped xenon. *Proc. Fifth Lunar Sci. Conf.* **2**, 2023–2031.

Liu, L. G. (1974). Silicate perovskite from phase transformations of pyrope-garnet at high pressures and temperatures. *Geophys. Res. Lett.* **1**, 277–280.

(1975). Post-oxide phases pf olivine and pyroxene and mineralogy of the mantle. *Nature* **258**, 510–512.

(1976). The high-pressure phases of MgSiO₃. *Earth Planet. Sci. Lett.* **31**, 200–208.

(1977a). Mineralogy and chemistry of the Earth's mantle above 1000 km. *Geophys. J.* **418**, 53–62.

(1977b). High-pressure NaAlSiO₄: The first silicate calcium ferrite isotype. *Geophys. Res. Lett.* **4**, 183–186.

and A. E. Ringwood (1975). Synthesis of a perovskite-type polymorph of CaSiO₃. *Earth Planet. Sci. Lett.* **28**, 209–211.

Lodochnikov, V. N. (1939). Some general problems connected with magma producing basaltic rocks. *Zap. Mineral. O-va.* **64**, 207–223.

Loubet, M., N. Shimizu, and C. J. Allegre (1975). Rare earth elements in alpine peridotites. *Contrib. Mineral. Petrol.* **53**, 1–12.

Lovering, J. F. and D. A. Wark (1977). Marker events in the early evolution of the solar system: Evidence from rims on the Ca- and Al-rich inclusions in carbonaceous chondrites. *Proc. Eighth Lunar Sci. Conf.* **1**, 95–112.

Lubimova, E. A. (1958). Thermal history of the earth. *Geophys. J.* **1**, 115–134.

(1966). Sources of intraplanetary heat. In A. P. Vinogradov (ed.): *Chemistry of the Earth's Crust, Vol. 1.* Israel Program for Scientific Translations, Jerusalem, pp. 27–36.

McAdoo, D. C. and J. A. Burns (1973). Further evidence for collisions among asteroids. *Icarus* **18**, 285–293.

McCallum, I. S., F. P. Okamura, and S. Ghose (1975). Mineralogy and petrology of sample 67075 and the origin of lunar anorthosites. *Earth Planet. Sci. Lett.* **20**, 36–53.

McCrea, W. H. (1960). The origin of the solar system. *Proc. R. Soc. London, Ser. A* **256**, 245–266.

and I. P. Williams (1965). Segregation of materials in cosmogony. *Proc. Roy. Soc. London, Ser. A* **287**, 143–164.

(1972). Origin of the solar system. In H. Reeves (ed.): *The Origin of the Solar System.* Proceedings of a Symposium held in Nice. CNRS, Paris, pp. 2–17.

MacDonald, G. J. F. (1959). Calculations on the thermal history of the earth. *J. Geophys. Res.* **64**, 1967–2000.

(1962). On the internal constitution of the inner planets. *J. Geophys. Res.* **67**, 2945–2974.

(1964). Tidal friction. *Rev. Geophys.* **2**, 467–541.

McElroy, M. B., Y. Yung, and A. O. Nier (1976). Isotopic composition of nitrogen: Implications for the past history of Mars' atmosphere. *Science* **194**, 70–72.

McElhinny, W. (1973). *Paleomagnetism and Plate Tectonics.* Cambridge Univ. Press, London.

McQueen, R. G. and S. P. Marsh (1960). Equations of state for nineteen metallic elements from shock wave measurements to two megabars. *J. Appl. Physics* **31**, 1253–1269.

McQueen, R. G. and S. P. Marsh (1966). In S. P. Clark (ed.): *Handbook of Physical Constants. Geol. Soc. Am. Mem. 97.*

S. P. Marsh and J. N. Fritz (1967). Hugoniot equation of state of twelve rocks. *J. Geophys. Res.* **72**, 4999–5036.

Mao, N. (1974). Velocity–density systematics and iron content of the mantle. *EOS* **55**, 416.

Mason, B. (1966). Composition of the earth. *Nature* **211**, 616–618.

(ed.) (1971). *Handbook of Elemental Abundances in Meteorites.* Gordon and Breach, New York.

Mestel, L. (1972). Magnetohydrodynamics, hydrodynamics, dynamics of the solar system in the different models. In H. Reeves (ed.): *Origin of the Solar System.* Proceedings of a Symposium held at Nice. CRNS, Paris, pp. 21–27.

Metzger, A. E., J. Trombka, R. Reedy, and J. Arnold (1974). Element concentrations from lunar orbital gamma ray experiments. *Proc. Fifth Lunar Sci. Conf.* **2**, 1067–1078.

E. Haines, R. Parker, and R. Radocinski (1977). Thorium concentrations in the lunar surface. I: Regional values and crustal content. *Proc. Eighth Lunar Sci. Conf.* **1**, 949–999.

Meyer, C., Jr., D. S. McKay, D. H. Anderson, and P. Butler, Jr. (1975). The source of sublimates on the Apollo 15 green and Apollo 17 orange glass samples. *Proc. Sixth Lunar Sci. Conf.* **2**, 1673–1699.

Miller, S. L. (1953). A production of amino acids under possible primitive earth conditions. *Science* **117**, 528–529.

Mitler, H. E. (1975). Formation of an iron-poor Moon by partial capture. *Icarus* **24**, 256–268.

Mizutani, H., T. Matsui, and H. Takeuchi (1972). Accretion process of the moon. *The Moon* **4**, 476–489.

Moorbath, S., R. K. O'Nions, R. J. Pankhurst, N. H. Gale, and V. R. McGregor (1972). Further rubidium–strontium age determinations on the very early Precambrian rocks of the Godthaab District, West Greenland. *Nature* **240**, 78–82.

Morgan, J. W. (1971). Uranium. In B. Mason (ed.): *Handbook of Elemental Abundances in Meteorites.* Gordon and Breach, New York, pp. 529–548.

Morris, G. B., R. W. Raitt, and G. G. Shor (1969). Velocity anisotropy and delay-time maps of the mantle near Hawaii. *J. Geophys. Res.* **74**, 4300–4316.

Mueller, R. F. (1964). Phase equilibria and the crystallization of chondritic meteorites. *Geochim. Cosmochim. Acta* **28**, 189–207.

Mufson, S. C. and H. List (1975). Mass loss from the infrared star CIT6. *Astrophys. J.* **202**, 183–190.

Munk, W. H. and D. Davies (1964). The relationship between core accretion and the rotation rate of the Earth. In H. Craig, S. Miller, and G. Wasserburg (eds.): *Isotopic and Cosmic Chemistry.* North Holland, Amsterdam, pp. 341–346.

Murthy, V. R. and H. Hall (1970). The chemical composition of the earth's core: Possibility of sulphur in the core. *Phys. Earth Planet. Inter.* **2**, 276–282.

N. M. Evenson and H. T. Hall (1971). A model of early lunar differentiation. *Nature* **234**, 267.

Mutch, T. A. and J. W. Head (1975). The geology of Mars: A brief review of some recent results. *Rev. Geophys. Space Phys.* **13**, 411–416.

—— and R. S. Saunders (1976). The geologic development of Mars: A review. *Space Sci. Rev.* **19**, 3–57.

Mysen, B. and I. Kushiro (1976). Partitioning of iron, nickel and magnesium between metal, oxide and silicates in Allende meteorite as a function of f_{O_2}. *Carnegie Inst. Washington Yearb.* **75**, 678–684.

Nakamura, Y., G. Latham, D. Lammlein, M. Ewing, F. Duennebier, and J. Dorman (1974). Deep lunar interior inferred from recent seismic data. *Geophys. Res. Lett.* **1**, 137–140.

—— F. Duennebier, G. Latham, and J. Dorman (1976a). Structure of the lunar mantle. *J. Geophys. Res.* **81**, 4818–4824.

—— G. V. Latham, J. Dorman, and F. Duennebier (1976b). Seismic structure of the moon: A summary of current status. *Proc. Seventh Lunar Sci. Conf.* **3**, 3113–3121.

Naldrett, A. J. and G. D. Mason (1968). Contrasting Archaean ultramafic igneous bodies in Dundonald and Clerque Townships, Ontario. *Can. J. Earth Sci.* **5**, 111–143.

Narita, S., T. Nakano, and C. Hayashi (1970). Rapid contraction of protostar to the shape of quasi-hydrostatic equilibrium. III. *Prog. Theor. Phys. Jpn.* **43**, 942–964.

Nesbitt, R. W. (1972). Skeletal crystal forms in the ultramafic rocks of the Yilgarn Block, Western Australia; evidence for an Archaean ultramafic liquid. *Geol. Soc. Aust. Spec. Pub.* **3**, 331–350.

—— and S. S. Sun (1976). Geochemistry of Archaean spinifex-textured peridotites and magnesian and low-magnesian tholeiites. *Earth Planet. Sci. Lett.* **31**, 443–453.

Ness, N. F., K. Behannon, R. Leeping, and Y. Whang (1976). Observations of Mercury's magnetic field. *Icarus* **28**, 479–488.

Neugebaur, G., E. Becklin, and H. Hyland (1971). Infrared sources of radiation. *Ann. Rev. Astron. Astrophys.* **9**, 67–102.

Niemeyer, S. and D. Leich (1976). Atmospheric rare gases in lunar rock 60015. *Proc. Seventh Lunar Sci. Conf.* **1**, 587–597.

Nunes, P. D., M. Tatsumoto, and D. Unruh (1975). U–Th–Pb systematics of anorthositic gabbros 78155 and 77017—implications for early lunar evolution. *Proc. Sixth Lunar Sci. Conf.* **2**, 1431–1444.

Nuttli, O. W. (1969). Travel times and amplitudes of *S* waves from nuclear explosions in Nevada. *Bull. Seismol. Soc. Am.* **59**, 385–398.

Nyquist, L. E., B. Bansal, H. Wooden, and H. Wiesman (1977). Sr-isotopic constraints on the petrogenesis of Apollo 12 basalts. *Proc. Eighth Lunar Sci. Conf.* **2**, 1383–1415.

O'Keefe, J. A. (1966). The origin of the moon and the core of the earth. In B. G. Marsden and A. G. W. Cameron (eds.): *The Earth-Moon System*, Plenum Press, New York, pp. 224–233.

—— (1969). Origin of the Moon. *J. Geophys. Res.* **74**, 2758–2767.

—— (1970). The origin of the Moon. *J. Geophys. Res.* **75**, 6565–6574.

—— (1972a). The origin of the Moon: Theories involving joint formation with the Earth. *Astrophys. Space Sci.* **16**, 201–211.

(1972b). Geochemical evidence for the origin of the Moon. *Naturwissenschaften* **59**, 45–52.

(1972c). Inclination of the Moon's orbit: The early history. *Irish Astron. J.* **10**, 241–250.

and E. Sullivan (1978). Fission origin of the Moon: Cause and timing. *Icarus* **35**, 272–283.

and H. C. Urey (1977). The deficiency of siderophile elements in the Moon. *Phil. Trans. R. Soc. London, Ser. A* **285**, 569–575.

Olinger, B. (1976). The compression of stishovite. *J. Geophys. Res.* **81**, 5341–5343.

Olsen, E., L. Fuchs, and W. Forbes (1973). Chromium and phosphorus enrichment in the metal of Type II carbonaceous chondrites. *Geochim. Cosmochim. Acta* **37**, 2037–2042.

Onuma, N., R. N. Clayton, and T. K. Mayeda (1972). Oxygen isotope cosmothermometer. *Geochim. Cosmochim. Acta* **36**, 169–188.

Öpik, E. J. (1955). The origin of the moon. *Irish Astron. J.* **3**, 245–248.

(1961). Tidal deformations and the origin of the Moon. *Astron. J.* **66**, 60–67.

(1963a). Selective escape of gases. *Geophys. J.* **7**, 490–509.

(1963b). Survival of comet nuclei and the asteroids. *Adv. Astron. Astrophys.* **2**, 219–262.

(1967). Evolution of the Moon's surface, 1. *Irish Astron. J.* **8**, 38–52.

(1969). The Moon's surface. *Ann. Rev. Astron. Astrophys.* **7**, 473–526.

(1972). Comments on lunar origin. *Irish Astron. J.* **10**, 190–238.

Orowan, E. (1969). Density of the Moon and nucleation of the planets. *Nature* **222**, 867.

Ostriker, J. P. (1972). Hydrodynamics of the collapse; rotation and contraction. In H. Reeves (ed.): *The Origin of the Solar System*. Proceedings of a Symposium held in Nice, CNRS, Paris, 154–162.

Oversby, V. M. and A. E. Ringwood (1971). Time of formation of the earth's core. *Nature* **234**, 463–465.

Owen, T., K. Biemann, D. Rushneck, J. Biller, D. Howarth, and A. Lafleur (1977). The composition of the atmosphere at the surface of Mars. *J. Geophys. Res.* **82**, 4635–4639.

Oxburgh, E. R. and D. L. Turcotte (1970). Thermal structure of island arcs. *Bull. Geol. Soc. Am.* **81**, 1665–1688.

Papanastassiou, D. and G. J. Wasserburg (1973). Rb–Sr ages and initial strontium in basalts from Apollo 15. *Earth Planet. Sci. Lett.* **17**, 324–337.

and G. J. Wasserburg (1975). Rb–Sr study of a lunar dunite and evidence for early lunar differentiates. *Proc. Sixth Lunar Sci. Conf.* **2**, 1467–1489.

Phillips, R. J. and R. S. Saunders (1975). The isostatic state of Martian topography. *J. Geophys. Res.* **80**, 2893–2898.

Pidgeon, R. T. (1978). Big Stubby and the early history of the Earth. In R. Zartman (ed.): *Fourth International Conference, Geochronology, Cosmochronology, Isotope Geology*, pp. 334–335. U.S. Geological Survey Open-File Report 78–701.

Podolak, M. and A. G. W. Cameron (1974). Models of the giant planets. *Icarus* **22**, 123–148.

Poldervaart, A. (1955). Chemistry of the earth's crust. *Geol. Soc. Am. Spec. Paper* **62**, 119–144.

Pollack, J. B. and R. Reynolds (1974). Implications of Jupiter's early contraction history for the composition of the Galilean satellites. *Icarus* **21**, 248–253.

    D. Colburn, R. Kahn, J. Hunter, W. Camp, C. Carlston, and M. Wolf (1977a). Properties of aerosols in the Martian atmosphere as inferred from Viking lander imaging data. *J. Geophys. Res.* **82**, 4479–4496.

    A. Grossman, R. Moore, and H. Graboske (1977b). A calculation of Saturn's gravitational contraction history. *Icarus* **30**, 111–128.

Prentice, A. J. R. (1974). The formation of planetary systems. In J. P. Wild (ed.): *In the Beginning* . . . Copernicus 500th Birthday Symposium, Australian Academy of Sciences, Canberra, pp. 15–47.

Press, F. (1970). Earth models consistent with geophysical data. *Phys. Earth Planet. Inter.* **3**, 3–22.

Prior, G. T. (1920). Classification of meteorites. *Min. Mag.* **19**, 51–63.

    (1953). *Catalogue of Meteorites.* British Museum, London, 2nd Ed.

Rader, L. F., W. Swadley, C. Hoffman, and H. Lipp (1963). New chemical determinations of zinc in basalts and rocks of similar composition. *Geochim. Coschim. Acta* **27**, 695–716.

Raitt, R. (1963). The crustal rocks. In M. N. Hill (ed.): *The Sea, Vol. 3*, Interscience, New York, Chapter 6.

Rammensee W. and H. Wänke (1977). On the partition coefficient of tungsten between metal and silicate and its bearing on the origin of the moon. *Proc. Eighth Lunar Sci. Conf.* **1**, 399–409.

Ramsey, W. H. (1948). On the constitution of the terrestrial planets. *Mon. Not. Roy. Astron. Soc.* **108**, 406–413.

    (1949). On the nature of the earth's core. *Mon. Not. R. Astron. Soc.* **5**, 409–426.

Reasenberg, R. (1977). The moment of inertia and isostasy of Mars. *J. Geophys. Res.* **82**, 369–375.

Reed, G. W., R. Allen, and S. Jovanovic (1977). Volatile metal deposits on lunar soils—relation to volcanism. *Proc. Eighth Lunar Sci. Conf.* **3**, 3917–3930.

Reid, A. F. and A. E. Ringwood (1974). New dense phases of geophysical significance. *Nature* **252**, 681–682.

    and A. E. Ringwood (1975). High pressure modification of $ScAlO_3$ and some geophysical implications. *J. Geophys. Res.* **80**, 3363–3370.

    A. D. Wadsley, and A. E. Ringwood (1967). High pressure $NaAlGeO_4$, a calcium ferrite isomorph and model structure for silicates at depth in the earth's mantle. *Acta Crystallogr.* **23**, 736–739.

Reid, A. M., C. Meyer, Jr., R. S. Harmon, and R. Brett (1970). Metal grains in Apollo 12 igneous rocks. *Earth Planet. Sci. Lett.* **9**, 1–5.

Reynolds, R. T. and A. L. Summers (1969). Calculations on the composition of the terrestrial planets. *J. Geophys. Res.* **74**, 2494–2511.

Ringwood, A. E. (1958). Constitution of the mantle, III: Consequences of the olivine–spinel transition. *Geochim. Cosmochim. Acta* **15**, 195–212.

(1959). On the chemical evolution and densities of the planets. *Geochim. Cosmochim. Acta* **15**, 257–283.

(1960). Some aspects of the thermal evolution of the earth. *Geochim. Cosmochim. Acta* **20**, 241–259.

(1962). Mineralogical constitution of the deep mantle. *J. Geophys. Res.* **67**, 4005–4010.

(1963). The origin of the high-temperature minerals in carbonaceous chondrites. *J. Geophys. Res.* **68**, 1141–1143.

(1966a). The chemical composition and origin of the earth. In P. M. Hurley (ed.): *Advances in Earth Sciences.* 287–356. MIT Press, Cambridge.

(1966b). Chemical evolution of the terrestrial planets. *Geochim. Cosmochim. Acta* **30**, 41–104.

(1966c). Genesis of chondritic meteorites. *Rev. Geophys.* **4**, 113–174.

(1969). Composition and evolution of the upper mantle. In P. Hart (ed.): *The Earth's Crust and Upper Mantle,* 1–17. Am. Geophys. U. Geophys. Monogr. 13.

(1970a). Phase transformations and the constitution of the mantle. *Phys. Earth Planet. Inter.* **3**, 109–155.

(1970b). Origin of the Moon: The precipitation hypothesis. *Earth Planet. Sci. Lett.* **8**, 131–140.

(1970c). Petrogenesis of Apollo 11 basalts and implications for lunar origin. *J. Geophys. Res.* **75**, 6453–6479.

(1971). Core–mantle equilibrium: Comments on a paper by R. Brett. *Geochim. Cosmochim. Acta* **35**, 223–230.

(1972). Some comparative aspects of lunar origin. *Phys. Earth Planet. Inter.* **6**, 366–376.

(1974). The early evolution of planets. In J. P. Wild (ed): *In the Beginning . . . .* Copernicus 500th Birthday Symposium, Australian Academy of Sciences, Chapter 3, 48–85.

(1975a). *Composition and Petrology of the Earth's Mantle.* McGraw-Hill, New York.

(1975b). Some aspects of the minor element chemistry of lunar mare basalts. *The Moon* **12**, 127–157.

(1976). Limits on the bulk composition of the moon. *Icarus* **28**, 325–349.

(1977a). Basaltic magmatism and the composition of the Moon I: Major and heat producing elements. *The Moon* **16**, 389–423.

(1977b). Composition of the core and implications for origin of the Earth. *Geochem. J.* **11**, 111–135.

(1978a). Origin of the Moon. *Lunar and Planetary Science* **9**, 961–963. Lunar and Planetary Institute, Houston.

(1978b). Water in the solar system. In A. McIntyre (ed.): *Water—Planets, Plants and People.* Austral. Acad. Sci., Canberra, pp. 18–34.

and D. L. Anderson (1977). Earth and Venus: A comparative study. *Icarus* **30**, 243–253.

and S. P. Clark (1971). Internal constitution of Mars. *Nature* **234**, 89–92.

and E. Essene (1970). Petrogenesis of Apollo 11 basalts, internal constitution and origin of the moon. *Proc. Apollo 11 Lunar Sci. Conf.* **1**, 769–799.

and S. E. Kesson (1976). A dynamic model for mare basalt petrogenesis. *Proc. Seventh Lunar Sci. Conf.* **2**, 1697–1722.

and S. E. Kesson (1977). Basaltic magmatism and the bulk composition of the Moon, II. Siderophile and volatile elements in Moon, Earth and chondrites: Implications for lunar origin. *The Moon* **16**, 425–464.

and A. Major (1966). Synthesis of Mg_2SiO_4–Fe_2SiO_4 solid solutions. *Earth Planet. Sci. Lett.* **1**, 241–245.

and A. Major (1968). High pressure transformations in pyroxene II. *Earth Planet. Sci. Lett.* **5**, 76–78.

and A. Major (1970). The system Mg_2SiO_4–Fe_2SiO_4 at high pressures and temperatures. *Phys. Earth Planet. Inter.* **3**, 89–108.

and A. Major (1971). Synthesis of majorite and other high pressure garnets and perovskites. *Earth Planet. Sci. Lett.* **12**, 411–418.

Ross, C. J., M. D. Foster, and A. T. Myers (1954). Origin of dunites and olivine-rich inclusions in basaltic rocks. *Am. Mineral.* **39**, 693–737.

Ross, J. E. and L. H. Aller (1976). The chemical composition of the sun. *Science* **191**, 1223–1229.

Rubey, W. W. (1951). Geologic history of sea water. *Bull. Geol. Soc. Am.* **62**, 1111–1147.

(1955). Development of the hydrosphere and atmosphere with special reference to the probable composition of the early atmosphere. In A. Poldervaart (ed.): *Crust of the Earth*. Geol. Soc. Am. Spec. Paper **62**, pp. 631–650.

Rubincam, D. P. (1975). Tidal friction and the early history of the moon's orbit. *J. Geophys.* **80**, 1537–1548.

Runcorn, S. K. (1962). Palaeomagnetic evidence for continental drift and its geophysical cause. In S. K. Runcorn (ed.): *Continental Drift*. Academic Press, New York, pp. 1–39.

(1965). Changes in the convection pattern in the earth's mantle and continental drift: Evidence for a cold origin of the earth. *Phil. Trans. R. Soc. London, Ser. A* **258**, 228–251.

(1977). Early melting of the moon. *Proc. Eighth Lunar Sci. Conf.* **1**, 463–469.

Ruskol, E. L. (1960). The origin of the moon, 1. Formation of a swarm of bodies around the earth. *Sov. Astron. AJ.* **4**, 657–668.

(1963). On the origin of the moon, 2. The growth of the moon in the circumterrestrial swarm of satellites. *Sov. Astron. AJ.* **7**, 221–227.

(1972a). The origin of the moon, 3. Some aspects of the dynamics of the circumterrestrial swarm. *Sov. Astron. AJ.* **15**, 646–654.

(1972b). On the possible differences in the bulk composition of the Earth and Moon forming in the circumterrestrial swarm. In S. K. Runcorn and H. C. Urey (eds.): *The Moon*. Reidel, Dordrecht, Holland, pp. 426–428.

(1973). On the tidal changes of the orbital inclinations of the Uranus satellites relative to its equatorial plane. *Astron. Vestn.* **7**, 150–158.

(1977). The origin of the Moon. In J. Pomeroy and N. Hubbard (eds.): *Proceedings of the Soviet-American Conference on Cosmochemistry of the Moon and Planets*, NASA SP-370, Washington, pp. 815–822.

Ryder, G. and J. A. Wood (1977). Serenitatus and Imbrium impact melts: Implications for large-scale layering in the lunar crust. *Proc. Eighth Lunar Sci. Conf.* **1**, 655–668.

Safronov, V. S. (1954). On the growth of planets in the protoplanetary cloud. *Astron. Zh.* **31**, 499–510.

(1959). On the primeval temperature of the earth. *Bull. Acad. Sci. USSR, Geophys, Ser.* **1**, 139–143.

(1964). The primary inhomogeneities of the Earth's mantle. *Tectonophysics* **1**, 217–221.

(1972a). *Evolution of the Protoplanetary Cloud and Formation of the Earth and Planets.* Translated from the Russian, Israel Program for Scientific Translations, Tel Aviv.

(1972b). On the mass transport in the model of the solar system by A. G. W. Cameron. In H. Reeves (ed.): *On the Origin of the Solar System.* CNRS, Paris, pp. 361–366.

and E. V. Zvjagina (1969). Relative sizes of the largest bodies during the accumulation of planets. *Icarus* **10**, 109–115.

Sagan, C. (1962). Structure of lower atmosphere of Venus. *Icarus* **1**, 151–169.

(1967). Origins of the atmospheres of the Earth and planets. In S. K. Runcorn (ed.): *International Dictionary of Geophysics.* Pergamon, London.

Sato, M., H. Hickling, and J. McLane (1973). Oxygen fugacity values of Apollo 12, 14, and 15 lunar samples and reduced state of lunar magmas. *Proc. Fourth Lunar Sci. Conf.* **1**, 1061–1079.

Sato, Y. (1977). Pressure–volume relationship of stishovite under hydrostatic compression. *Earth Planet. Sci. Lett.* **34**, 307–312.

Schatzman, E. (1962). A theory of the origin of magnetic activity during star formation. *Ann. Astrophys.* **25**, 18–29.

(1967). Cosmogony of the solar system and origin of the deuterium. *Ann. Astrophys.* **30**, 963–974.

Schilling, J. G. (1973). Iceland mantle; plume: Geochemical study of Reykjanes Ridge. *Nature* **243**, 565–571.

Schmidt, O. Y. (1944). Meteoritic theory of the origin of the earth and plantes. *Dokl. Akad. Nauk. SSSR* **45**, 245–263.

(1950). *Four Lectures on the Origin of the Earth.* 2nd Ed. pp. 65–66. (In Russian.)

(1958). *A Theory of the Origin of the Earth: Four Lectures.* Foreign Languages Publishing House, Moscow, pp. 58–59. (Lawrence and Wishart, London, 1959.)

Schnetzler, C. C. and J. A. Philpotts (1971). Alkali, alkaline earth and rare earth element concentrations in some Apollo 12 soils, rocks and separated phases. *Proc. Second Lunar Sci. Conf.* **2**, 1101–1122.

Schonfeld, E. (1975). A model for the lunar anorthostic gabbro. *Proc. Sixth Lunar Sci. Conf.* **2**, 1375–1386.

(1977a). Comparison of orbital chemistry with crustal thickness and lunar sample chemistry. *Proc. Eighth Lunar Sci. Conf.* **1**, 1149–1162.

(1977b). Martian volcanism (abstract). *Lunar Sci.* **8**, 843–845.

and M. J. Bielefield (1977). Correlation between geochemical features and photogeological features in Mare Crisium (abstract). In *Conference on Luna 24.* Lunar Science Institute, Houston, Texas, pp. 167–169.

Schreiber, E. and O. L. Anderson (1970). Properties and composition of lunar materials: Earth analogues. *Science* **168**, 1579–1580.

Shoemaker, E. (1972). Cratering history and early evolution of the moon. In C. Watkins (ed.): *Lunar Science III*. Lunar Science Institute, Houston, Texas, pp. 696–698.

Siegfried, R. W. and S. C. Solomon (1974). Mercury: Internal structure and thermal evolution. *Icarus* **23**, 192–205.

Sill, G. T. (1972). Sulphuric acid in the Venus clouds. *Commun. Lunar Planet. Lab., Univ. Arizona* **171**, 191–198.

Silver, L. T. (1972). Lead volatilization and volatile transfer processes on the Moon. In C. Watkins (ed.): *Lunar Science III*. Lunar Science Institute, Houston, Texas, pp. 701–703.

Sinclair, W. S., W. Sjogren, J. Williams, and A. Ferrari (1976). The lunar moment of inertia derived from combined Doppler and laser ranging data. In C. Watkins (ed.): *Lunar Science VII*. Lunar Science Institute, Houston, Texas, pp. 817–819.

Singer, S. F. (1968). The origin of the moon and geophysical consequences. *Geophys. J.* **15**, 205–226.

——— (1970). Origin of the moon by capture and its consequences. *Trans. Am. Geophys. U.* **51**, 637–641.

Smith, J. V. (1974). Origin of the Moon by disintegrative capture with chemical differentiation followed by sequential accretion. In: *Lunar Science V*, (abstract) Lunar Science Institute, Houston, Texas, pp. 718–720.

Solomon, S. C. (1976). Some aspects of core-formation in Mercury. *Icarus* **28**, 509–521.

——— and J. Chaiken (1976). Thermal expansion and thermal stress in the moon and terrestrial planets: Clues to early thermal history. *Proc. Seventh Lunar Sci. Conf.* **3**, 3229–3243.

Sonett, C. P., D. Colburn, D. Dyal, C. Parkin, B. Smith, G. Schubert, and K. Schwartz (1971). The lunar electrical conductivity profile. *Nature* **230**, 359–362.

Steele, I. M. and J. V. Smith (1975). Minor elements in lunar olivine as a petrologic indicator. *Proc. Sixth Lunar Sci. Conf.* **1**, 451–467.

Steinhart, J. S. and R. P. Meyer (1961). Explosion studies of continental structure. *Carnegie Inst. Washington Publ.* **622**, 409–520.

Steuber, A. M. and G. C. Goles (1967). Abundances of Na, Mn, Cr, Sc, and Co in ultramafic rocks. *Geochim. Cosmochim. Acta* **31**, 75–93.

Stevenson, D. J. (1977). Hydrogen in the earth's core. *Nature* **268**, 130–131.

——— (1978). The outer planets and their satellites. In: *The Origin of the Solar System*, John Wiley, New York, pp. 395–431.

——— and E. Salpeter (1976). Interior models of Jupiter. In T. J. Gehrels (ed.): *Jupiter*. Proceedings of the IAU Symposium No. 30. Arizona University Press, Tucson.

Stolper, E. (1977). Experimental petrology of eucritic meteorites. *Geochim. Cosmochim. Acta* **41**, 587–611.

——— (1979). Trace elements in shergottite meteorites: Implications for the origin of planets. *Earth Planet. Sci. Lett.* **42**, 239–242.

Strom, R. G. (1977). Origin and relative age of lunar and Mercurian intercrater plains. *Phys. Earth Planet. Inter.* **15**, 156–172.

——— N. Trask and J. Guest (1975a). Tectonism and volcanism on Mercury. *J. Geophys. Res.* **80**, 2478–2507.

Strom, S. E., K. M. Strom, and G. L. Grasdalen (1975b). Young stellar objects and dark interstellar clouds. *Ann. Rev. Astron. Astrophys.* **13**, 187–216.

Studier, M., R. Hayatsu, and E. Anders (1965). Organic compounds in carbonaceous chondrites. *Science* **149**, 1455–1459.

R. Hayatsu and E. Anders (1968). Origin of organic matter in early solar system, 1. Hydrocarbons. *Geochim. Cosmochim. Acta* **32**, 151–174.

R. Hayatsu and E. Anders (1972). Origin of organic matter in early solar system, 5. Further studies of meteoritic hydrocarbons and a discussion of their origin. *Geochim. Cosmochim. Acta* **36**, 189–215.

Sun, S. S. and R. W. Nesbitt (1977). Chemical heterogeneity of the Archaean mantle, composition of the earth and mantle evolution. *Earth Planet. Sci. Lett.* **35**, 429–448.

Surkov, Y. A. (1977). Geochemical studies of Venus and Venera 9 and 10 automatic interplanetary stations. *Proc. Eighth Lunar Sci. Conf.* **3**, 2665–2685.

and G. A. Fedoseyev (1977). Radioactivity of the moon, planets and meteorites. In J. Pomeroy and N. J. Hubbard (eds.): *Proc. Soviet-American Conference on Cosmochemistry of the Moon and Planets.* Moscow, June 1974. NASA, Washington D.C. pp. 201–218.

Tatsumoto, M. (1978). Isotopic composition of lead in oceanic basalt and its implication to mantle evolution. *Earth Planet. Sci. Lett.* **38**, 63–87.

R. Knight, and B. Doe (1971). U–Th–Pb systematics of Apollo 12 lunar samples. *Proc. Second Lunar Sci. Conf.* **2**, 1521–1546.

C. Hedge, R. Knight, D. Unruh, and B. Doe (1972). U–Th–Pb, Rb–Sr and K measurements on some Apollo 15 and 16 samples. In J. W. Chamberlain and C. Watkins (eds.): *The Apollo 14 Lunar Samples.* The Lunar Science Institute, Houston, Texas, pp. 391–395.

P. Nunes, R. Knight, C. Hedge and D. Unruh (1973). U–Th–Pb, Rb–Sr and K measurements of two Apollo 17 samples. *EOS* **54**, 614–615.

Taylor, C. R. and J. Chipman (1943). Equilibria of liquid iron and simple basic and acid slags in a rotating iron furnace. *AIME Trans.* **154**, 228–245.

Taylor, H. P., Jr. (1968). The oxygen isotope geochemistry of igneous rocks. *Contrib. Miner. Petrol.* **19**, 1–71.

M. Duke, L. Silver, and S. Epstein (1965). Oxygen isotope studies of minerals in stony meteorites. *Geochim. Cosmochim. Acta.* **29**, 489–512.

Taylor, S. R. (1973a). Chemical evidence for lunar melting and differentiation. *Nature* **245**, 203–205.

(1973b). Tektites: A post-Apollo view. *Earth Sci. Rev.* **9**, 101–123.

(1977). Island arc models and the composition of the continental crust. In M. Talwani and W. Pitman (eds.): *Island Arcs, Deep Sea Trenches and Backarc Basins.* Am. Geophys. Un. Monogr. **19**, American Geophysical Union, Washington D.C., pp. 325–335.

and P. Jakeš (1974). The geochemical evolution of the moon. *Proc. Fifth Lunar Sci. Conf.* **2**, 1287–1305.

and A. E. Bence (1975). Evolution of the lunar highlands crust. *Proc. Sixth Lunar Sci. Conf.* **1**, 1121–1142.

Tera, F. and G. J. Wasserburg (1974). U–Th–Pb systematics on lunar rocks and inferences about lunar evolution and the age of the moon. *Proc. Fifth Lunar Sci. Conf.* **2**, 1571–1599.

and G. J. Wasserburg (1976). Lunar ball games and other sports (abstract). In C. Watkins (ed.): *Lunar Science VII*. The Lunar Science Institute, Houston, Texas, pp. 858–860.

Ter Haar, D. (1950). Further studies of the origin of the solar system. *Astrophys. J.* **111**, 179–190.

and A. G. L. Cameron (1963). Historical review of theories of the origin of the solar system. In R. Jastrow and A. G. L. Cameron (eds.): *Origin of the Solar System*, Academic, New York, pp. 4–37.

Thompson, R. I., P. Struttmatter, E. Erikson, F. Witteborn, and D. Strecker (1977). Observation of preplanetary discs around MWC 349 and LkH α 101. *Astron. J.* **218**, 170–180.

Tittmann, B. R., L. Ahlberg, and J. Curnow (1976). Internal friction and velocity measurements. *Proc. Seventh Lunar Sci. Conf.* **3**, 3123–3132.

Toksöz, M. N. and A. T. Hsui (1978). Thermal history and evolution of Mars. *Icarus* **34**, 537–547.

A. Dainty, S. Solomon, and K. Anderson (1974). Structure of the moon, *Rev. Geophys. Space Phys.* **12**, 539–567.

A. T. Hsui and D. Johnston (1976). Evolution of the moon revisited. In C. Watkins (ed.): *Lunar Science VII*. The Lunar Science Institute, Houston, Texas, pp. 867–869.

Toulmin, P., A. Baird, B. Clark, K. Keil, H. Rose, R. Christian, P. Evans, and W. Kelliher (1977). Geochemical and mineralogical interpretation of the Viking inorganic chemical results. *J. Geophys. Res.* **82**, 4625–4634.

Trunin, R. F., V. Lonshakova, G. Simakov, and N. Galden (1965). Investigation of rocks under high pressures and temperatures of shock compression. *Izv. Akad. SSSR, Fiz. Zemle.* **9**, 1–12.

Turekian, K. and S. P. Clark (1969). Inhomogeneous accumulation of the earth from the primitive solar nebula. *Earth Planet. Sci. Lett.* **6**, 346–348.

Unruh, D. and M. Tatsumoto (1977). Evolution of mare basalts: The complexity of the U–Th–Pb system. *Proc. Eighth Lunar Sci. Conf.* **2**, 1673–1696.

(1978). Implications from Luna 24 sample 24170 to U-Pb evolution in the lunar mantle. In: *Mare Crisium: The View from Luna 24*, Pergamon, New York, pp. 679–694.

Urey, H. C. (1952). *The Planets*. Yale University Press, New Haven, Connecticut.

(1954). On the dissipation of gas and volatilized elements from proto-planets. *Astrophys. J. Suppl.* **1**, 147–173.

(1956). Diamonds, meteorites and the origin of the solar system. *Astrophys. J.* **124**, 623–637.

(1957a). Boundary conditions for the origin of the solar system. In L. H. Ahrens, F. Press, K. Rankama, and S. Runcorn, (eds.): *Physics and Chemistry of the Earth*, *Vol. 2*. Pergamon Press, London, pp. 46–76.

(1957b). *Meteorites and the Origin of the Solar System*. 41st Guthrie Lecture, Yearbook of the Physical Society, London, pp. 14–29.

(1958). The early history of the solar system as indicated by the meteorites. *Proc. Chem. Soc.* 67–78.

(1959). Primitive planetary atmospheres and the origin of life. In: *The Origin of Life on the Earth, Vol. 1.* Symposium of International Union of Biochemistry (Moscow, 1957). MacMillan, New York, pp. 1–16.

(1960a). Chemical evidence relative to the origin of the solar system. *Mon. Not. R. Astron. Soc.* **131**, 199–223.

(1960b). On the chemical evolution and densities of the planets. *Geochim. Cosmochim. Acta* **18**, 151–153.

(1962a). Evidence regarding the origin of the earth. *Geochim. Cosmochim. Acta* **26**, 1–13.

(1962b). Origin and history of the moon. In Z. Kopal (ed.): *Physics and Astronomy of the Moon.* Academic, New York, Chapter 13, 481–523.

(1963). The origin and evolution of the solar system. In D. P. LeGalley (ed.): *Space Science.* John Wiley, New York, Chapter 4, 123–168.

and H. Craig (1953). The composition of the stone meteorites and the origin of meteorites. *Geochim. Cosmochim. Acta* **4**, 36–82.

Vaniman, D. and J. Papike (1977). The Apollo 17 drill core: Chemistry and stratigraphy of monomineralic fragments and the discovery of a new very low-Ti (VLT) mare basalt. In C. Watkins (ed.): *Lunar Science VIII.* Lunar Science Institute, Houston, Texas, pp. 952–954.

Vereshchagin, L. F., E. M. Yakovlev, and Y. A. Timofeev (1975). Possibility of transition of hydrogen into the metallic state. *JETP Letters* **21**, 85–86.

Verhoogen, J. (1956). Temperatures within the earth. In L. Ahrens, K. Rankama and S. Runcorn (eds.): *Physics and Chemistry of the Earth, Vol. 1.* Pergamon, London, pp. 17–43.

Viljoen, R. P. and M. J. Viljoen (1969a). Evidence for the existence of a mobile extrusive peridotitic magma from the Komati Formation of the Onverwacht Group. *Geol. Soc. S. Afr. Spec. Pub*[1]. **2**, 87–112.

and M. J. Viljoen (1969b). Evidence for the composition of the primitive mantle and its products of partial melting, from a study of the rocks of the Barberton Mountain Land. *Geol. Soc. S. Afr. Spec. Publ.* **2**, 275–295.

Vilminot, J. C. (1965). Les enclaves de peridotite et de pyroxenolite a spinelle dans le basalte du Rocher du Lion. *Bull. Soc. Fr. Mineral. Cristallogr.* **88**, 109–118.

Vinogradov, A. P. (1959). The chemical evolution of the earth. *Chtenie im. V.I. Vernadskogo* I. Izd. AN. SSR., Moscow.

(1961). The origin of the material of the earth's crust. Communication 1. *Geochemistry USSR*, 1–32 (English Translation).

(1967). Gas regime of the Earth. In A. P. Vinogradov (ed.): *Chemistry of the Earth's Crust, Vol. II.* 1–19. Israel Program for Technical Translations, Jerusalem.

Vollmer, R. (1977). Terrestrial lead evolution and formation time of the Earth's core. *Nature* **270**, 144–147.

von Weizsäcker, C. F. (1944). Über die Enstehung des Planetensystems. *Z. Astrophys.* **22**, 319–355.

Wai, C. M. and J. Wasson (1977). Nebula condensation of moderately volatile elements and their abundances in ordinary chondrites. *Earth Planet. Sci. Lett.* **36**, 1–13.

G. W. Wetherill, and J. T. Wasson (1968). The distribution of trace quantities of germanium between metal, silicate and sulfide phases. *Geochim. Cosmochim. Acta* **32**, 1269–1278.

Walker, D. and J. F. Hays (1977). Plagioclase flotation and lunar crust formation. *Geology* **5**, 425–428.

J. Longhi, and J. F. Hays (1975a). Differentiation of a very thick magma body and its implications for the source regions of mare basalts. *Proc. Sixth Lunar Sci. Conf.* **1**, 1103–1120.

J. Longhi, E. M. Stolper, T. L. Grove, and J. F. Hays (1975b). Origin of titaniferous lunar basalts. *Geochim. Cosmochim. Acta* **39**, 1219–1236.

R. J. Kirkpatrick, J. Longhi, and J. F. Hays (1976). Crystallization history and origin of lunar picritic basalt 12002: Phase equilibria, cooling rate studies, and physical properties of the parent magma. *Geol. Soc. Am. Bull.* **87**, 646–656.

Walker, J. C., K. Turekian, and D. Hunten (1970). An estimate of the present-day deep-mantle degassing rate from data on the atmosphere of Venus. *J. Geophys. Res.* **75**, 3558–3561.

Wang, C. Y. (1972). A simple earth model. *J. Geophys. Res.* **77**, 4318–4329.

Wang, H. and G. Simmons (1972). FeO and SiO_2 in the lower mantle. *Earth Planet. Sci. Lett.* **14**, 83–86.

T. Todd, D. Richter, and G. Simmons (1973). Elastic properties of plagioclase aggregates. *Proc. Fourth Lunar Sci. Conf.* **3**, 2663–2671.

Wänke, H., K. Palme, H. Baddenhausen, G. Dreibus, E. Jagoutz, H. Kruse, B. Spettel, F. Teschke, and R. Thacker (1974). Chemistry of Apollo 16 and 17 samples: Bulk composition, late stage accumulation and early differentiation of the moon. *Proc. Fifth Lunar Sci. Conf.* **2**, 1307–1335.

H. Palme, H. Baddenhausen, G. Dreibus, E. Jagoutz, H. Kruse, C. Palme, B. Spettel, F. Teschke, and R. Thacker (1975). New data on the chemistry of lunar samples: Primary matter in the lunar highlands and the bulk composition of the moon. *Proc. Sixth Lunar Sci. Conf.* **2**, 1313–1340.

H. Palme, H. Kruse, H. Baddenhausen. M. Cendales, G. Dreibus, H. Hofmeister, E. Jagoutz, C. Palme, B. Spettel, and R. Thacker (1976). Chemistry of lunar highland rocks: A refined evaluation of the composition of the primary matter. *Proc. Seventh Lunar Sci. Conf.* **3**, 3479–3499.

H. Baddenhausen, K. Blum, M. Cendales, G. Dreibus, H. Hofmeister, H. Kruse, E. Jagoutz, C. Palme, B. Spettel, R. Thacker, and E. Vilcsek (1977). On the chemistry of lunar samples and achondrites. Primary matter in the lunar highlands: A re-evaluation. *Proc. Eighth Lunar Sci. Conf.* **2**, 2191–2213.

G. Dreibus and H. Palme (1978a). Are the siderophile elements found in the lunar highlands of truly meteoritic origin? In: *Lunar and Planetary Science, Vol. 9.* Lunar and Planetary Institute, Houston, Texas, pp. 1202–1204.

G. Dreibus and H. Palme (1978b). Primary matter in the lunar highlands: The case of the siderophile elements. In: *Proc. Ninth Lunar Planet. Sci. Conf.* **1**, 83–110.

Ward, W. R. and A. G. W. Cameron (1978). Disc evolution within the Roche limit. In: *Lunar and Planetary Science, Vol. 9*. Lunar and Planetary Institute, Houston, Texas, pp. 1205–1207.

Warren, P. H. and J. T. Wasson (1978). Compositional-petrographic investigation of pristine nonmare rocks. *Proc. Ninth Lunar Planet. Sci. Conf.* **1**, 185–218.

Wasson, J. T. (1972). Formation of ordinary chondrites. *Rev. Geophys. Space Phys.* **10**, 711–759.

    (1974). *Meteorites*. Springer-Verlag, Berlin.

    and C. C. Chou (1974). Fractionation of moderately volatile elements in ordinary chondrites. *Meteoritics* **9**, 69–84.

    and C. M. Wai (1976). Explanation for the very low Ga and Ge concentrations in some iron meteorite groups. *Nature* **261**, 114–116.

Wedepohl, K. (ed.) (1974). *Handbook of Geochemistry II*. Springer-Verlag, New York.

Weidenschilling, S. J. (1975). Mass loss from the region of Mars and the asteroid belt. *Icarus* **26**, 361–366.

    (1976). Accretion of the terrestrial planets II. *Icarus* **27**, 161–170.

    (1977). The distribution of mass in the planetary system and solar nebula. *Astrophys. Space Sci.* **51**, 153–158.

    (1978). Iron/silicate fractionation and the origin of Mercury. *Icarus* **35**, 99–111.

Werner, M. W., E. E. Becklin, and G. Neugebaur (1977). Infrared studies of star formation. *Science* **197**, 723–732.

Wetherill, G. W. (1975). Late heavy bombardment of the moon and terrestrial planets. *Proc. Sixth Lunar Sci. Conf.* **2**, 1539–1561.

    (1976a). The role of large bodies in the formation of the Earth and Moon. *Proc. Seventh Lunar Sci. Conf.* **3**, 3245–3257.

    (1976b). Where do the meteorites come from? A re-evaluation of the Earth-crossing Apollo asteroids as sources of chondritic meteorites. *Geochim. Cosmochim. Acta* **40**, 1297–1317.

    (1977). Evolution of the Earth's planetesimal swarm subsequent to the formation of the Earth and Moon. *Proc. Eighth Lunar Sci. Conf.* **1**, 1–16.

    (1977). Fragmentation of asteroids and delivery of fragments to Earth. In A. H. Delsemme (ed.): *Proceedings of the IAU Colloquium 39: Relationships between comets, minor planets and meteorites*. Univ. of Toledo press, Toledo, Ohio.

    and J. G. Williams (1978). Origin of differentiated meteorites. In *Proceedings of the Second International Conference on the Origin and Distribution of Elements*. Paris (in press).

Wiik, H. B. (1956). The chemical composition of some stony meteorites. *Geochim. Cosmochim. Acta* **2**, 91–117.

Wilhelms, D. E. (1976). Mercurian volcanism questioned. *Icarus* **28**, 551–558.

Williams, D. L. and R. P. Von Herzen (1974). Heat loss from the Earth; new estimate. *Geology* **2**, 327–328.

Wise, D. U. (1963). An origin of the moon by rotational fission during formation of the earth's core. *J. Geophys. Res.* **68**, 1547–1554.

    (1969). Origin of the Moon from the Earth: Some new mechanisms and comparisons. *J. Geophys. Res.* **74**, 6034–6045.

Wiskerchen, M. J. and C. P. Sonett (1977). A lunar metal core? *Proc. Eighth Lunar Sci. Conf.* **3**, 515–535.

Wlotzka, F., B. Spettel, and H. Wänke (1973). On the composition of metal from Apollo 16 fines and the meteoritic component. *Proc. Fourth Lunar Sci. Conf.* **2**, 1483–1491.

Wood, J. A. (1970). Petrology of the lunar soil and geophysical implications. *J. Geophys. Res.* **32**, 6497–6513.

——— (1975). Lunar petrogenesis in a well-stirred magma ocean. *Proc. Sixth Lunar Sci. Conf.* **1**, 1087–1102.

——— and H. Y. McSween (1977). Chondrules as condensation products. In A. H. Delsemme (ed.): *Proceedings of the IAU Colloquium No. 39: Relationships Between Comets, Minor Planets and Meteorites.* Univ. of Toledo Press, Toledo, Ohio, pp. 365–373.

——— and H. E. Mitler (1974). Origin of the moon by a modified capture mechanism, or: Half a loaf is better than a whole one (abstract). *Lunar Science, Vol. V.* Lunar Science Institute, Houston, Texas, pp. 851–853.

Wu, H. and L. Broadfoot (1977). The extreme ultraviolet albedos of the planet Mercury and the Moon. *J. Geophys. Res.* **82**, 759–761.

Young, A. T. (1973). Are the clouds of Venus sulphuric acid? *Icarus* **18**, 564–582.

Index